Signals and Communication Technology

For further volumes:
http://www.springer.com/series/4748

Waldemar Rebizant · Janusz Szafran
Andrzej Wiszniewski

Digital Signal Processing in Power System Protection and Control

Springer

Prof. Waldemar Rebizant
Wroclaw University of Technology
Wybrzeze Wyspianskiego 27
50-370 Wroclaw
Poland
e-mail: waldemar.rebizant@pwr.wroc.pl

Prof. Andrzej Wiszniewski
Wroclaw University of Technology
Wybrzeze Wyspianskiego 27
50-370 Wroclaw
Poland
e-mail: andrzej.wiszniewski@pwr.wroc.pl

Prof. Janusz Szafran
Wroclaw University of Technology
Wybrzeze Wyspianskiego 27
50-370 Wroclaw
Poland
e-mail: janusz.szafran@pwr.wroc.pl

ISSN 1860-4862
ISBN 978-0-85729-801-0 e-ISBN 978-0-85729-802-7
DOI 10.1007/978-0-85729-802-7
Springer London Dordrecht Heidelberg New York

British Library Cataloguing in Publication Data
A catalogue record for this book is available from the British Library

Cover design: eStudio Calamar, Berlin/Figueres

Printed on acid-free paper

Springer is part of Springer Science+Business Media (www.springer.com)

Contents

Symbols and Abbreviations

Symbols

$\underline{a}$	Complex rotation operator (by $2\pi/3$)
a_c, a_s	Coefficients of the orthogonal filters F_c, F_s with sine, cosine windows
a_k, b_k	Constant factors in a difference equation nominator and denominator
$a(k)$, $b(k)$	Filter equation coefficients
a_n, b_n, a_0	Coefficients of the Fourier series
A	Signal magnitude
A, $\overline{A}$	Fuzzy set and its complement
$A_s(z)$, $A_s(j\Omega)$	Z and Fourier transfer functions of the sine window filter
$A_c(z)$, $A_c(j\Omega)$	Z and Fourier transfer functions of the cosine window filter
A, B	Probabilistic thresholds (SPRT test)
A, B, C	Coefficients of the filter design procedure with bilinear transformation
A_k, s_k	Residuum values and poles of the transfer function partial expansion
B_s	Flux density value leading to CT saturation
$B_{\max}$	Maximal expected flux density level in the CT core
c_n	Coefficients of the complex Fourier series
C	Capacitance
d	A/D converter range resolution
$d_{CC}(n)$, $d_{SS}(n)$, $d_{SC}(n)$, $d_{CS}(n)$	Filter coefficients in transient state
$\mathrm{d}f/\mathrm{d}t$	Rate of frequency change
D_X	Range of possible values of the decision random variable X
$e(t)$, $e(n)$	Error signal, continuous and discrete

$\boldsymbol{e}_N$	Vector of estimation error samples
E	Signal energy
f	Frequency
f_1, ω_1	Frequency and angular frequency of the fundamental harmonic component
f_C	Cut-off frequency of a filter
f_k, ω_k	Frequency and angular frequency of the *k*th harmonic component
f_m	Measured frequency
f_S	Sampling frequency: $f_S = 1/T_S$
$f(\boldsymbol{X}\vert H_i)$	Probability density function of the decision vector $\boldsymbol{X}$ for hypothesis H_i
F_c, F_s	Orthogonal filters
$F(\mathrm{j}\omega)$	Fourier transform of a time function
$F(s)$	Laplace transform of a time function
$F(z)$	Z transform of the signal f
F_{1C}, F_{1S}	Gains of filters of odd (S) and even (C) symmetry of impulse response for fundamental frequency
G, B, Y	Conductance, susceptance and admittance
$h(t)$	Continuous impulse response of a system
$h_a(t)$	Impulse response of an analog filter
$h(n)$	Discrete impulse response of a system
$\boldsymbol{h}_N$	Matrix of LSE signal model elements for different sampling instants
$H(\mathrm{j}\omega)$	Fourier transfer function
$H(s)$	Laplace transfer function
$H_a(s)$	Laplace transfer function of an analog filter
$H(z)$	Z Transfer function
$H_i(\Omega)$, $W(\Omega)$	Frequency responses of ideal filter and rectangular window, respectively
$i(n)$	Current sample of current signal
$i(t)$	Instantaneous value of current at the instant t
$i_m(n)$	Magnetizing current at given time instant
i_1, i_2, $i_{2\mathrm{e}}$	Primary, secondary and estimated secondary CT currents
I, U	Magnitudes of current and voltage signals
I_{km}, U_{km}	Magnitudes of the *k*th harmonic of current and voltage signals
I_a	Initial value of the decaying DC component
I_d	Differential current magnitude
I_{hm}	Phasor magnitude of transient oscillatory decaying current component
$I_{L\_\max}$, $I_{\text{pick-up}}$, $I_{f\_\min}$	Maximum load current, relay pick-up and minimum fault current

I_{L1}, I_{L2}, I_{L3}	Magnitudes of current fundamental components in phases $L1$, $L2$ and $L3$
I_s	Stabilization current magnitude
I_0, I_1, I_2	Zero, positive and negative symmetrical components of three phase currents
IN	Amount of information
j	Imaginary operator
k, l, m	Number of samples of delay, integers
L	Inductance
$L(\omega)$	Logarithmic frequency response
m	Number of bits of an A/D converter
M	A/D converter range
$M_{0.5}$	Number of reference samples during half period of the signal
n	Current number of sample
n_{ANN}	ANN size (total number of neurons)
N	Number of samples in the signal period / in the filter data window
N_u, N_i	Transformation ratio of a VT/CT
N_0	Number of sine/cosine filter coefficients within one cycle of frequency Ω_0
N_1	Number of samples in the period of the fundamental frequency
p_m, p_c	Probability of mutation and crossover
P, Q, S	Active, reactive and apparent powers
P_0, P_1, P_2	Zero, positive and negative symmetrical components of active power
$\boldsymbol{P}_N$	LSE transformation matrix
Q	Quality index, fitness value
R, X, Z	Resistance, reactance and impedance
$S(m)$	Secondary CT current integral for observing CT saturation end
$\underline{\boldsymbol{S}}$	Symmetrical components complex transformation matrix
$\boldsymbol{S}_R$, $\boldsymbol{S}_I$	Real and imaginary parts of the transformation matrix $\underline{\boldsymbol{S}}$
t	Current moment at the time scale
t_a, k_a	Regulator time constant and gain values (AVR)
t_e	Exciter time constant (AVR)
t_f, k_f	Feedback connection time constant and gain (AVR)
t_r	Time constant of the input filter (AVR)
t_{s1}, t_{s2}	Time instants of beginning and ending of CT saturation
T, T_0	Duration of the signal period

T_a	Time constant of current decaying DC component
T_h	Time constant of transient oscillatory decaying current component
T_s	Saturation time constant of a CT
T_S	Time between two successive samples (sampling period)
T_w	Duration of the data window
T_1	Duration of the fundamental frequency period
$u(n)$	Current sample of voltage signal
$v(t)$, $u(t)$	Instantaneous value of voltage
U_{exc}	Excitation voltage (AVR output),
U_{L1}, U_{L2}, U_{L3}	Magnitudes of voltage fundamental components in phases $L1$, $L2$ and $L3$
U_0, U_1, U_2	Zero, positive and negative symmetrical components of three phase voltages
V_{ref}	Reference voltage
w_i	Weighting factors for particular criteria
$\boldsymbol{W}$, $\boldsymbol{B}$	Matrices of neuron/ANN weights and bias values
$W_0(z)$, $W_0(j\Omega)$	Z and Fourier transfer functions of the 0-order Walsh filter
$W_1(z)$, $W_1(j\Omega)$	Z and Fourier transfer functions of the 1-order Walsh filter
$W_2(z)$, $W_2(j\Omega)$	Z and Fourier transfer functions of the 2-order Walsh filter
x, y	Input and output signals of a block or unit
$x(t)$	Continuous signal x
$x^*(t)$	Sampled signal x
$x(n)$	Current sample of signal x
$x(n\text{-}k)$	Sample of signal x delayed by k sampling periods
x_C, x_S	Orthogonal components of signal x (direct, quadrature)
x_{F1C}, x_{F1S}	Orthogonal components of signal x obtained from orthogonal filters
$\underline{\boldsymbol{x}}_{012}$	Vector of symmetrical components' phasors
$\underline{\boldsymbol{x}}_{L1L2L3}$	Vector of phase signals' phasors
$x_{012}^{C(S)}$, $x_{L1L2L3}^{C(S)}$	Orthogonal components of symmetrical/phase signals
X_{kC}, X_{kS}	Correlation coefficients for kth harmonic components
$X(j\omega)$	Fourier transform of the signal x
$X^*(j\omega)$	Fourier transform of the sampled signal x
$X(k)$	FFT of signal x at discrete frequency points k
$X_{\min}$	Smallest expected signal value
X_{1m}	Magnitude of the fundamental harmonic component

$\boldsymbol{x}_N$	Vector of signal samples for LSE procedure
$\hat{\boldsymbol{X}}$	Approximate vector of signal model parameters determined with LSE method
$\delta(t)$, $\delta(n)$	Continuous and discrete Dirac delta function
δ_i	Total support for the ith decision, fuzzy support value
Δ	Discrimination threshold
ΔI	Current difference
$\Delta\varphi$	Difference of current phase angles
ε	Error value
ε_0, ε_1	Assumed values of the first and second order error probabilities
φ	Phase angle of the phasor: $\varphi = \text{arctg}\,(x_S/x_C)$; $\underline{X} = X\text{e}^{\text{j}\varphi}$
ϕ	Phase difference between two phasors
Φ	Magnetic flux
Φ_S	Saturation flux in the ferromagnetic core
Φ_r	Residual flux in the ferromagnetic core
ψ	Flux linkage in the CT core
ψ_c, ψ_s	Arguments of the orthogonal transfer functions G_c, G_s
γ_S	Phase difference between two successive samples: $\gamma_S = \omega_1 T_S$
λ	Group delay value
$\mu_A(x)$	Membership function of fuzzy set A
$\mu_{\overline{A}}(x)$	Membership function of a complement for fuzzy set A
ν_{ik}	Support value for the kth decision from the ith criterion, result of fuzzy comparison
ω	Angular frequency
ω_a, ω_d	Angular frequencies in transfer functions of analog and digital filters
ω_{aC}, ω_{dC}	Cut-off angular frequency of an analog/digital filter
ω_C	Cut-off angular frequency of a filter
ω_{dCH}, ω_{dCL}	High and low cut-off angular frequencies of a digital band pass or band rejection filter
ω_S	Angular sampling frequency: $\omega_S = 2\pi f_S$
ω_0	Angular frequency of a signal with the period T_0
Ω	Relative angular frequency: $\Omega = \omega/f_S$
Ω_1	Relative fundamental angular frequency: $\Omega_1 = \omega_1/f_S$
Ω_1, Ω_2	Sub-regions of the decisive space
Ω_k	kth multiple of the relative fundamental angular frequency Ω_1

Ω_0	Natural relative angular frequency of a sine/cosine filter
τ	Continuous time as an independent variable
η	Learning rate (ANN)
Θ_k	SPRT test index at the instant k
ϑ	Temperature

Abbreviations

A/D	Analog to digital (converter)
ADALINE	Adaptive linear model (of a neuron)
AI	Artificial intelligence
ANFIS	Adaptive network fuzzy inference system
ANN	Artificial neural network
AVR	Automatic voltage regulator
BP	BackPropagation (algorithm)
COG	Center of gravity (defuzzification)
CT	Current transformer
CVT	Capacitive voltage transformer
DFT	Discrete Fourier transform
EMTP-ATP	Electromagnetic transients program-Alternative transients program
EP	Evolutionary programming
ES	Evolutionary strategy
FACTS	Flexible AC transmission systems
FAM	Fuzzy associative memory
FFT	Fast Fourier transform
FIR	Finite impulse response (filter)
FIS	Fuzzy inference system
FL	Fuzzy logic
FSC	Ferroresonance suppressing circuit
GA	Genetic algorithm
GPS	Global positioning system
HVDC	High voltage direct current
IED	Intelligent electronic device
IIR	Infinite impulse response (filter)
I>	Overcurrent protection
LAN	Local area network
LSE	Least squares estimation technique
MF	Membership function
MOM	Mean of maxima (defuzzification)
MSE	Mean square error (technique, value)
MSPRT	Multi-hypotheses sequential probability ratio test
NIM	Network interface module
OS	Out-of-step (protection)
PDF	Probability density function

PMU	Phasor measurement unit
PSP	Power system protection
RBF	Radial basis function (ANN)
RTU	Remote terminal unit
SAS	Substation automation system
SCADA	Supervisory control and data acquisition
SOM	Self-organizing map (ANN)
SPRT	Sequential probability ratio test
SVM	Support vector machine (ANN)
VDU	Visual display unit
WAMPS	Wide area measurement and protection systems
WAN	Wide area network
VT	Voltage transformer
Z<	Underimpedance (distance) protection

Chapter 1
Introduction

Electric power system is a backbone of contemporary technical civilization and today's economy. One cannot imagine functioning without stable, uninterruptable supply of electric energy that is needed both in industry and in most areas of our private lives. The users of electric energy treat it sometimes as granted, considering it as a primary good and even as one of the human rights.

Electric energy is produced in power plants and then transmitted over long distances and distributed to the final consumers, industrial and domestic. The electric network is a vast structure embracing a lot of generators in power plants or distributed energy sources, hundreds of thousand kilometers of lines, thousands of transformers, busbars, and other equipment. This structure has to be maintained, controlled and protected against possible faults and other abnormal phenomena. Thus a great variety of protection and control devices aimed at smooth and stable running of power systems have been developed and installed in modern power networks. Most of them are nowadays produced in digital technology, performing their functions through numerical calculations and signal analysis.

Many books related to power system protection and control describe the basic theory as well as applications of protection principles, which result from understanding of physical nature of encountered phenomena, to protection of particular elements of power networks and to maintaining system integrity. Therefore the authors are not going to cover the same field again. Instead, deeper insight into functioning of digital protection and control devices is proposed, which includes both hardware and software details. This book is meant to provide thorough information on signal processing methods and algorithms that are used for measurement of protection criteria values, being a basis for final decision making in protective relays.

This book is aimed at bridging the gap between theory of protection and control and practical applications of protection equipment. Understanding the ways of protection functioning is crucial not only for equipment developers and manufacturers but also for their users that are going to install, set and operate the

W. Rebizant et al., *Digital Signal Processing in Power System Protection and Control,* Signals and Communication Technology,
DOI: 10.1007/978-0-85729-802-7_1, © Springer-Verlag London Limited 2011

protection devices in an appropriate manner. The information provided here can be useful for protection engineers working at various levels of electric network, in utilities, as well as for students of electrical engineering, especially electrical power engineering. It may also be helpful for all others willing to get acquainted and to apply the filtering, measuring and decision making algorithms for purpose other than protection and control, everywhere fast and on-line signal analysis is needed for proper functioning of the apparatus.

This book starts from a short overview of abnormal states in power systems and enumeration of principles and criteria for their recognition and elimination (Chap. 2). Next functional development of protection devices and systems is outlined, with brief description of protection functional blocks and place of protective relays in broad protection and control systems of power networks (Chap. 3). Fundamentals of system and signal analysis are provided in Chap. 4, including all basic transformations needed in analysis of continuous and discrete systems, which is meant as a foundation for further more sophisticated algorithms.

The digital algorithms for signal filtering with infinite and finite impulse response are presented in Chaps. 5 and 6, respectively, which is followed by methods and algorithms that can be applied for additional correction or reconstruction of distorted secondary signals of voltage and current transformers. Measurement algorithms of most commonly used protection criteria values are described in Chaps. 8 and 9, whereas in the latter their dynamics and features in frequency domain are discussed and some adaptive measurement algorithms are introduced. Finally, decision making methods with deterministic and statistical approach as well as multiple criteria and adaptive decision schemes are outlined in Chap. 10.

Chapters 11–15 are devoted to basic theory and applications of chosen Artificial Intelligence techniques for protection and control. Fuzzy Logic based schemes, Artificial Neural Networks, Expert Systems and Genetic Algorithms with their advantages and drawbacks are presented. In Chap. 15 the techniques mentioned are compared and their useful combinations (hybrids) are shown that are hoped to eliminate disadvantages and magnify good features of particular techniques.

Theory provided in the book is illustrated with lot of computational examples as well as with examples of the algorithms' application to power system protection. Special attention is paid to applications of the intelligent approaches and paradigms that seem to have a great potential for improving operation of protective devices.

The authors wish respectable readers pleasant reading and a lot of fun by discovering the beauty of presented mathematical apparatus, signal processing algorithms, adaptive methods and intelligent techniques!

Chapter 2
Abnormal States in Power Systems and Criteria for Their Recognition

2.1 Introduction

The protection equipment is applied with the aim of minimizing the effects of faults and other abnormal phenomena on the operation of electrical power systems. An electrical power system is considered as all the plants required to generate, transmit and distribute electrical power, including generators, power transformers, lines and cables, circuit-breakers, etc.

Faults in power systems can be a result of external or internal influences, whereas the most common reasons of abnormal states are:

- overvoltage due to lightning,
- short-circuits due to mechanical destruction or bridging of isolation (e.g. during road works, caused by falling trees, animals, birds, etc.),
- thermal overload (overcurrents),
- aging of isolation,
- maintenance/staff mistakes,
- climatic disasters (floods, earthquakes, heavy snowfalls, icing, strong wind, etc.).

The consequences of a fault may be:

- damage to the plant due to dynamic and thermal effects of the fault current,
- loss of supply to loads,
- danger to human lives,
- loss of system stability,
- possibility of cascading events leading to blackouts.

The protection devices may react to the events with tripping, alarm or signaling, depending on the expected consequences of the event, see Table 2.1. As one can notice, not always the protection devices are about tripping, sometimes an alarm or signaling is enough, depending on the situation at hand. Nevertheless, the

W. Rebizant et al., *Digital Signal Processing in Power System Protection and Control,* Signals and Communication Technology,
DOI: 10.1007/978-0-85729-802-7_2,

Table 2.1 Selected phenomena and possible protection reaction

Exposure	Fault	Damage	Operation	Task
Short-circuit	Yes	Yes	Response with trip command	Limitation of damage range
Ground fault	Yes	No	Response with signaling	Indication and location of fault
Overload Unbalance	No	No	Acting with command/ signal	Avoiding of possible damage
Overvoltage Undervoltage Underfrequency	No	No	Acting with command/ signal	Avoiding of possible damage

abnormal phenomena have to be promptly detected and classified, which is performed by measuring of certain criteria values. In the following subsections the dangerous phenomena for most important power system components are outlined, the protection criteria for detecting faults are enumerated and basic requirements for protective devices are discussed.

2.2 Faults and Abnormal Phenomena in Power Networks

The protection relays should differentiate the normal operating conditions from the abnormal ones, including short-circuits and other impacts that can be dangerous to protected equipment itself and may impair safe operation of the power system. Under normal operating conditions the power system is characterized by:

- almost symmetrical three-phase voltages and currents,
- operational currents remaining below the pre-set levels (including some permissible overload),
- voltage level within permissible range around the nominal value,
- frequency of the signals equal or very close to the nominal 50/60 Hz,
- harmonics content within permissible limits,
- power quality satisfying customer and standard requirements.

Of course, the power system is rather seldom in an ideal state, thus voltage and current signals may vary due to variations of load and generation as well as switching operations.

It must be noted here that some phenomena occurring in power system under normal conditions may also be seen as abnormal for the sake of current/voltage signal parameters being far distant from the nominal values. However, their reason is not always a fault and the protection should not operate in such cases. Examples of such situations are:

- charging of the transmission line after switching in,
- starting of large induction motors,

- inrush currents due to energization of power transformers,
- stationary overexcitation (overfluxing) of power transformers due to increased voltage and/or decreased frequency,
- asymmetric load,
- frequency excursions and power swings, etc.

In such circumstances one may observe temporary increase of the currents beyond acceptable limits, appearance of negative and zero sequence components, increase of harmonic contents, etc., which would suggest that a fault has probably occurred. Additional problems may appear during current transformers saturation caused by external faults, which significantly falsifies information delivered to the protection devices. Therefore appropriate set of criteria rather than a single one is often applied, which leads to multi-criteria solutions characterized by increased stability against unwanted operations.

The protection devices should issue a tripping/warning decision in cases of faults within the protected plant or zone. It should be said that the faults may be of various kind, type, intensity and possible consequences. The most dangerous are short-circuits, since the currents flowing in the circuit are then a high multiple of the nominal level. The fault current level depends on many parameters including: voltage level, operation mode of the network star point (isolated, resistively grounded or compensated), fault resistance, fault type, and so on.

For particular power system elements the following kinds of faults and abnormal phenomena, for which the protection relays should react, could be listed [6]:

- transmission/distribution lines:
  - single-phase-to-ground faults,
  - phase-to-phase faults,
  - phase-to-phase-to-ground faults,
  - three-phase faults,
  - high-impedance faults,
  - intermittent arcing faults,
  - overload;
- busbars:
  - phase-to-ground short-circuits,
  - phase faults;
- power transformers:
  - ground faults,
  - phase faults at the terminals,
  - short-circuits between turns,
  - short-circuits between windings,
  - tap-changer failures,
  - transformer tank oil leaks,

- overexcitation,
- overload;

- synchronous generators:
  - ground faults,
  - phase faults between windings,
  - inter-turn faults,
  - duplicate ground faults,
  - stator/rotor overload,
  - loss of excitation,
  - pole slipping,
  - under-/overfrequency,
  - overvoltage,
  - asymmetric load,
  - motoring;

- HV three-phase motors:
  - phase faults between the stator windings or on the motor feeder,
  - phase faults on a stator winding,
  - stator inter-turn short-circuits,
  - overload,
  - phase imbalance,
  - voltage dips,
  - asynchronous operation (in case of synchronous machines).

Each of the above enumerated fault types can be recognized through determination (measurement) of certain features, called criteria, resulting from processing of information contained in voltages, currents and other signals. Below a list of most frequently used protection criteria are given. It is not the aim of this book to thoroughly discuss their application for particular power system elements nor to discuss the rules for relay settings and their coordination. The main focus is directed on the methods and detailed algorithms for criteria values measurement, which is described in further chapters.

2.3 Criteria Signals

Overcurrent

An overcurrent is detected when the maximum load current permissible for an item of an electrical plant is exceeded. Overcurrent protection devices monitor the current being conducted by the protected unit and issue a tripping command for the circuit-breaker when the current exceeds the setting (the so called relay pick-up current). According to the operating speed one can distinguish instantaneous and time-delayed (definite time or inverse time) overcurrent relays.

Over- and Undervoltage

Over- and undervoltage protections react when the voltages measured exceed the permissible limits.

Frequency

The deviation of frequency from its rated value is an indication of power imbalance in the system.

Differential Principle

Difference of current magnitudes or phase angles at the terminals of protected plant is a clear sign of internal fault.

Impedance

Impedance is measured to detect faults on transmission systems or underexcitation or out-of-step conditions of generators.

Symmetrical Components

In some applications they are more suitable for protection purposes than the phase values. The basic types of faults detected by monitoring of symmetrical components: phase-to-phase and ground faults in solid and low-resistance grounded systems, ground faults in medium voltage systems (ungrounded or inductively grounded networks), asymmetric system configuration, asymmetric load and open-circuit phase conductor.

Power Direction

Discrimination of power flow direction is used in combination with overcurrent units when the overcurrent criterion on its own is insufficient to preserve selectivity of tripping. Typical applications are for making the relays selective for ring and parallel lines.

Other Criteria

Other criteria used by protective devices are:

- temperature (typical application: transformer oil temperature),
- rate of oil flow, accumulation of gas (for detecting internal transformer faults),
- harmonics in the neutral current and voltage (for detecting ground faults in inductively grounded systems),
- harmonics in generator current (for detecting internal generator faults),
- transient current and voltage signals, traveling waves (e.g. in transmission lines protection and fault location).

A very basic overview of how the protection criteria are applied for protecting against the most frequent abnormal phenomena can be found in Table 2.2. No further discussion is provided here, for more specific rules of protecting particular power system elements the reader should refer to some other books, e.g. [1–6].

2.4 Requirements for Protective Devices

Protection devices/relays cannot unfortunately prevent the faults that happen, they act only after fault appearance. The protection devices are designed and expected to:

Table 2.2 Faults and criteria for their detection

Type of fault	Variable used for detection	
Phase fault in general	Phase current magnitude	I
	Current difference	ΔI
	Difference of current phase angle	$\Delta\varphi$
	Phase voltage magnitude	U
	Power direction	P
	Impedance	Z
Asymmetrical faults	As above plus zero and negative sequence components of	
	Current	I_0, I_2
	Voltage	U_0, U_2
	Power	P_0, P_2
Ground faults	Zero sequence components of	
	Current	I_0
	Voltage	U_0
	Power	P_0
Asymmetric configuration	Negative sequence component of current	I_2
Asymmetric load		
Interrupted conductor		
Overload	Phase current magnitude	I
	Temperature	ϑ
Real power deficit	Frequency	f
	Rate of frequency change	$\mathrm{d}f/\mathrm{d}t$
Real power excess	Frequency	f

- detect eventual faults/abnormal conditions possibly fast and selectively,
- help avoiding extensive damage of the faulted element through its isolation from supply,
- enable further operation of the rest of power system,
- prevent treats to environment, people and animals,
- eliminate or limit voltage dips/sags,
- prevent developing outages and blackouts.

Having the above in mind one can formulate the following general requirements the protection relays have to fulfill:

Selectivity—an ability to isolate only the defective plant from the rest of the system. Selectivity can be achieved by:

- time grading, i.e. the protection device nearest the fault trips the fastest and all the others—slower (overcurrent and distance protection),
- magnitude and/or phase comparison of the currents at both sides of the protected unit (pilot wire and differential protection),
- determination of fault power flow direction at both sides of the protected unit (directional comparison protection).

Reliability—high security against underfunction (lack of tripping in case of fault) and overfunction (false tripping). Reliability is an ability of a protective device to fulfill its purpose throughout its operational life. The related notions are here:

- dependability: the assurance that the protection device will perform its function and selectively trip the protected item of a primary plant in the event of a fault,
- security: the assurance that the protection device will not trip unless there is a fault on the protected item,
- availability: the ratio of the time that a protection device is actually serviceable to the total time it is in operation.

Sensitivity—ability to react even to relatively small deviations of the monitored variables due to a disturbance in the system.

Operating speed—ability to isolate the faulted element within shortest possible period of time. The protection trip time is the time between fault inception and trip command. One should understand that the total fault clearing (TFC) time is a sum of the protection reaction time and circuit-breaker operating time (including arc extinguishing). The TFC time amounts: 60–100 ms in EHV/HV networks, up to 200 ms in 110 kV networks, and 0.1–1.0 s in MV networks. With modern protection devices the operating time is typically one period of the power system frequency.

Most of the commonly applied protective relays perform their job in the "traditional way" (relay's modus operandi) that can be summarized as follows:

- collecting the local information (currents, voltages, etc.),
- extracting (measurement) certain features such as magnitude, phase difference, impedance, etc.
- comparing the measured criteria values with the operating characteristics or thresholds,
- applying certain delays,
- issuing trip command.

The above algorithm (even though realized in digital technology) reproduces the well known protection principles and cannot by-pass certain constraints of those principles. Even though the digital protection devices are better and better, offering fast and accurate signal processing algorithms, possibilities of almost free shaping and setting of decision characteristics, as well as number of additional auxiliary functions, there are situations and plants where classical approach and methods may not guarantee proper relay operation. A number of difficulties can still be faced, among which the following seem to be the most significant:

- Limited recognition abilities of the known protection principles, e.g.:
  - impedance is not a perfect measure of the distance to a fault on a transmission line since the same impedance value may be seen by the relay for both internal and external faults—consequences: faults close to the line remote terminals (in front of and behind) are difficult to distinguish, thus only 80–90% of the line length is efficiently protected,
  - 2nd harmonic ratio is not a perfect indicator of the magnetizing inrush in a transformer since the 2nd harmonic may occur during internal faults as well—consequences: either false tripping or delayed operation may occur depending on the threshold;

- Trade-offs in relay setting, e.g.:
  - the 2nd harmonic restraint may be set low to increase the security during inrush conditions, but this would increase delay or result in missing operations during internal faults,
  - the bias characteristic of transformer differential relay may be set sensitive but this may result in false tripping during thorough faults and saturation of the CTs;
- Limited speed of measurement:
  - any relay asserts the tripping signal based on the symptoms of abnormalities,
  - those symptoms need certain time to occur,
  - extra time is needed for accurate enough measurement of those symptoms,
  - there is no perfect measuring algorithm that solves the conflict between the speed and the accuracy.
- New complex devices in today's power systems create new problems or even make some of the traditional relaying principles to fail (HVDC, series compensation, non-linear loads, FACTS).

In all above mentioned cases the normal operation and overload/fault regions may overlap and thus it is difficult or even impossible to set a threshold that would separate the operation and blocking areas completely. Besides, even if the steady-state loci of measured criteria values are situated within proper decision areas, they may be seen in wrong part of the decision space during measurement transients. Therefore, new intelligent ideas are needed to improve operation of protection devices for such situations.

The protection improvement possibilities may include:

- application of adaptive protection schemes—adjusting of the protection functions in real time (criterion values measurement and decision-making algorithms can be adequately adapted),
- processing larger amount of information (multi-criteria approach, decision-making with probabilistic algorithms),
- application of Artificial Intelligence techniques (Artificial Neural Networks, Fuzzy Logic, Expert Systems, etc.).

Both the traditional and intelligent approaches and algorithms of signal processing and reasoning in protective relays are presented in the following chapters of this book.

References

1. Clemens C, Rothe K (1991) Schutztechnik in Elektroenergiesystemen. VDE Verlag, Berlin
2. Elmore WA (1994) Protective relaying—theory and applications. Marcel Dekker Inc., New York

3. Hewitson LG, Brown M, Balakrishnan R (2004) Practical power system protection. Newnes, Burlington
4. Horowitz SH, Phadke AG (1988) Power system relaying. Research Studies Press, Taunton
5. Johns AT, Salman SK (1995) Digital protection for power systems. P. Peregrinus Ltd., on behalf of IEE, London
6. Ungrad H, Winkler W, Wiszniewski A (1995) Protection techniques in electrical energy systems. Marcel Dekker Inc., New York

Chapter 3
Hardware and Functional Development of Protection Devices and Systems

3.1 Protection Generations

The age of protection began some 105 years ago, when control apparatus were for the first time connected to current and voltage transformers, which reduced the primary currents and voltages to secondary levels. The secondary currents and voltages could be processed to generate TRIP or NO-TRIP decisions (Fig. 3.1).

The first relays were simple overcurrent and undervoltage apparatus (1905). Both were electromagnetic or magnetoelectric in nature and resembled the measuring apparatus. The relays were actuated either by r.m.s. values or mean values of rectified signals. They had moving parts and their operation delay was the time needed to close the output contacts.

Next step (around 1915) was introduction of inverse time overcurrent relays, either thermal, or with rotating discs. Some 5 years later the differential principle was adopted, which was a great step forward in protection of the power system apparatus. The differential relays compared two currents from the protected plant terminals.

From differential relays there was only one step to distance relays (1930), which compared currents and voltages, their r.m.s. or mean values. They had the delay being a function of the measured impedance $\underline{Z} = \underline{U}/\underline{I}$. All those relays were magnetoelectric or electromagnetic, all had moving parts and we consider them as the first generation of protective devices.

The big change was caused by the invention of transistors (1947). They were introduced to protective relays around 1955. It enabled to build the static devices (without moving parts, except the output element) that began the era of second generation of protective relays. The static relays had the operational criteria the same as electromagnetic ones; however, their decisions (to trip or not to trip) were based on the modified ways of signal processing. Because of that they offered a number of advantages:

W. Rebizant et al., *Digital Signal Processing in Power System Protection and Control,* Signals and Communication Technology,
DOI: 10.1007/978-0-85729-802-7_3, © Springer-Verlag London Limited 2011

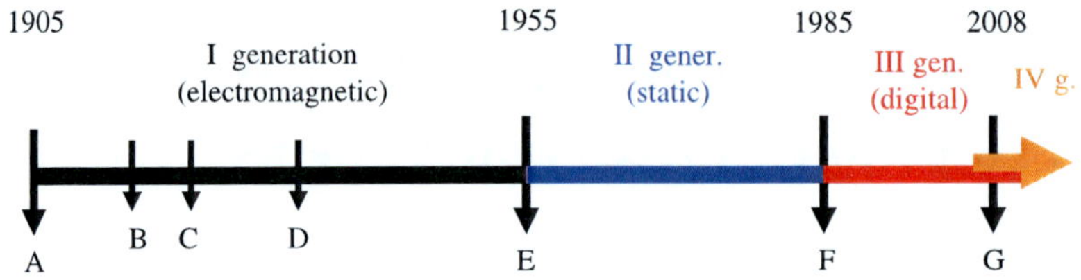

Fig. 3.1 History of protection technology. *A* overcurrent and undervoltage relays with CTs and VTs, *B* inverse time overcurrent relays, *C* differential relays and directional relays, *D* distance relays with time graded characteristics, *E* static relays with filters and comparators, *F* digital relays measuring phasors, *G* first wide-area measurement protective systems

- improvement of cooperation with CTs and VTs,
- reduced dimensions and modularity,
- facilitation of testing, maintenance and repairs,
- complex operational characteristics,
- increased speed of operation.

The third generation of protective devices started with the widespread application of inexpensive digital processors and memories (around 1985). Within the digital relay processing of relay input signals are realized in five basic steps:

- input of analog signals, which usually are currents and voltages derived from CTs and VTs, enter the antialiasing low-pass filters that remove components of high frequency, which could irreversibly corrupt the digital signals;
- from the antialiasing filter the signals—still in the analog form—enter the A/D converter and at the output of that block they have forms of trains of samples;
- the digital signals enter the block of initial processing, where they are filtered, orthogonalized, etc.;
- the results of initial processing enter the block of digital measurements, in which the criteria signals, their specific parameters and mutual relations are calculated;
- the final block generates the protection decisions, which are based on comparing of calculated criteria values with the pre-set thresholds or other characteristics.

One has to remark that the relays of previous generations did not actually measure particular signals. Their operations were based on comparators, which only decided, if the given quantity was smaller or greater than the operational threshold. In cases of digital relays the signals are actually measured and the comparison with the thresholds come afterwards [5–9].

Starting from around 1985 the digital relays became dominant in the offer of manufacturers. It is so, because they offer a number of advantages:

- integration of functions,
- further reduction of power derived from secondary terminals of CTs and VTs,
- reduction of secondary cabling,
- complex algorithms, which process digital signals using values of the samples,
- increased speed of operation,

Table 3.1 Enhancements in processing power of digital systems

Year	Memory RAM/EPROM	Data bus width (bits)	Calculation speed (MIPS)
1986	64k/128k	16	0.5
1992	256k/512k	16	1.0
1999	512k/4 MB +4 MB D-RAM	32	35
2004	16 MB Flash +12 MB D-RAM	32	80

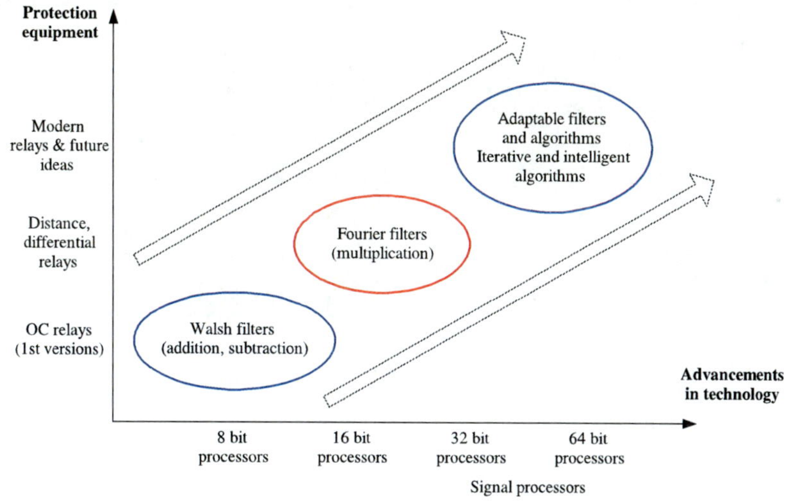

Fig. 3.2 Development of protection equipment vs. advancements in technology

- facility of communication,
- self-testing capabilities.

Realization of more and more complex measurements and other protection functions have become possible thanks to advancements in digital technology. Better and more powerful processors (Table 3.1) enable implementation of more sophisticated methods and algorithms (Fig. 3.2), including also self-testing and communication with the environment. The integration of more and more protection functions in one cubicle has become a common practice. Examples of feeder protection relays ($Z<$, $I>$ directional) of three generations can be seen in Fig. 3.3.

More and more efficient digital processing of relaying signals as well as continuous progress in telecommunication enabled coming of the era of 4th generation of protective devices. It is the era of Wide-Area Measurement and Protection Systems (WAMPS), which combine sophisticated digital processing with the fast and reliable exchange of information via telecommunication links [1, 3, 14].

The 4th generation has the following properties:

- wide area measuring of signals and transfer of the results to the decision-making points,

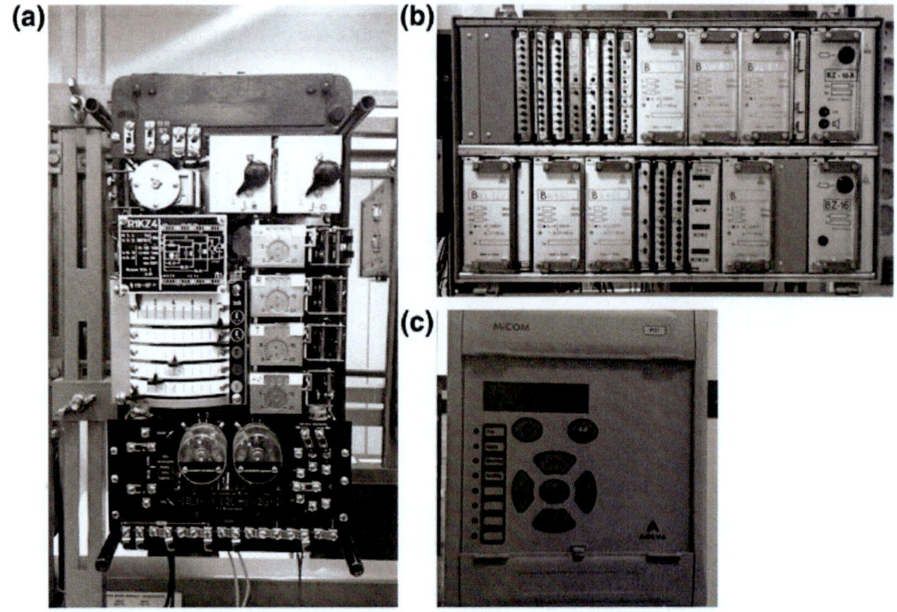

Fig. 3.3 Comparison of three generations of protective equipment ($Z<$, $I>$ relays): **a** electromechanical (R1KZ4, without housing), **b** static, **c** digital (MICOM P127) (pictures taken by the author in protection lab at the WUT, Wroclaw, Poland)

- integration of protection, control, monitoring and measurements,
- adaptability to the existing conditions,
- intelligent decisions, estimation of actual conditions and possible consequences of wrong decisions.

The basic scheme of the 4th generation protection device is shown in Fig. 3.4. Vital place in its operation plays the communication, which can be realized via wires, high-frequency radio signals or optical fibers. It is essential that protection devices of various vendors communicate and can understand each other. The latter issue is solved by introduction of common communication standards (e.g. the IEC-61850 for communication within a substation) [4].

The 4th generation came in timely, because the philosophy of relaying has recently slightly changed. Previously the principal requirement was to assure reliable and fast protection of given power system apparatus. Therefore the relays were object-oriented.

Nowadays the priorities of relaying are slightly shifted. The first one is protection of the power system against developing disturbances, which could lead to blackouts. Although the need to protect given power system component is still fundamental, however, the relays ought to be system-oriented. Therefore, undesirable tripping may be considered as dangerous as the delayed tripping of the fault.

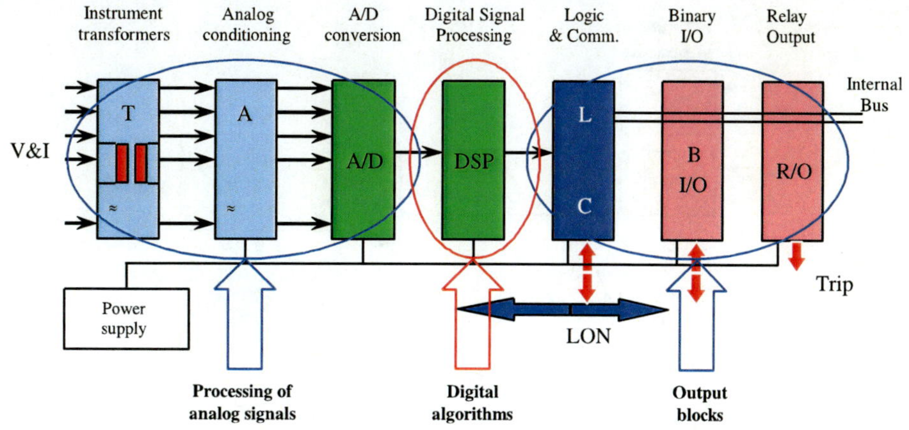

Fig. 3.4 Structure of a digital relay of the 4th generation

Within this philosophy the system-wide protection and control has emerged, with the following (exemplary) functions:

- adaptive protection system with adjustment to various network topologies,
- wide-area differential protection embracing network objects over a number of voltage levels,
- global (centralized) decision-making based on exchange of information,
- short-circuit detection and location at the relaying center,
- application of adaptive protection settings,
- additional functions (monitoring of system stability, connection with substation automation systems, etc.).

An example of such a structure for system stability monitoring and control is shown in Fig. 3.5. The information from Phasor Measurement Units installed at selected places in the network is processed to evaluate the safety margin. In case of low level of this margin certain control actions can be performed, with the aim of reversing the process and maintaining system stability [2, 11].

All the advantages of 3rd and 4th generation of protective devices require efficient digital processing of the input signals. Both theoretical basics and specific algorithms for signal conditioning, calculation of criteria values and decision-making are provided in the following sections of this book.

3.2 Functional Blocks of a Digital Protection Device

3.2.1 Analog Antialiasing Filtering

Before sampling the input signals go through the antialiasing low-pass filters that are used to pass all the components with frequencies lower than f_k (highest

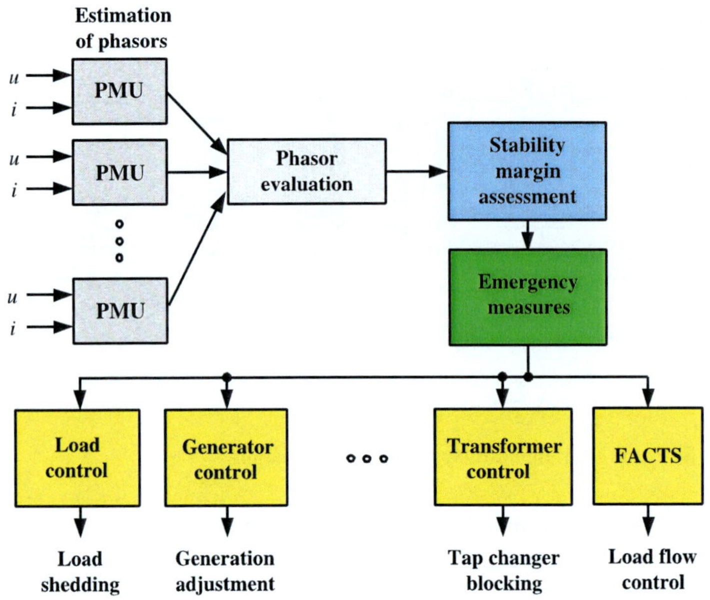

Fig. 3.5 Example of WAMPS functions for maintaining system integrity

frequency of the component important for protection operation) and to eliminate frequencies higher than that. The cut-off frequency of the filter f_C ought to meet the following requirements [15]:

$$f_k < f_C \leq \frac{f_S - f_k}{3}, \qquad (3.1)$$

where f_S is a sampling frequency.

This would make all the components with frequencies lower than f_k pass with the minimal distortion, while the components with frequencies greater than that would be suppressed. The high-side limitation for the cut-off frequency results from the repercussions of the sampling process, discussed in the next section.

A general picture of the low-pass filter frequency response is presented in Fig. 3.6. The filter should have possibly flat characteristic in both pass and reject regions (possibly small values of the errors ε_1 and ε_2). The analog input filters are in most cases implemented in form of cascade RC circuits. Required slope of the transition part of filter spectrum is obtained by choosing appropriate filter order, which means that respective number of the first order RC four-poles are connected in series.

Considering that the relay operation is based on the level of the fundamental frequency component ($f_k = 50$ Hz) of the voltage or current signal and the sampling frequency is $f_S = 400$ Hz, the required filter cut-off frequency is $f_C = 117$ Hz. If the relay (e.g. transformer protection) has to evaluate the level of the fifth harmonic ($f_k = 250$ Hz), and the sampling frequency is $f_S = 1600$ Hz (note that the sampling rate resulting from (3.1) should be higher than 1000 Hz, plus some reasonable margin), the filter cut-off frequency should be $f_C = 450$ Hz.

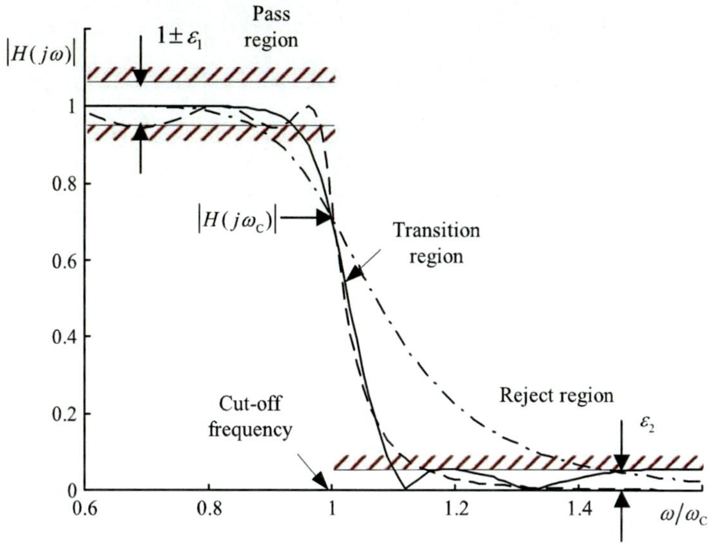

Fig. 3.6 Frequency response of a low-pass filter

For the first order RC four-pole its transfer function in frequency domain amounts:

$$H(j\omega) = \frac{\frac{1}{\omega C}}{\frac{1}{\omega C} + R} = \frac{1}{\omega RC + 1}. \tag{3.2}$$

From the very definition of cut-off frequency:

$$|H(j\omega_C)| = \frac{1}{\sqrt{2}} \tag{3.3}$$

it results that

$$RC = \frac{1}{\omega_C}, \tag{3.4}$$

thus, for such a filter one can easily select the values of R and C. The frequency response of the filter is shown in Fig. 3.7. One can see that the characteristic is far from ideal, which implies utilization of higher order filters (min. 2nd order), having both higher slope in the transient region and better filtering properties in pass and rejection areas, see Fig. 3.8 for the third order filter.

3.2.2 Sampling Process and A/D Conversion

If the input signal is a current in transient state caused by a short-circuit, it may be represented in the form:

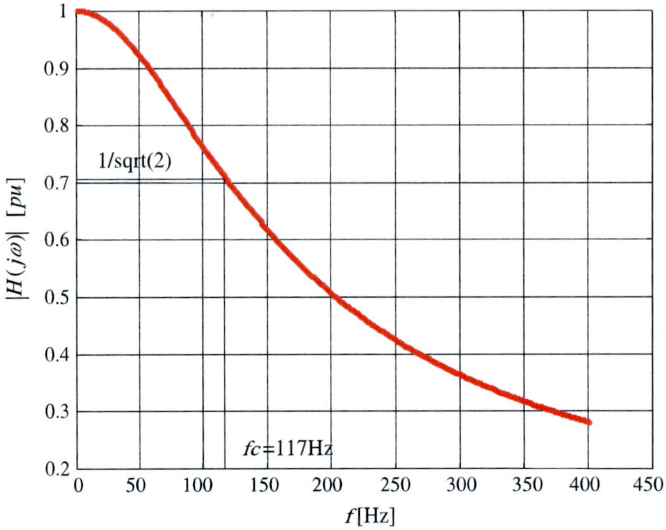

Fig. 3.7 Frequency response of the 1st order RC filter

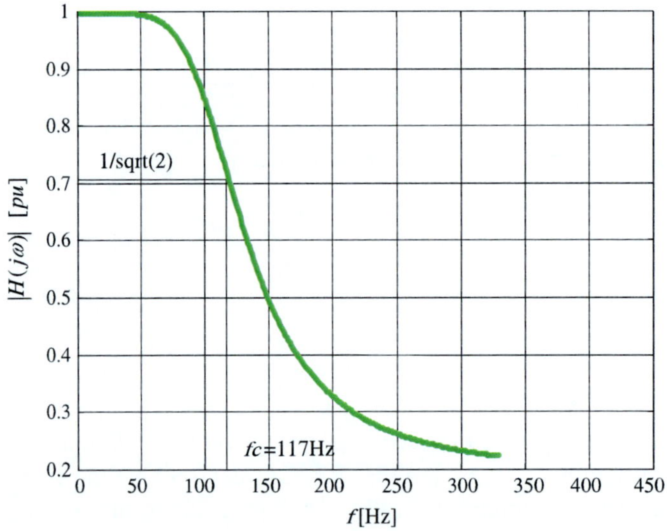

Fig. 3.8 Frequency response of the 3rd order RC filter

$$i(\tau) = I_{1m}\cos(\omega_1\tau - \alpha_1) + I_a \exp\left(-\frac{\tau}{T_a}\right) + \sum_{k=2}^{p} I_{km}\cos(k\omega_1\tau - \alpha_k)$$
$$+ \sum_{h=1}^{r} I_{hm}\left[\exp\left(-\frac{\tau}{T_h}\right)\right]\cos(\omega_h\tau - \beta_h). \quad (3.5)$$

One may observe that the signal model consists of several terms, including the fundamental frequency, decaying DC, higher frequency and transient oscillatory decaying components, where:

I_{1m} is a magnitude of the fundamental frequency component, I_a, T_a are initial value and time constant of the decaying DC component, I_{km} are magnitudes of higher frequency components, I_{hm}, T_h are magnitudes and time constants of transient oscillatory decaying components.

The useful information is usually contained in the fundamental frequency component, and sometimes in the selected higher frequency components (2nd, 3rd or 5th). Therefore, all the other components are to be rejected. Particularly dangerous are components with very high frequencies (close to the sampling frequency) since they may cause irreversible corruption of the digital signal.

Therefore, selection of the sampling frequency f_S is a compromise. It must not be too low, to enable reproduction of the components that are vital for the relaying decisions. On the other hand, it must not be too high, to avoid unnecessary burden for the digital processing.

Basically, the minimum sampling frequency results from the Shannon–Kotielnikov theorem that defines conditions for possibility of signal reconstruction after sampling. According to that, there should be at least two samples of the signal taken within the period of the signal component that should be represented in digital form without loss of information about frequency [10, 15].

If the component that should be reproduced correctly has the frequency f_k then the sampling frequency ought to be:

$$f_S \geq 2f_k. \tag{3.6}$$

In real installations the sampling frequency is seldom lower than 800 Hz (16 samples per one period of the fundamental frequency 50 Hz component, 4 samples per one period of the 4th harmonic, etc.). Contemporary digital protection relays offer sampling rates up to several kHz. Special solutions, where higher frequency components are used for generating the trip decision, may have sampling rates in the range of many hundreds of kHz [12, 13].

The sampling process leading to extraction of the sampled values in time domain has also significant consequences in frequency domain. One can prove that the spectrum of an analog signal becomes duplicated after signal sampling. The spectrum of sampled signal is a sum of copies of original spectrum shifted left and right by a multiple of sampling frequency (or angular frequency), according to:

$$X^*(j\omega) = \frac{1}{T_S}\sum_{k=-\infty}^{\infty}\left\{\int_{-\infty}^{\infty} x(t)\exp[-j(\omega - k\omega_S)t]\mathrm{d}t\right\} = \frac{1}{T_S}\sum_{k=-\infty}^{\infty} X[j(\omega - k\omega_S)]. \tag{3.7}$$

This feature of digital spectrum is illustrated in Fig. 3.9. One can conclude that the analog signal can be reconstructed from its samples (by using an appropriate

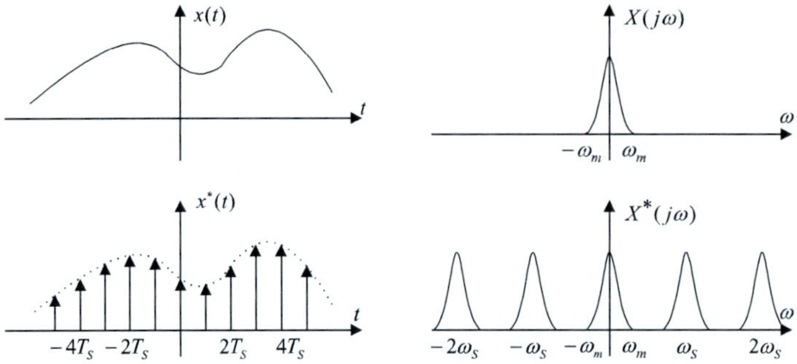

Fig. 3.9 An illustration of sampling process in time and frequency domains

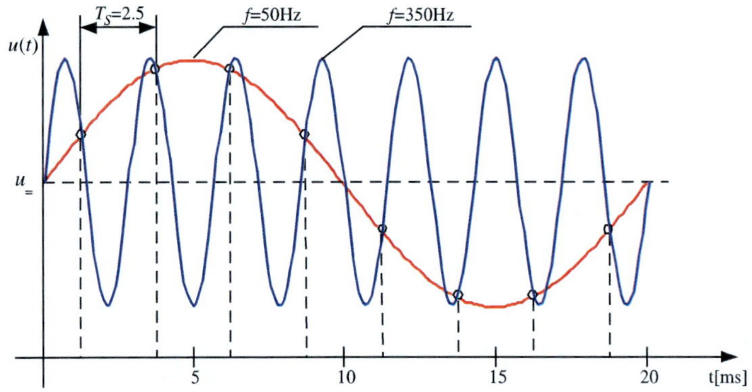

Fig. 3.10 Sampling of the fundamental and (sampling-fundamental) frequency signals

low-pass filter) only when the copies of original spectrum are separated one from another, which coincides with the Shannon–Kotielnikov sampling theorem.

Another consequence of sampling can be seen in Fig. 3.10. It can be observed that for the sampling period $T_S = 2.5$ ms, i.e. for the sampling frequency $f_S = 400$ Hz (8 samples per cycle of the fundamental 50 Hz component), both the 50 Hz signal and its reflected frequency component ($400 - 50 = 350$ Hz) after sampling give exactly the same sampled values, in other words—they become indistinguishable. If the signal contains both frequency components, then depending on their phases the result of sampling would be a weighted sum/difference of both components. This is why the analog filtering is necessary—to filter out the all components of frequencies higher than half of the sampling rate, and thus eliminate the aliasing effects.

After sampling the signal becomes discretized in time, however, the values remain still analog. For further processing they have to be represented in digital form with finite number of digits, which is a role of an A/D converter. In modern protective relays one A/D converter is used even if many channels for numerous signals have to be processed. The A/D converter is then switched consecutively to subsequent channels with use of a multiplexer and analog memory (usually a capacitor bank), as shown in Fig. 3.11.

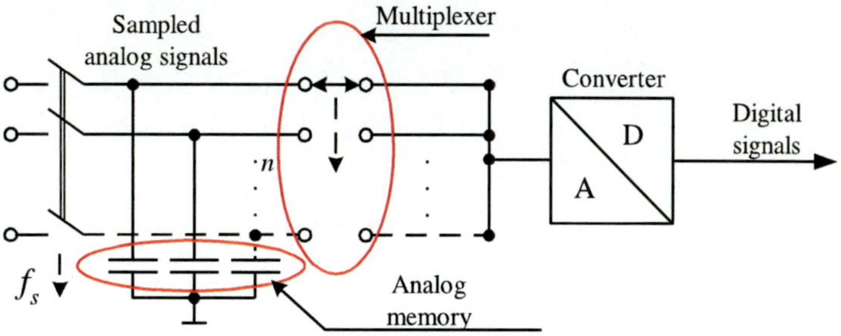

Fig. 3.11 A/D converter with analog memory and multiplexer

The A/D converters are characterized by two basic parameters:

- number of bits m (called also the word length of A/D converter, $m = 8\ldots16$),
- converter range M.

The converter resolution is a derivative of the parameters m and M. The grain of conversion is defined as:

$$d = \frac{M}{2^m - 1}. \tag{3.8}$$

From (3.8) results that higher resolution (lower grain) is obtained for higher number of bits m. When the signal changes in wider range the resolution for the same number of bits gets lower. That is why the A/D converters for current signals must have higher number of bits ($m = 12\ldots16$), since fault currents are significantly higher than normal operation (load) levels.

If one wants to maintain the relative error of A/D conversion ε (related to the smallest expected signal value X_{min}) at given level, then the minimum required number of bits should obey the relationship:

$$m \geq \log_2\left(\frac{0.5M}{\varepsilon X_{min}}\right) + 1. \tag{3.9}$$

3.2.3 Digital Signal Processing

The protection operation related to determination of state of the protected plant (faulty or healthy) and issuing the final decision is based on digital processing of sampled input signals, with the aim of calculation of the respective criteria values.

A great number of protection criteria measurement algorithms have been developed so far, that can be roughly grouped into four families, as illustrated in Fig. 3.12. One can see that the first step of signal processing is usually filtering out of signal components that are hoped to bear the information about the protected plant state. The other components are treated as noise and should be rejected or at

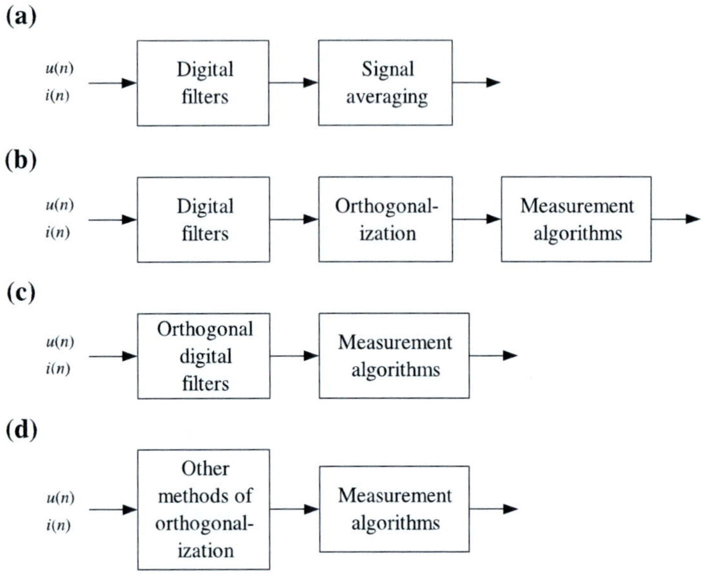

Fig. 3.12 Various ways of protection criteria measurement

least significantly suppressed. Then the criteria values can be calculated either by signal averaging or by using of measurement algorithms based on orthogonal components of the signals [15].

The orthogonalization process is meant as calculation/extraction of two mutually orthogonal components of the sinusoidal signal. It can be performed by special methods of orthogonalization (e.g. by time delay) or by taking two orthogonal filters. The correlation method, discrete Fourier transformation, Kalman filtering, and others can also be used for the purpose.

The filter types, their design and analysis are thoroughly described in Chaps. 5 and 6. The details related to measurement procedures and algorithms can be found in Chaps. 8 and 9.

Measured values are further compared with operational limits/protection thresholds or characteristics to reach the final protection decision, which is described in Chap. 10.

3.3 Hierarchical Structure of Protection and Control

Contemporary protection relays are multifunctional universal devices, realizing a lot of additional functions (metering, event/fault recording…), offered as standard even with smallest relays. Power quality monitoring is only partly covered by protection relays (registration of voltage dips longer than 10 ms and harmonics up to the 5th or 10th order—at least; more PQ functions expected to arrive in coming future). Complete protection of a power system component can now be provided by only a few integrated relays (e.g. 3 relays, each with about 15 protection functions, needed for protection of a larger generating unit).

Most of the present-day relays are combined protection and control units. The protection devices provide control of many circuit breakers, function of key-operated switching authority, feeder control diagram, status indication of feeder devices at graphic display, measured-value acquisition, signal and command indications, P, Q, cos φ and meter-reading calculation, measured-value recording, event logging, switchgear statistics, switchgear interlocking, etc.

There exist a clear tendency towards:

- *Substation Integration*—integration of protection, control and data acquisition functions into a minimal number of platforms to reduce capital costs, reduce panel and control room space and eliminate redundant equipment and databases.
- *Substation Automation*—deployment of substation and feeder operating functions and applications ranging from SCADA and alarm processing to integrated volt/var control in order to optimize the management of capital assets and enhance operation and maintenance efficiencies with minimal human intervention.

One can observe a process of converting the substation structure of *traditional protection and substation control* being a conglomeration of often totally different devices (relays, meters, switchboards and RTUs), constituting a centralized solution with extensive parallel wiring, control with mimic display, pushbuttons, position indicators, interposing relays, local/remote switch, into a modern *coordinated protection and substation control* system characterized by:

- decentralized protection and control structure,
- a few multifunctional, intelligent devices of uniform design,
- a few serial links used instead of extensive parallel wiring,
- control of the substation takes place with menu-guided procedures at a central Visual Display Unit workplace.

Today's trends in substation automation technology is related to substation (local) control as well as remote control. Modern integrated protection and control substations apply a lot of Intelligent Electronic Devices (IEDs) that can build an open communication systems, operated centrally with a PC, with communication to higher level control centre [9].

In wider sense the structure of power system control is a hierarchical structure with vertical connections, depicted schematically in Fig. 3.13. Modern protection and control systems within substations offer also horizontal links, utilizing the common data bus for extensive data exchange between feeder/object level devices (Fig. 3.14).

The substation IEDs are either LAN enabled or are connected to the data bus through the Network Interface Module devices. A lot of information is stored in the Data Concentrator. The substation data can be remotely accessed from outside, time referenced (GPS) and used in SCADA applications. The operational and

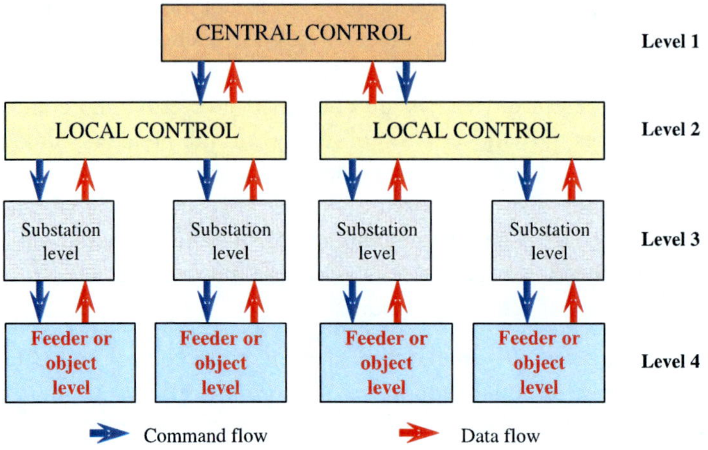

Fig. 3.13 Hierarchical structure of protection and control

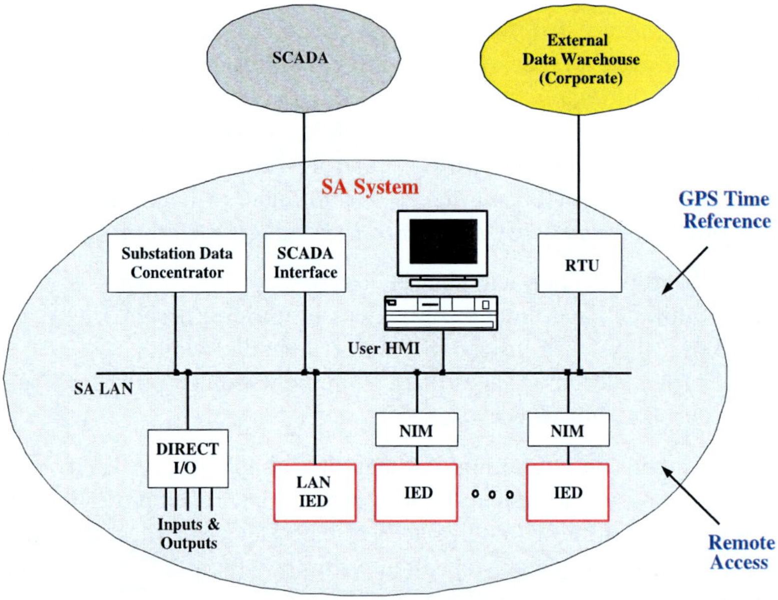

Fig. 3.14 System functional architecture diagram

non-operational data can also be transferred to the corporate Data Warehouse that enables users to access substation data, providing necessary firewall to substation control and operation functions.

Summarizing this chapter one can say that:

- modern media and cost pressure are the diving force for system integration and automation in substations,

- expected further progress in data acquisition, processing and storage capability will allow further upgrade of protection functions and seamless monitoring of load, fault events and switchgear state,
- wide-band communication LANs and Internet technology will make the information available at any place of the enterprise,
- access for operation and diagnosis will surely be soon possible from any space via mobile communication,
- if applicable standards (e.g. IEC 81650) and tools are available the idea of "plug and play" in SA systems may soon come to reality.

References

1. Adamiak MG, Novosel D, Kasztenny B, Madani V, Sykes J, Phadke AG (2006) Wide area protection and control—today and tomorrow. IEEE Publ No 0-7803-9193-4/06
2. Benmouyal G, Schweitzer EO, Guzman A (2002) Synchronized phasor measurement in protective relays for protection, control, and analysis of electrical power systems. In: Proceedings of the western protective relay conference, Spokane, Washington
3. Bertsch J, Karlsson D (2005) Wide-area protection and power system utilization. Proc IEEE 93:997–1003
4. Brand KP, Brunner C, Wimmer W (2004) Design of IEC 61850 based substation automation systems according to customer requirements. In: Proceedings of CIGRE Session, Paris: B5-103
5. Clemens C, Rothe K (1991) Schutztechnik in Elektroenergiesystemen. VDE Verlag, Berlin
6. Elmore WA (1994) Protective relaying—theory and applications. Marcel Dekker Inc., New York
7. Horowitz SH, Phadke AG (1988) Power system relaying. Research Studies Press, Taunton
8. Johns AT, Salman SK (1995) Digital protection for power systems. P. Peregrinus Ltd., on behalf of IEE, London
9. Lin X, Li Z, Wu K, Weng H (2009) Principles and implementations of hierarchical region defensive systems of power grid. IEEE Trans Power Deliv 24:30–37
10. Marks JR (1991) Introduction to Shannon sampling and interpolation theory. Springer, New York
11. Phadke AG, Thorp JS (2008) Synchronized phasor measurements and their applications. Springer Science & Business Media, LLC, New York
12. Power engineering guide (2010) 6th edn. Siemens AG, Erlangen, Germany
13. Protective relays application guide (1987) 3rd edn. AREVA T&D Protection and Control, UK
14. Seethalekshmi K, Singh SN, Srivastava SC (2008) Wide area protection and control: present status and key challenges. In: Proceedings of the 15th national power systems conference, IIT Bombay, pp 169–175
15. Ungrad H, Winkler W, Wiszniewski A (1995) Protection techniques in electrical energy systems. Marcel Dekker Inc., New York

Chapter 4
Fundamentals of System Analysis and Synthesis

4.1 Introduction

Synthesis of digital power system protection systems is possible with application of many advanced and complex mathematical tools. Understanding of digital systems, digital signal processing, analysis and synthesis of digital filters require transformations between time and frequency domain as well as transitions between continuous and discrete data and systems. All of them belong to the family of methods being a part of signal and system theory and use Fourier, Laplace and *Z* transforms as well as Fourier series as the most important mathematical basis.

The aim of this chapter is an introduction and short description of these fundamental tools to make studying following parts of this book easier for the reader. The chapter presents the very basic information, provided without any proof and mathematical precision, since it is impossible within the book devoted to applications. Even this background is quite wide, including: Fourier series, Fourier transform, Laplace transform, *Z* transform, Fourier transform with discrete time as well as discrete Fourier transform (DFT).

Different methods of signals and systems descriptions are related to some Fourier transformations and resulting options—Laplace and *Z* transforms. The usage of given transform depends on signal models, parameters and technical applications at hand. The following transforms are the most important in signals and systems analysis and synthesis:

(a) *Fourier series* (complex Fourier series) that can be used to represent continuous periodic signals belonging to class of power signals. A power signal is such one that has infinite energy

$$E = \int_{-\infty}^{\infty} x^2(t)\mathrm{d}t \Rightarrow \infty \quad \text{but finite power} \quad P = \lim_{T\to\infty} \frac{1}{2T} \int_{-T}^{T} x^2(t)\mathrm{d}t < \infty.$$

W. Rebizant et al., *Digital Signal Processing in Power System Protection and Control,* Signals and Communication Technology,
DOI: 10.1007/978-0-85729-802-7_4, © Springer-Verlag London Limited 2011

(b) *Fourier integral* that can be applied to analyze aperiodic (decaying) signals which are also called energy signals, i.e. which have limited energy

$$E = \int_{-\infty}^{\infty} x^2(t)\mathrm{d}t < \infty.$$

The Fourier integral can also be applied to periodic signals being power signals when impulse Dirac function is used. It is said that Fourier integrals of these signals exist in distribution meaning. The so called sifting feature of delta function is utilized here, which can be shortly expressed in the following way:

$$\int_{-\infty}^{\infty} x(t)\delta(t-\tau)\mathrm{d}t = x(\tau).$$

(c) *Laplace transform* that is applied for continuous and sub-continuous signals as well as for solving of the differential and partial differential equations which arise in many engineering problems and describe the majority of linear time invariant dynamical systems. The Laplace transform is closely related to the Fourier transform that is equivalent to evaluating the bilateral Laplace transform with complex argument $s = j\omega$. This relationship between the Laplace and Fourier transforms is often used to determine the frequency spectrum of a signal or dynamical system.

(d) *Z transform* that delivers frequency domain representation of a discrete time domain signal, being a sequence of real or complex numbers. The Z transform can be considered as a discrete equivalent of the Laplace transform.

(e) *Fourier transform of the discrete signal* that is applied for analysis of signals with non-zero values only at discrete (usually equi-distant) time instants. The signals can be treated either directly as discrete or as being an effect of continuous signals sampling. This transformation is in a sense equivalent to the complex Fourier series that is used for analysis of periodic signals, while similarity refers here to periodicity in frequency domain.

(f) *Discrete Fourier Transform* and its fast version—*Fast Fourier Transform* that can be applied for discrete signals with limited number of samples $0 < n < N$. Utilization of DFT is possible for aperiodic signals only and its result is also a periodic function in frequency domain. Unlike traditional Fourier transform where discrete time domain signal is transformed into continuous frequency domain function, in case of the DFT both signal and its transform have discrete representation in time and frequency domains, respectively.

4.2 Fourier Series

Let there be given a periodic time function (signal) having a period of T_0, i.e. such that $x(t + kT_0) = x(t)$, $k \in C$. The function can be represented by infinite series called Fourier series according to the following equation [2, 9]:

$$x(t) = \sum_{n=0}^{\infty} [a_n \cos(n\omega_0 t) + b_n \sin(n\omega_0 t)], \tag{4.1}$$

where

$$a_n = \frac{2}{T_0} \int_0^{T_0} x(t) \cos(n\omega_0 t) \mathrm{d}t,$$

$$b_n = \frac{2}{T_0} \int_0^{T_0} x(t) \sin(n\omega_0 t) \mathrm{d}t,$$

$$a_0 = \frac{1}{T_0} \int_0^{T_0} x(t) \mathrm{d}t,$$

$$\omega_0 = \frac{2\pi}{T_0}.$$

It can be noted here that above signal equivalent is obtained with minimization of mean square error and that the set of functions sin, cos is a complete set of orthogonal functions. There are also other sets of orthogonal functions, for instance Walsh, Haar and others which can be used as well for analysis of periodic signals.

The formula of Fourier series (4.1) can be expressed multifold. If one uses the Euler substitution:

$$\cos(\alpha) = \frac{1}{2}[\exp(j\alpha) + \exp(-j\alpha)],$$

$$\sin(\alpha) = \frac{1}{2j}[\exp(j\alpha) - \exp(-j\alpha)],$$

where $j = \sqrt{-1}$ is an imaginary operator, then after simple rearrangements the complex Fourier series is yielded, in the form:

$$x(t) = \sum_{n=0}^{\infty} c_n \exp(jn\omega_0 t), \tag{4.2}$$

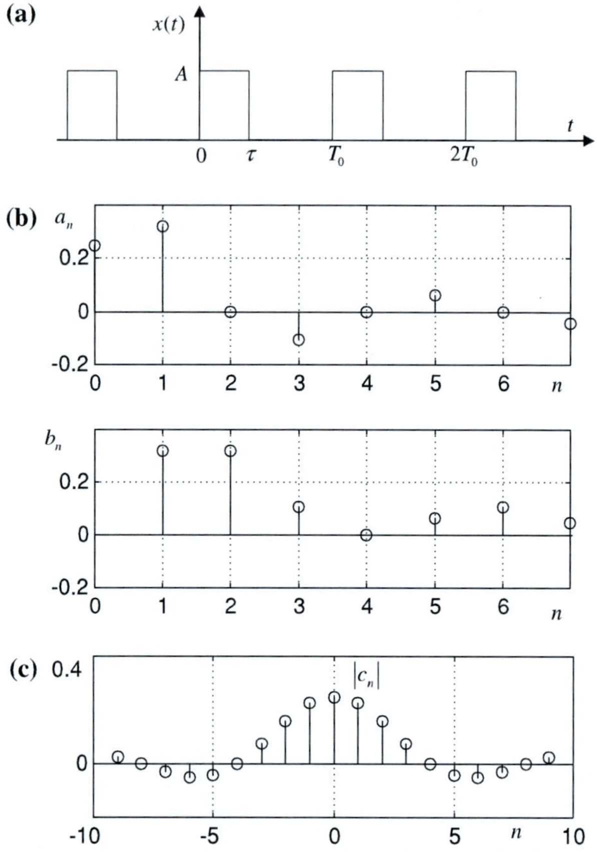

Fig. 4.1 Periodic function **a**, its Fourier series coefficients **b** and absolute values of the complex Fourier series coefficients **c**

where

$$c_n = \frac{1}{2}(a_n - jb_n) = \frac{1}{T_0}\int_0^{T_0} x(t)\exp(-jn\omega_0 t)\mathrm{d}t.$$

Example 4.1 Determine the coefficients of Fourier series as well as complex Fourier series for a periodic time function shown in Fig. 4.1a.

Solution From Eq. 4.1 one obtains:

$$a_n = \frac{2}{T_0}\int_0^{\tau} A\cos(n\omega_0 t)\mathrm{d}t = \frac{A}{n\pi}\sin\left(2\pi n\frac{\tau}{T_0}\right),$$

$$b_n = \frac{2}{T_0}\int_0^{\tau} A\sin(n\omega_0 t)\mathrm{d}t = \frac{A}{n\pi}\left[1 - \cos\left(2\pi n\frac{\tau}{T_0}\right)\right],$$

$$a_0 = \frac{2}{T_0}\int_0^{\tau} A\,\mathrm{d}t = A\frac{\tau}{T_0}.$$

Graphical presentation of the calculated coefficients as a function of relative angular frequency for the signal magnitude A equal unity and $\tau/T_0 = 0.25$ is shown in Fig. 4.1b.

Further, according to Eq. 4.2 one obtains:

$$c_n = \frac{1}{T_0}\int_0^{T_0} A\,\exp(-jn\omega_0 t)\mathrm{d}t = \exp\left(-jn\pi\frac{\tau}{T_0}\right)\frac{A}{n\pi}\sin\left(n\pi\frac{\tau}{T_0}\right).$$

It is seen that the coefficients are complex numbers defining in an integrated way both magnitudes and arguments of particular signal components. The absolute values of c_n represent magnitudes of given frequency components:

$$|c_n| = \frac{A}{n\pi}\left|\sin\left(n\pi\frac{\tau}{T_0}\right)\right|,$$

with

$$\lim_{n\to 0}|c_n| = A\frac{\tau}{T_0}$$

that are shown graphically as a function of frequency in Fig. 4.1c. The graph was prepared for $A = 1$ and $\tau/T_0 = 0.25$, similarly as in previous example. Introducing calculated values c_n to (4.2) one gets the complex Fourier series of the considered signal:

$$x(t) = \frac{A}{\pi}\sum_{n=-\infty}^{\infty}\frac{1}{n}\sin\left(n\pi\frac{\tau}{T_0}\right)\exp[jn\omega_0(t-\tau/2)].$$

4.3 Fourier Transform

Non-periodic signals can be represented in frequency domain as well. One of many transitions between Fourier series and Fourier transform (integral) rely on representing non-periodic functions as periodic ones with their period approaching infinity. This transition causes that instead of samples in frequency domain (as in Fourier series) one obtains continuous spectra. Without discussing any particular assumptions concerning the transition it can be stated that Fourier transformation concerns energy signals, i.e. satisfying the condition:

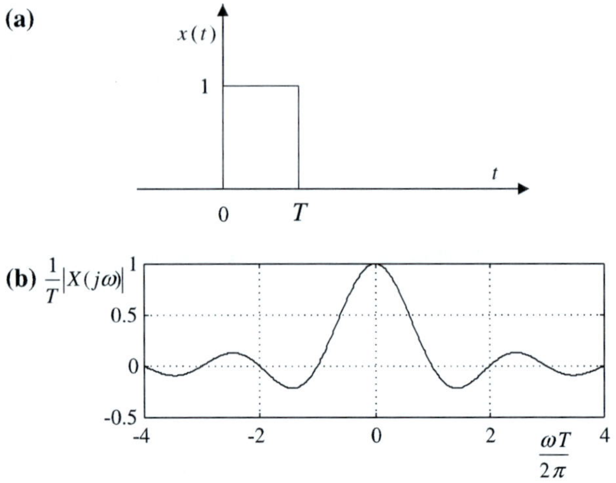

Fig. 4.2 Signal **a** and its Fourier transform **b**

$$\int_{-\infty}^{\infty} |x(t)| \mathrm{d}t < \infty.$$

Whatever the way to get Fourier transform a pair of Fourier integrals allows to go from time to frequency domain and the other way round, [5, 8]. The integrals describing simple and inverse transformations are as follows:

$$X(j\omega) = \int_{-\infty}^{\infty} x(t) \exp(-j\omega t) \mathrm{d}t, \tag{4.3}$$

$$x(t) = \frac{1}{2\pi} \int_{-\infty}^{\infty} X(j\omega) \exp(j\omega t) \mathrm{d}\omega. \tag{4.4}$$

Example 4.2 Applying Fourier transform determine spectrum of the signal $x(t)$ that is equal 1.0 for the time range from zero to T and equal 0.0 otherwise (Fig. 4.2a). Draw the magnitude of this spectrum as a function of frequency.

Solution Application of the formula (4.3) yields:

$$X(j\omega) = \int_{0}^{T} \exp(-j\omega t) \mathrm{d}t = T \frac{\sin(p)}{p} \exp(-jp),$$

where

$$p = \frac{\omega T}{2}$$

and the magnitude of this spectrum is equal:

$$|X(j\omega)| = T\left|\frac{\sin(p)}{p}\right|$$

and has zero points at $p = k\pi$, i.e. for angular frequencies $\omega_k = 2\pi k/T$ (see Fig. 4.2b).

As it was mentioned above all periodic signals can be represented and analyzed with Fourier series, while the aperiodic ones—with use of Fourier integral. For practical applications it would be convenient and favorable if the same technique could be utilized for possibly wide class of signals. Since the Fourier series must not be applied for non-periodic signals, one may eventually try applying the Fourier integral for periodic (power) signals. Such an extension can be reached through appropriate substitution of the complex exponential function (component of the Euler's equations for sin, cos) with the delta Dirac function. Taking advantage of the sifting feature of delta Dirac function one may arrive at:

$$\int_{-\infty}^{\infty} \delta(\omega - \omega_0)\exp(j\omega t)\mathrm{d}\omega = \exp(j\omega_0 t), \tag{4.5}$$

and by comparison of (4.5) with (4.4) one gets:

$$F^{-1}\{2\pi\delta(\omega - \omega_0)\} = \exp(j\omega_0 t), \text{ i.e. } F\{\exp(j\omega_0 t\} = 2\pi\delta(\omega - \omega_0),$$

where $F\{\}$ denotes simple Fourier transform (4.3) and $F^{-1}\{\}$ stands for inverse transform (4.4).

Example 4.3 Determine the Fourier transform of the periodic function $A\ \cos(\omega_0 t)$.

Solution With application of the Euler's substitutions one obtains:

$$x(t) = A\ \cos(\omega_0 t) = \frac{A}{2}[\exp(j\omega_0 t) + \exp(-j\omega_0 t)].$$

Now, applying the formula (4.5) to both exponential components one gets the sought Fourier transform:

$$F\{x(t)\} = X(j\omega) = \pi A[\delta(\omega - \omega_0) + \delta(\omega + \omega_0)].$$

The signal and its transform are presented in Fig. 4.3.

4.4 Laplace Transform

The continuous signal x in time domain can be transformed into frequency domain by calculation of the integral:

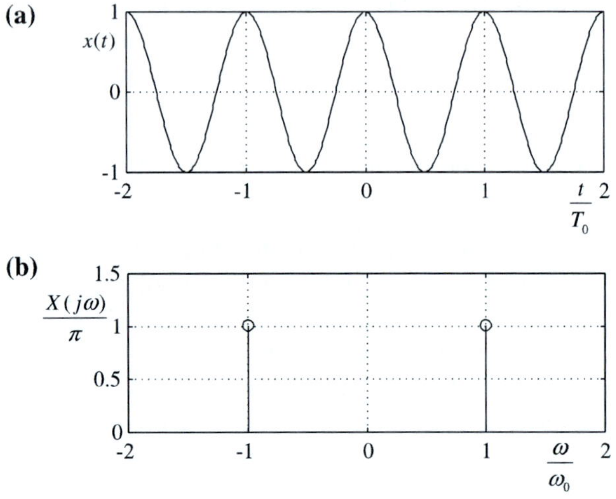

Fig. 4.3 Periodic signal **a** and its Fourier transform **b**

$$X(\sigma + j\omega) = \int_{-\infty}^{\infty} x(t) \exp[-(\sigma + j\omega)t]\mathrm{d}t. \tag{4.6}$$

The inverse transformation can be done according to:

$$x(t) = \frac{1}{2\pi j} \int_{-\infty}^{\infty} X(\sigma + j\omega) \exp[(\sigma + j\omega)t]\mathrm{d}\omega. \tag{4.7}$$

Substituting $s = \sigma + j\omega$ (i.e. $\mathrm{d}s = j\,\mathrm{d}\omega$) one obtains, [2, 3]:

$$X(s) = \int_{-\infty}^{\infty} x(t) \exp(-st)\mathrm{d}t \tag{4.8}$$

and

$$x(s) = \frac{1}{2\pi j} \int_{\sigma - j\infty}^{\sigma + j\infty} X(s) \exp(st)\mathrm{d}s, \tag{4.9}$$

which represent the Laplace transform and its inverse.

The main features of the Laplace transform that facilitate calculation for more complex signals are collected in Table 4.1, whereas the Laplace transforms of selected most commonly analyzed signals are given in Table 4.2.

Example 4.4 Applying the definition (4.8) and/or the theorems from Table 4.1 calculate the Laplace transforms of the signals: (a) $x(t) = 1(t)$, (b) $x(t) = \cos(\omega t)1(t)$.

Table 4.1 Fundamental features of Laplace transform

$y(t)$	$Y(s)$
$x(t)\exp(\alpha t)$	$X(s-\alpha)$
$x(t-t_0)$	$X(s)e^{-st_0}$
$\frac{dx(t)}{dt}$	$sX(s)-x(0)$
$\int_{-\infty}^{t} x(\tau)d\tau$	$\frac{1}{s}X(s)$
$x(t)*g(t)$	$X(s)G(s)$

Table 4.2 Important time functions and their Laplace and Z transforms

$x(t)$	$X(s)$	$X(z)$
$k\delta(t)$	k	k
$1(t)$	$\frac{1}{s}$	$\frac{z}{z-1}$
$t1(t)$	$\frac{1}{s^2}$	$\frac{zT_S}{(z-1)^2}$
$\frac{t^2}{2}1(t)$	$\frac{1}{s^3}$	$\frac{z(z+1)T_S^2}{(z-1)^2}$
$\exp(-\alpha t)1(t)$	$\frac{1}{s+\alpha}$	$\frac{z}{z-\exp(-\alpha T_S)}$
$t\exp(-\alpha t)1(t)$	$\frac{1}{(s+\alpha)^2}$	$\frac{zT_S\exp(-\alpha T_S)}{[z-\exp(-\alpha T_S)]^2}$
$\sin(\beta t)1(t)$	$\frac{\beta}{s^2+\beta^2}$	$\frac{z\sin(\beta T_S)}{z^2-2z\cos(\beta T_S)+1}$
$\cos(\beta t)1(t)$	$\frac{s}{(s^2+\beta^2)}$	$\frac{z[z-\cos(\beta T_S)]}{z^2-2z\cos(\beta T_S)+1}$
	$\frac{1-\exp(-sT_S)}{s}X(s)$	$\frac{z-1}{z}Z\left\{\frac{X(s)}{s}\right\}$

Solutions

(a) From the definition (4.8) one obtains

$$X(s)=\int_{-\infty}^{\infty} 1(t)e^{-st}\,dt=\int_{-\infty}^{\infty} e^{-st}\,dt=-\frac{1}{s}e^{-st}\Big|_0^{\infty}=0-\left(-\frac{1}{s}\right)=\frac{1}{s}.$$

(b) Expanding the cos function and using the features from Table 4.1 one may derive:

$$X(s)=L\left\{\frac{1}{2}\left(e^{j\omega t}+e^{-j\omega t}\right)1(t)\right\}=\frac{1}{2}\left(\frac{1}{s-j\omega}+\frac{1}{s+j\omega}\right)=\frac{s}{s^2+\omega^2}.$$

Table 4.3 Fundamental features of Z transform

$y(t)$	$Y(z)$
$\exp(\alpha t)x(t)$	$X\{z\exp(-\alpha T_S)\}$
$tx(t)$	$-zT_S\frac{\mathrm{d}}{\mathrm{d}z}\{X(z)\}$
$x(t-mT_S)$	$z^{-m}X(z)+z^{-m}\sum_{k=1}^{m}x(-kT_S)z^k$
$x(t-mT_S)1(t-mT_S)$	$z^{-m}X(z)$
$x(t+mT_S)$	$z^mX(z)-z^m\sum_{k=0}^{m-1}x(kT_S)Z^{-k}$

4.5 Z Transform for Sampled Data Signals and Systems

Among many methods of impulse modulation of continuous signal the most important and the most popular is impulse modulation of magnitude realized with constant sampling period T_S. A mathematical model of such sampled signal can be an infinite sum of signal samples where each sample is described by its peak value and position along time axis:

$$x^*(t) = x(t)\sum_{n=0}^{\infty}\delta(t-nT_S) = \sum_{n=0}^{\infty}x(nT_S)\delta(t-nT_S). \tag{4.10}$$

Laplace transform of above sum is easy to be calculated if one remembers that delay of given signal by T_S in time domain is equivalent to multiplication of its Laplace transform by the function $\exp(-sT_S)$. Then one obtains:

$$L\{x^*(t)\} = \sum_{n=0}^{\infty}x(nT_S)\exp(-nsT_S). \tag{4.11}$$

Substituting $z = \exp(sT_S)$ one gets:

$$L\{x^*(t)\}_{z=\exp(sT_S)} = Z\{x(t)\} = X(z)$$

which further leads to the well known definition of Z transform, [3, 4, 6, 7]:

$$X(z) = \sum_{n=0}^{\infty}x(nT_S)z^{-n}. \tag{4.12}$$

The definition (4.12) allows for calculating Z transform of many time functions. Some important examples are given in Table 4.2.

In many cases of more complex signals the fundamental features of Z transform resulting from its definition may be useful. They are given in Table 4.3.

Example 4.5 Applying the definition (4.12) calculate Z transform of the step signal $x(t) = 1(t)$.

Solution

$$X(z) = \sum_{n=0}^{\infty} z^{-n} = \frac{1}{1 - z^{-1}} = \frac{z}{z - 1},$$

where for assuring series convergence $|z^{-1}| < 1$ or $|z| > 1$ or should hold.

Example 4.6 Basing on the result of Example 4.5 and the Z transform features (Table 4.3) determine Z equivalents of the signals: (a) $y(t) = t1(t)$; (b) $y(t) = \exp(at)1(t)$; (c) $y(t) = \cos(\omega_0 t)1(t)$.

Solutions

(a)

$$Y(z) = z\{t1(t)\} = -zT_S \frac{\mathrm{d}}{\mathrm{d}z}\left[\frac{z}{z-1}\right] = -zT_S \frac{(z-1) - z}{(z-1)^2} = \frac{zT_S}{(z-1)^2}.$$

(b)

$$Y(z) = Z\{\exp(at)1(t)\} = \frac{z}{z-1}\bigg|_{z = z\exp(-aT_S)} = \frac{z}{z - \exp(aT_S)}.$$

(c) Using an appropriate decomposition of the cosine function and previous result (example b) one may derive:

$$\begin{aligned} Y(z) &= Z\{\cos(\omega_0 t)1(t)\} = 0.5Z\{1(t)\exp(j\omega_0 t) + 1(t)\exp(-j\omega_0 t)\} \\ &= 0.5\left[\frac{z}{z - \exp(j\omega_0 T_S)} - \frac{z}{z - \exp(-j\omega_0 T_S)}\right] = \frac{z[z - \cos(\omega_0 T_S)]}{z^2 - 2z\cos(\omega_0 T_S) + 1}. \end{aligned}$$

The opposite task to presented above is calculation of the signal samples for given Z transform. There are various possibilities to realize the task. Three most frequently used are the following: the equivalent of power series, the fractional decomposition and the method of residua.

If Z transform can be represented as a power series given by the equation:

$$X(z) = a_0 + a_1 z^{-1} + a_2 z^{-2} + \cdots + . \tag{4.13}$$

then comparing it with the sum (4.8) results in equalities:

$$u(0) = a_0, \quad u(T_S) = a_1, \quad u(2T_S) = a_2, \text{ etc.} \tag{4.14}$$

The sampled signal can thus be written in the form:

$$u^*(t) = a_0\delta(t) + a_1\delta(t - T_S) + a_2\delta(t - 2T_S) + \cdots +, \tag{4.15}$$

which is the solution of the problem. It can be added here that there are many methods to find power series and one of them is division of numerator by denominator.

Example 4.7 The following Z transform of a signal is given:

$$X(z) = \frac{z}{z^2 - 3z + 2}.$$

Applying power series expansion through division of polynomials calculate the few first samples of the signal $x(nT_S)$.

Solution Dividing the nominator by denominator polynomials one obtains:

$$X(z) = 0 + z^{-1} + 3z^{-2} + 7z^{-3} + \cdots +$$

and thus:

$$x^*(t) = 0\delta(t) + 1\delta(t - T_S) + 3\delta(t - 2T_S) + 7\delta(t - 3T_S) + \cdots +$$

Another method for determination of reverse Z transfer is based on *fractional decomposition* and calculation of inverse transforms of all resulting fractions. One can observe that since the nominator order of $X(z)$ cannot be higher than the order of denominator $\left(\lim\limits_{z\to\infty}[X(z)] = x(0)\right)$, one should fractionalize the function $X(z)/z$, which yields the following decomposition:

$$\frac{X(z)}{z} = \sum_{k=1}^{r} \frac{a_k}{z - z_k} \quad \text{or} \quad X(z) = \sum_{k=1}^{r} \frac{a_k z}{z - z_k} \tag{4.16}$$

and thus:

$$x(nT_S) = \sum_{k=1}^{r} a_k z_k^n, \quad n \geq 0. \tag{4.17}$$

Example 4.8 With use of the fractional decomposition approach calculate the values of signal samples for the Z transform

$$X(z) = \frac{z}{z^2 - 3z + 2}.$$

Solution

$$\frac{X(z)}{z} = \frac{1}{z^2 - 3z + 2} = \frac{A}{z - 1} = \frac{B}{z - 2},$$

thus

$$A(z - 2) + B(z - 1) = 1, \quad \text{i.e. } A = -1, B = 1,$$

and

$$X(z) = \frac{-z}{z - 1} + \frac{z}{z - 2}.$$

Finally:

$$x(nT_S) = -1^n + 2^n.$$

This example was the same as the previous one, whereas the solution obtained is now given in analytical form. One can easily check that the signal samples calculated from the above equation and the ones from Example 4.7 are identical.

The last of presented methods is the *method of residua*—the method which gives the result being the sum of all elements for given single and multiple poles of Z transform. The elements of time response are calculated for single poles according to equation:

$$x(nT_S) = \sum_{k=1}^{r} x_k(nT_S) = \sum_{k=1}^{r} (z - z_k)X(z)z^{n-1}|_{z=z_k}, \tag{4.18}$$

while the elements for multiple poles are equal:

$$x_k(nT_S) = \frac{1}{(q-1)!}\left\{\frac{\mathrm{d}^{q-1}}{\mathrm{d}z^{q-1}}\left[(z-z_k)^q X(z)z^{n-1}\right]\right\}|_{z=z_k}. \tag{4.19}$$

Example 4.9 Applying the method of residua determine the signal values at sample instances for the Z transform as in Examples 4.7 and 4.8.

Solution The discussed Z transform has its poles at $z = 1$ and $z = 2$, therefore the signal samples are the following:

$$\begin{aligned} x(nT_S) &= (z-1)\frac{z}{(z-1)(z-2)}z^{n-1}|_{\text{for } z=1} + (z-2)\frac{z}{(z-1)(z-2)}z^{n-2}|_{\text{for } z=2} \\ &= -1^n + 2^n. \end{aligned}$$

This solution is identical as in Example 4.8, where it was obtained with the fractional decomposition method.

4.6 Fourier Transform of Sampled Data

The spectrum of sampled signal given by (4.10) can be derived from the very definition of Fourier transform. Since the sampled signal is non-zero at time instants $t = nT_S$ only, the integral (4.3) becomes a sum:

$$X^*(j\omega) = X(j\omega)|_{t=nT_S} = \sum_{n=-\infty}^{n=\infty} x(n)\exp(-jn\omega T_S). \tag{4.20}$$

The same result can be obtained coming out from the Z transform by applying appropriate substitution:

$$X^*(j\omega) = X(z)|_{z=\exp(j\omega T_S)} = \sum_{n=-\infty}^{n=\infty} x(n)\exp(-jn\omega T_S). \tag{4.21}$$

It is worth noticing that since $\exp(j\omega T_S) = \exp[j(\omega T_S + 2k\pi)] = \exp[j(\omega + k\omega_S)T_S]$, the substitution and the spectrum of sampled signal $X^*(j\omega)$ are periodic in frequency domain with the period equal to the sampling angular frequency. This feature is essential for determination of the sampling frequency for the signal at hand, which is thoroughly explained in Chap. 2.

The inverse transformation (from the frequency to discrete time domain) is expressed by:

$$x(n) = \frac{1}{2\pi}\int_{-\pi/T_S}^{\pi/T_S} X^*(j\omega)\exp(jn\omega T_S)\mathrm{d}\omega. \tag{4.22}$$

4.7 Discrete Fourier Transform

A starting point for determination of the equations of direct and inverse DFT can be Z transform of sampled signal $x(n)$ having limited number of N elements with all remaining elements equal to zero. The Z transform of such a signal is then given by the equation:

$$X^*(z) = \sum_{n=0}^{N-1} x(n)z^{-n}. \tag{4.23}$$

If numerical calculations are made, there is no possibility to determine the transform for all frequencies but only for certain limited, chosen set of values. It is known from sampling theorem that one turn on Z plane is equivalent to passing frequency range equal to sampling frequency. Taking that into account and substituting:

$$z = \exp(j\omega T_S) = \exp(j\Omega) \tag{4.24}$$

for a set of equally spaced angular frequencies along the unit circle:

$$\Omega_k = k\frac{2\pi}{N}, \quad k = 1\ldots N \tag{4.25}$$

and calculating frequency response at these frequencies yields the transformation in the form:

$$X(k) = \sum_{n=0}^{N-1} x(n)\exp\left(-jnk\frac{2\pi}{N}\right) \text{ where } k = 0, 1, \ldots, N-1. \tag{4.26}$$

The inverse transform is then given by:

$$x(n) = \frac{1}{N}\sum_{k=0}^{N-1} X(k)\exp\left(jnk\frac{2\pi}{N}\right) \text{ where } n = 0, 1, \ldots, N-1 \tag{4.27}$$

allowing to get the time series $x(n)$ from given frequency samples $X(k)$.

A pair of above equations describes DFT. It allows to pass between time and frequency domains and has many applications to analysis of signals and systems, to analysis and synthesis of digital filters, and others, [1, 4, 8].

Example 4.10 Determine the DFT of the signals: (a) $x(n) = 1$ for $0 \le n \le N-1$, (b) $x(n) = \cos(n(2\pi/N))$ for the same range of samples as in example (a).

Solution

(a)

$$X(k) = \sum_{n=0}^{N-1} \exp\left(-jnk\frac{2\pi}{N}\right) = \frac{1-\exp(-jk2\pi)}{1-\exp(-jk2\pi/N)}$$

and thus

$$X(k) = \begin{cases} N & \text{for } k = 0 \\ 0 & \text{for other } k \end{cases}.$$

(b) The considered signal can be decomposed as follows:

$$x(n) = \cos\left(n\frac{2\pi}{N}\right) = 0.5\left\{\exp\left(jn\frac{2\pi}{N}\right) + \exp\left(-jn\frac{2\pi}{N}\right)\right\} = 0.5\{x_1(n) + x_2(n)\}.$$

Now, calculating the DFT for both exponential components one obtains:

$$X_1(k) = DFT\{x_1(n)\} = \sum_{n=0}^{N-1} \exp\left[jn\frac{2\pi}{N}(k-1)\right] = \frac{1-\exp[-j(k-1)2\pi]}{1-\exp\left[-j(k-1)\frac{2\pi}{N}\right]},$$

which yields:

$$X_1(k) = \begin{cases} N & \text{for } k = 1 \\ 0 & \text{for other } k \end{cases} \quad \text{where } 0 \le k \le N-1$$

and analogously:

$$X_2(k) = \frac{1-\exp[-j(k+1)2\pi]}{1-\exp\left[-j(k+1)\frac{2\pi}{N}\right]}.$$

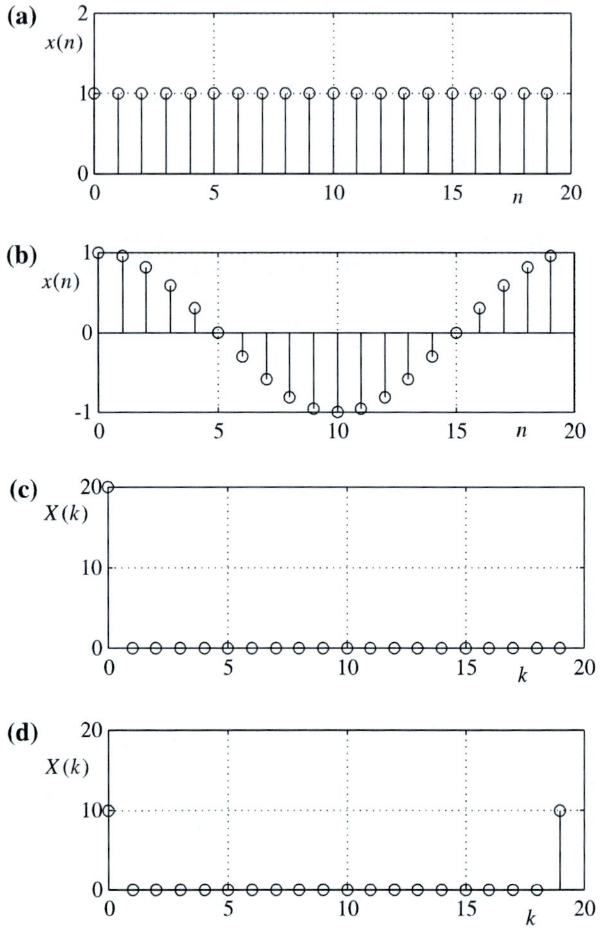

Fig. 4.4 Analyzed signals (**a**, **b**) and their DFTs (**c**, **d**) for $N = 20$

Thus

$$X_2(k) = \begin{cases} N & \text{for } k = N-1 \\ 0 & \text{for other } k \end{cases}.$$

and finally, the sought transform is equal:

$$X(k) = DFT\left\{\cos\left(n\frac{2\pi}{N}\right)\right\} = \frac{N}{2}[\delta(k-1) + \delta(k-N+1)].$$

The considered signals along with calculated DFT functions are presented in Fig. 4.4.

Example 4.11 Determine the inverse DFT for a signal with the following DFT:

$$X(k) = \begin{cases} N & \text{for } k = 0 \\ 0 & \text{for other } k \end{cases}.$$

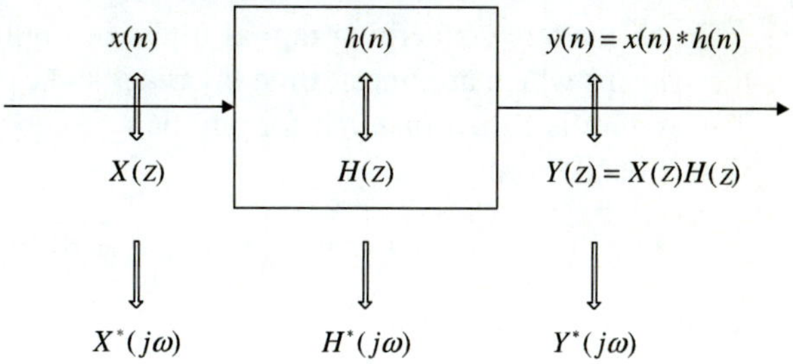

Fig. 4.5 Discrete system and signals and their relations in time and frequency domains

Solution From (4.27) the signal samples are:

$$x(n) = \frac{1}{N} \sum_{\text{for } k=0} N \exp\left(jnk\frac{2\pi}{N}\right) = 1,$$

which means that the signal is equal to unity for all n (step signal).

4.8 Description of Discrete Dynamic Systems in Time and Frequency Domains

4.8.1 Description of Discrete Systems in Time Domain

Discrete systems can be defined and described either in time or in frequency domains. These two domains are closely related according to adequate Fourier transform. To describe given system is then enough to define it in one of the domains [7].

Discrete dynamic system is defined in time domain by mathematical operation realized on its input series $x(n)$ to get the output series $y(n) = F\{x(n)\}$. This process is illustrated in the block scheme in Fig. 4.5.

Examples of such relationship could be the following equations:

$$y(n) = 0.5[x(n) + x(n-2)]$$

or

$$y(n) = [x(n) - 0.99y(n-1)].$$

Actually, there are certain simplifications in equations above and further, not important from the viewpoint of reaching their solutions. In these equations the sampling period T_S was omitted, thus instead of $x(nT_S)$ and $y(nT_S)$ the simplified $x(n)$ and $y(n)$ could be used. It may also be interpreted as a system or equation with sampling period $T_S = 1$. Such notation is usually much simpler and does not lead to misunderstanding in time domain.

In all matters concerning digital filters or digital signal processing in general, the systems which are linear, time invariant and causal are considered in this book. The system is called *linear* if for any pair of equations between input and output values of the form:

$$y_1(n) = F\{x_1(n)\} \quad y_2(n) = F\{x_2(n)\} \tag{4.28a}$$

the following relationship holds:

$$y(n) = F\{a_1x_1(n) + a_2x_2(n)\} = a_1F\{x_1(n)\} + a_2F\{x_2(n)\} = a_1y_1(n) + a_2y_2(n) \tag{4.28b}$$

It is easy to check that the system described by the equation $y(n) = 0.5[x(n) + x(n-2)]$ is linear but those described by the equations $y(n) = x^2(n)$ or $y(n) = |x(n)|$ are not.

The system is called *time invariant* when the shift of input signals by certain number of samples results in the same shift of output signals, i.e. when for $y_1(n) = F\{x_1(n)\}$ and $x_2(n) = x_1(n+m)$ one gets:

$$y_2(n) = F\{x_2(n)\} = F\{x_1(n+m)\} = y_1(n+m). \tag{4.29}$$

One can easily check that the systems described by equations $y(n) = x(n) + x(n-2)$ or $y(n) = x^2(n)$ are time invariant, however, the one given by equation $y(n) = nx(n)$ is not.

A system is called *causal* if for zero input signal for all instants before given reference instant, say $n = 0$, the corresponding output signal for those instants remains also zero. In other words in a causal system a non-zero output signal may appear not earlier but after a cause, which means that if $x(n) = 0$ for $n < 0$ then also $y(n) = 0$ for $n < 0$.

Discrete Convolution

It is easy to notice that, according to discrete signal definition as well as simple and inverse Z transform equations, any discrete signal can be written in the form:

$$x(n) = \sum_{k=0}^{\infty} x(k)\delta(n-k), \tag{4.30}$$

where

$$\delta(n) = \begin{cases} 1 & \text{for } n = 0 \\ 0 & \text{for other } n \end{cases}.$$

Then the output of any linear system can be expressed as:

$$y(n) = F\{x(n)\} = F\left\{\sum_{k=0}^{\infty} x(k)\delta(n-k)\right\} = \sum_{k=0}^{\infty} x(k)F\{\delta(n-k)\}. \tag{4.31}$$

Denoting the impulse response by $h(n)$, i.e. $h(n) = F\{\delta(n)\}$, one obtains the discrete convolution equation allowing to calculate the output signal as a function of the input signal and the system's impulse response:

$$y(n) = \sum_{k=0}^{\infty} x(k)h(n-k) \tag{4.32}$$

or symbolically

$$y(n) = x(n) * h(n). \tag{4.33}$$

If both the input signal and the system follow the causal conditions, i.e. $x(n) = 0$ and $h(n) = 0$ for $n < 0$, then the discrete convolution can be written in the form:

$$y(n) = \sum_{k=0}^{n} x(k)h(n-k) = \sum_{k=0}^{n} x(n-k)h(k) \tag{4.34}$$

Example 4.12 Applying the method of discrete convolution calculate step response of the system with the impulse response $h(n) = \exp(-n)$.

Solution In accordance with (4.32) one can derive:

$$y(n) = \sum_{k=0}^{n} \exp(-k) = \frac{1 - \exp(-n+1)}{1 - \exp(-1)}.$$

Difference Equation

Another way of description of digital transferring of an input series into an output one is the linear difference equation with constant coefficients:

$$y(n) = \sum_{k=1}^{M} a_k y(n-k) + \sum_{k=0}^{N-1} b_k x(n-k), \tag{4.35}$$

where a_k, b_k are constant factors.

To solve Eq. 4.35 M initial conditions are needed. The solution, i.e. the function of $y(n)$, can be reached multifold. One of the ways is application of the Z transform, the other is an iterative approach, that is easy when a few output samples are sought only, or when digital means are applied.

Example 4.13 Let there be given a difference equation of the form:

$$y(n) = x(n) + ay(n-1).$$

Solve the equation for the input signal that is equal unity for $n \geq 0$ and is equal zero otherwise. Assume initial condition $y(-1) = 0$.

Solution

$$\begin{aligned} n<0, &\quad y(n) = 0 \\ n = 0, &\quad y(0) = x(0) + ay(-1) = 1 \\ n = 1, &\quad y(1) = x(1) + ay(0) = 1 + a \\ n = 2, &\quad y(2) = x(2) + ay(1) = 1 + a(1 + a) \end{aligned}$$

and so on.

Iteratively one can calculate any required number of the signal $y(n)$ samples. Sometimes it is also possible to reach the solution in general form.

4.8.2 Discrete System Description in Frequency Domain

The alternative way of description of a linear discrete system is its frequency response. This is a supplementary possibility, since sometimes one may need responses in both time and frequency domains. On the other hand, one of those descriptions is enough, since one can get the other one using Fourier transform.

The frequency response of a linear system can be obtained using different methods of description in time domain, especially the difference equations and the impulse response (discrete convolution). In the first case from difference equation one can get the system transfer function first, while in the second case the impulse response is used directly.

Transfer Function Versus Difference Equation

Calculating Z transform of both sides of Eq. 4.35 and assuming zero initial conditions yields:

$$Y(z) = \sum_{k=1}^{M} a_k z^{-k} Y(z) + \sum_{k=0}^{N-1} b_k z^{-k} X(z) \tag{4.36}$$

and rearranging it the discrete transfer function is reached, in the form:

$$\frac{Y(z)}{X(z)} = H(z) = \frac{\sum_{k=0}^{N-1} b_k z^{-k}}{1 - \sum_{k=1}^{M} a_k z^{-k}}, \tag{4.37}$$

which is directly related to the difference equation given before. This transfer function can be used to find the system response to any input signal. This is done by multiplying the system discrete transfer function by given Z transform of the input signal, and calculating inverse transform of that product.

The other way of getting the discrete transfer function results from discrete convolution. It is known that convolution in frequency domain results in a product of transforms. Calculating Z transform of both sides of Eq. 4.32 gives:

$$Y(z) = H(z)X(z), \tag{4.38}$$

where $X(z)$, $Y(z)$ are Z transforms of the input and output signals, and $H(z)$ is a discrete transfer function of the system being Z transform of its impulse response:

$$H(z) = \sum_{k=0}^{\infty} h(k)z^{-k}. \tag{4.39}$$

Example 4.14 Determine the discrete transfer functions of the systems described by the following difference equations:

(a) $y(n) = ax(n)$
(b) $y(n) = x(n-1)$
(c) $y(n) = b_0x(n) + b_1x(n-1) + b_2x(n-2)$
(d) $y(n) = x(n) + a_1y(n-1) + a_2y(n-2)$

Solutions Calculating transforms of the difference equations (assuming zero initial conditions in all cases), after simple rearrangements the following transfer functions are obtained:

(a) $H(z) = a$
(b) $H(z) = z^{-1}$
(c) $H(z) = b_0 + b_1z^{-1} + b_2z^{-2}$
(d) $H(z) = \frac{1}{1-a_1z^{-1}-a_2z^{-2}}$

Frequency Response

For the input signal of a discrete system being a phasor expressed as $x(n) = \exp(jn\omega T_S)$, where $\exp(jn\omega T_S) = \cos(n\omega T_S) + j\sin(n\omega T_S)$, applying the convolution (4.32) yields:

$$\begin{aligned} y(n) &= \sum_{k=0}^{\infty} h(k)x(n-k) = \sum_{k=0}^{\infty} h(k)\exp[j(n-k)\omega T_S] \\ &= \exp(jn\omega T_S)\sum_{k=0}^{\infty} h(k)\exp(-jk\omega T_S), \end{aligned} \tag{4.40}$$

where $h(k)$ is a impulse response of the system.

The second part of (4.40) is just the frequency response, describing the output signal when the input is a phasor of given frequency. The frequency response can thus be given by:

$$H^*(j\omega) = \sum_{k=0}^{\infty} h(k)\exp(-jk\omega T_S). \tag{4.41}$$

One can notice that the same result can be reached by substituting $z = \exp(j\omega T_S)$ in the transfer function (4.39). In the same way one can obtain spectrum of a signal directly from its Z transform. It is illustrated further with another examples.

Whatever methods are used to present the frequency response one can notice that it can also be expressed in the form:

$$H^*(j\omega) = |H^*(j\omega)| \exp\{j \arg[H^*(j\omega)]\}, \tag{4.42}$$

where $|H^*(j\omega)|$ is a magnitude of $H^*(j\omega)$, and $\arg[H^*(j\omega)]$ is a argument of $H^*(j\omega)$.

Substituting that in Eq. 4.40 one obtains:

$$y(n) = |H^*(j\omega)| \exp\{j[n\omega T_S + \arg(H^*(j\omega))]\} \tag{4.43}$$

and for instance, for $x(n) = \cos(n\omega T_S)$ one gets:

$$y(n) = |H^*(j\omega)| \cos\{n\omega T_S + \arg[H^*(j\omega)]\}.$$

Sometimes, especially when magnitudes of frequency response for some frequencies differ substantially, the logarithmic frequency response defined as:

$$L(\omega) = 20\log(|H^*(j\omega)|) \tag{4.44}$$

can be used. The units of (4.44) are decibels [dB].

Example 4.15 Determine frequency responses of the systems described by the following difference equations:

(a) $y(n) = 0.5[x(n) + x(n-1)]$
(b) $y(n) = 0.5[x(n) - x(n-1)]$
(c) $y(n) = x(n) + 0.9y(n-1)$

Solutions From (4.31) results that the corresponding systems' impulse responses are:

(a) $h(n) = 0.5[\delta(n) + \delta(n-1)]$
(b) $h(n) = 0.5[\delta(n) - \delta(n-1)]$
(c) $h(n) = (0.9)^n$

Coming out either from the difference equations or from the impulse responses one can determine the transfer functions for all the cases, in the form:

(a) $H(z) = 0.5(1 + z^{-1})$
(b) $H(z) = 0.5(1 - z^{-1})$
(c) $H(z) = \frac{1}{1-0.9z^{-1}} = \frac{z}{z-0.9}$

Now, substituting $z = \exp(j\omega T_S)$ the frequency responses of the considered system are reached:

Fig. 4.6 Frequency responses of the systems from Example 4.15

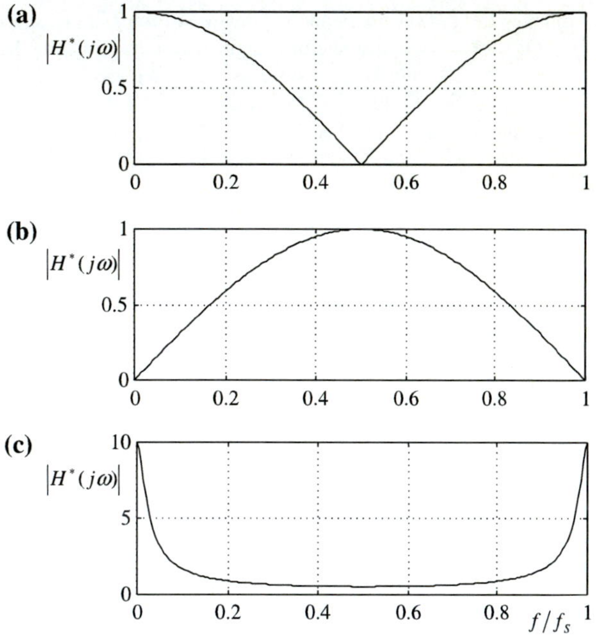

(a) $H^*(j\omega) = 0.5[1 + \exp(-j\omega T_S)] = \exp\left(-j\frac{\omega T_S}{2}\right)\cos\left(\frac{\omega T_S}{2}\right)$

(b) $H^*(j\omega) = 0.5[1 - \exp(-j\omega T_S)] = j\,\exp\left(-j\frac{\omega T_S}{2}\right)\sin\left(\frac{\omega T_S}{2}\right)$

(c) $H^*(j\omega) = \sum_{k=0}^{\infty}(0.9)^k \exp(-j\omega T_S) = \frac{1}{1-0.9\,\exp(-j\omega T_S)}$

One can observe that the first system has a low-pass filtering characteristic (in the frequency range up to half of the sampling rate), while the second one is a high-pass filter. The absolute value of the latter system is described by:

$$|H^*(j\omega)| = \frac{1}{\sqrt{[\cos(\omega T_S) - 0.9]^2 + \sin^2(\omega T_S)}}.$$

The absolute values of the calculated frequency responses are shown in Fig. 4.6.

References

1. Brigham E (1988) The fast Fourier transform and its applications. Prentice Hall Inc., Englewood Cliffs
2. Dyke PPG (1999) An introduction to Laplace transforms and Fourier series. Springer, London
3. Graf U (2004) Applied Laplace transforms and Z-transforms for scientists and engineers: a computational approach using a mathematica package. Birkhäuser, Basel
4. Hamming RW (1986) Numerical methods for scientists and engineers. McGraw-Hill, New York
5. James JF (2004) A student's guide to Fourier transforms: with applications in physics and engineering. Cambridge University Press, Cambridge

6. Jury EI (1973) Theory and application of the Z-transform method. Krieger Pub Co, Huntington
7. Ogata K (1995) Discrete time control systems, 2nd edn. Prentice Hall Inc., Upper Saddle River
8. Ukil A (2007) Intelligent systems and signal processing in power engineering. Springer, Berlin
9. Ungrad H, Winkler W, Wiszniewski A (1995) Protection techniques in electrical energy systems. Marcel Dekker Inc., New York

Chapter 5
Infinite Impulse Response Filters

5.1 IIR Filter Fundamentals

Digital filters allow to pass assumed spectrum part or particular component of the signal and prevent from passing the other signal components, sometimes called noise. Similarly to analog filters one can distinguish: low pass, high pass, band pass and band rejection digital filters [1–6]. The frequency response of these types of filters are shown in Fig. 5.1. In most cases it is required that the filters should have flat frequency response in pass band, steep transition region(-s) and close to zero rejection band. Phase frequency response is also important, especially useful are filters having this response linear. This is also valid in protection systems, where extracting orthogonal components is of importance.

Fundamental criteria describing requirements and performance of digital filters applied in protections are following [5]:

- accuracy in transmission of useful signal components and efficient rejection of noise,
- fast stabilizing of output signal after step change of input signal,
- linear phase shift frequency response,
- minimum burden during digital realization of given filter.

Some of the requirements are quite contradictory and that is why design of a digital filter is a matter of compromise.

All analog filters have infinite impulse response, i.e. their response is such at least theoretically. Digital filters, on the other hand, can be divided for:

- filters similar to analog ones having infinite impulse response called shortly IIR filters, (sometimes also called recursive filters)—described in detail in this chapter,

W. Rebizant et al., *Digital Signal Processing in Power System Protection and Control,* Signals and Communication Technology,
DOI: 10.1007/978-0-85729-802-7_5, © Springer-Verlag London Limited 2011

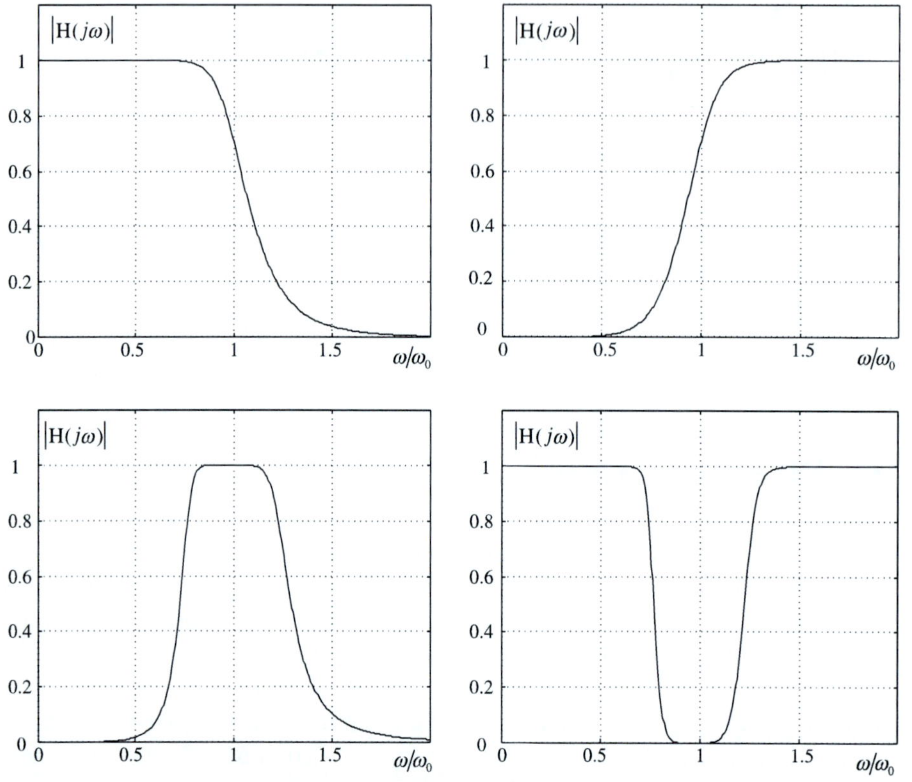

Fig. 5.1 Typical characteristics of lowpass, highpass, bandpass and band-rejection filters

- filters unknown in analog technique having finite impulse response called shortly FIR filters (sometimes also called non-recursive filters)—described in Chap. 6.

Typical algorithm of a digital IIR filter is given in the form [1, 3, 5]:

$$y(n) = \sum_{k=0}^{N-1} a(k)x(n-k) - \sum_{k=1}^{M} b(k)y(n-k) \tag{5.1}$$

where $y(n)$ is an output sample at discrete time instant n; $x(n)$ is an input sample at instant n, $a(k)$; $b(k)$ are filter coefficients.

Such a filter produces output as weighted sum of N recent input samples and M preceding outputs.

Digital filters can be described with use of a transfer function, which is determined using Z transform to both sides of Eq. 5.1. This is easy if one remembers that transfer functions are determined for zero initial conditions and signal delay by one step (sample) is equivalent to multiplication by z^{-1}. Consequently, the transfer function of IIR filter is given by following equation:

$$\frac{Y(z)}{X(z)} = H(z) = \frac{\sum_{k=0}^{N-1} a(k)z^{-k}}{1 + \sum_{k=1}^{M} b(k)z^{-k}}. \tag{5.2}$$

It is possible to get frequency response of the filters by substituting $e^{(j\omega T_S)}$ for z in Eq. 5.2. Then for IIR filter holds:

$$H(e^{j\omega T_s}) = H^*(j\omega) = \frac{\sum_{k=0}^{N-1} a(k)\mathrm{e}^{-jk\omega T_s}}{1 + \sum_{k=1}^{M} b(k)\mathrm{e}^{-jk\omega T_s}}. \tag{5.3}$$

There are three fundamental tasks concerning digital filters. These are: filter analysis, synthesis and realization. Filter analysis means checking its performance in time and frequency domains for known coefficients $a(k)$ and $b(k)$. Time domain filter performance concerns first of all adequately defined impulse response time or response for unit step (resulting from Eq. 5.1). In frequency domain one usually finds the magnitude and phase frequency responses. Synthesis is just an opposite process—assuming magnitude frequency response and some additional conditions concerning for instance phase shift frequency response or time response one should design the filter, i.e. calculate its coefficients $a(k)$ and $b(k)$. Filter realization means writing a program transmitting difference Eq. 5.1 to given digital system and starting its operation.

IIR digital filters are described in time domain with Eq. 5.1, using transfer function (5.2) or frequency response (5.3). As it was said, filter analysis in time domain means in general calculating either impulse or step response for known filter coefficients, whereas frequency domain analysis is understood as calculation of magnitude and phase shift as functions of frequency. Both calculations are easy for known filter coefficients and can be realized using handy calculator.

Filter synthesis is a little bit more difficult and means calculation of filter coefficients for assumed frequency response. There are many methods of synthesis. Most of them rely on normalized transfer functions of different types of standard filter approximations (e.g. Butterworth, Bessel, Tschebyshev) allowing to get required filter features. Next step is frequency transformation allowing to get required frequency response. The final step is transformation from analog to digital filter. Among several known methods of designing digital IIR filters below two selected ones will be presented: bilinear transformation and impulse time invariant methods [1, 5].

5.2 Synthesis of IIR Filters

5.2.1 Application of Bilinear Transformation

The starting point in filter synthesis is a transfer function of low pass normalized analog filter prototype. Next step is transformation of that filter to given type and given cut off frequency according to methods described below. The analog filter fulfilling the set of requirements is transferred from analog to digital version using bilinear transformation. The basis of this transformation is approximation of an integral by a sum within one sampling period. It leads to the following relationship between Laplace and Z transforms:

$$s \Rightarrow A\frac{1 - z^{-1}}{1 + z^{-1}} \tag{5.4}$$

called bilinear transformation. Important feature of this transformation is unambiguity (explicitness), i.e. given pole of analog filter is reflected by an equivalent pole of digital filter, stable analog filter gives stable digital filter and they are of the same orders.

From bilinear transformation also results relationship between angular frequencies in analog filter ω_a and digital filter ω_d. It can be obtained comparing magnitudes of both sides of (5.4):

$$\omega_a = A\frac{|1 - \exp(-j\omega_d T_S)|}{|1 + \exp(-j\omega_d T_S)|} = Atg(\omega_d T_S/2). \tag{5.5}$$

It is seen that certain "squeezing" of frequency in digital system appears, such that infinite frequency in analog system is equivalent to a half of sampling frequency in digital system. It results in a way also from the sampling theorem.

Synthesis of a low pass digital IIR filter

Design of low pass digital IIR filter which should have cut-off angular frequency ω_{d0} begins from selecting a transfer function of normalized (angular frequency equal to one) low pass filter fulfilling the requirements. One must first fix given approximation type and filter order, remembering that the aim is to design a digital filter having similar features (with respect to analog filter prototype) and given cut-off frequency. This digital filter can be obtained either directly from analog normalized filter or by transferring normalized filter to that of cut-off frequency ω_{dC} (sometimes such filter is on disposal from the very beginning). If one starts from normalized filter then the first step is substituting:

$$s \Rightarrow s_1/\omega_{aC}. \tag{5.6}$$

The obtained analog filter is transferred to digital domain using bilinear transformation:

$$s_1 \Rightarrow A\frac{1 - z^{-1}}{1 + z^{-1}}. \tag{5.7}$$

and one gets transfer function of sought digital filter.

One may also combine both transformations substituting for operator s:

$$s \Rightarrow \frac{A}{\omega_{aC}}\frac{1 - z^{-1}}{1 + z^{-1}}. \tag{5.8}$$

The ratio of constant coefficients results from (5.5):

$$\frac{A}{\omega_{aC}} = ctg(\omega_{dC}T_S/2). \tag{5.9}$$

and finally one gets the substitution in the form:

$$s \Rightarrow ctg(\omega_{dC}T_S/2)\frac{1 - z^{-1}}{1 + z^{-1}}. \tag{5.10}$$

The analog and resulting digital filters are the same orders.

If starting point is analog low pass filter having cut off frequency ω_{aC}, then the substitution is as follows:

$$s_1 \Rightarrow \omega_{aC}ctg(\omega_{dC}T_S/2)\frac{1 - z^{-1}}{1 + z^{-1}}. \tag{5.11}$$

It is seen that substitutions are equivalent and when $\omega_{aC} = 1$ (5.11) is the same as (5.10).

Synthesis of a high pass digital IIR filter

As before, the starting point is normalized transfer function of low pass filter prototype. It can be transferred to high pass filter having given cut-off angular frequency ω_{aC} by substituting:

$$s \Rightarrow \omega_{aC}/s_1. \tag{5.12}$$

Now, using bilinear transformation (5.8), i.e. substituting for operator s_1 formula as in this equation one can get transfer function of designed digital filter. It may be proved that frequency transformation from low pass filter to high pass filter and bilinear transformation can be combined giving:

$$s \Rightarrow \frac{\omega_{aC}}{A}\frac{1 + z^{-1}}{1 - z^{-1}}. \tag{5.13}$$

The ratio of constant coefficients can be calculated from (5.5), giving:

$$\frac{\omega_{aC}}{A} = tg(\omega_{dC}T_S/2). \tag{5.14}$$

Finally to get digital high pass filter having cut off frequency ω_{dC} from normalized analog low pass filter one should substitute in this normalized filter:

$$s \Rightarrow tg(\omega_{dC}T_S/2)\frac{1 + z^{-1}}{1 - z^{-1}}. \tag{5.15}$$

Synthesis of a band pass digital IIR filter

Here again the starting point is transfer function of a normalized low pass filter. The aim of synthesis is obtaining digital band pass filter having cut-off frequencies ω_{dCH} and ω_{dCL}, respectively. The way to reach the result is two-step. First, frequency transformation is used to transfer analog normalized low pass filter to analog band pass filter of cut-off frequencies as above:

$$s \Rightarrow \frac{1}{\omega_{dCH} - \omega_{dCL}}\left(s_1 + \frac{\omega_{dCH}\omega_{dCL}}{s_1}\right). \tag{5.16}$$

Then one should apply bilinear transformation:

$$s_1 \Rightarrow A\frac{1 - z^{-1}}{1 + z^{-1}}. \tag{5.17}$$

These two steps can be joined, giving the required substitution:

$$s \Rightarrow B\frac{1 - 2\alpha z^{-1} + z^{-2}}{1 - z^{-1}}, \tag{5.18}$$

where

$$B = ctg\left[\frac{(\omega_{dCH} - \omega_{dCL})T_S}{2}\right], \quad \alpha = \frac{\cos[0.5(\omega_{dCH} + \omega_{dCL})]}{\cos[0.5(\omega_{dCH} - \omega_{dCL})]}.$$

Synthesis of a band rejection digital IIR filter

Transfer function of digital band rejection digital IIR filter one obtains using the following substitution:

$$s \Rightarrow C\frac{1 - z^{-2}}{1 - 2\alpha z^{-1} + z^{-2}}, \tag{5.19}$$

where

$$C = tg\left[\frac{(\omega_{dCH} - \omega_{dCL})T_S}{2}\right]. \tag{5.20}$$

Example 5.1 Applying the bilinear transformation method that perform synthesis of a lowpass IIF filter, having the cutoff frequency $\omega_{dC} = 628$, taking the Butterworth II order low-pass approximation ($\omega_{aC} = 1$) as analog filter prototype.

$$H(s) = \frac{1}{s^2 + \sqrt{2}s + 1}$$

Assume sampling frequency $f_S = 1/T_S = 1000\,\text{Hz}$.

Solution The bilinear transformation method allows easy obtaining of the IIR filter transfer function and its difference equation. First one calculates the constant coefficient of bilinear transformation. For a lowpass digital filter this factor amounts to (according to (5.9)):

$$ctg(0.5\omega_{dC}T_S) = ctg(0.5 * 628 * 0.001) = 3.078$$

and in transfer function of analog filter one substitutes for s the expression:

$$s \Rightarrow 3.078\frac{1 - z^{-1}}{1 + z^{-1}}.$$

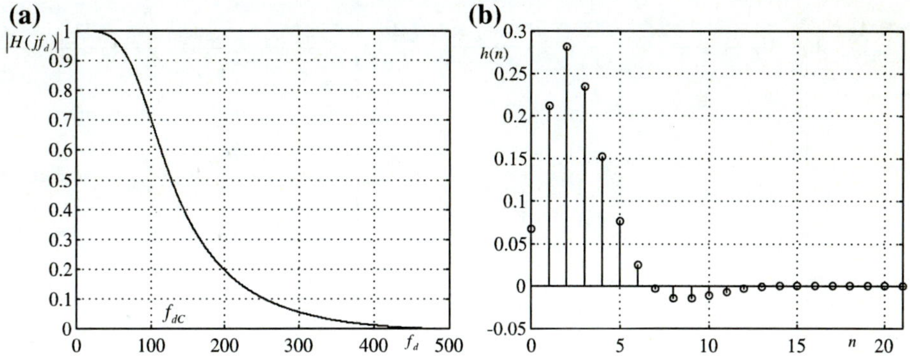

Fig. 5.2 Impulse response **a** and frequency response **b** of digital filter from Example 5.1

As a result the transfer function of digital filter equals:

$$H(z) = \frac{1}{\left(3.078\frac{1-z^{-1}}{1+z^{-1}}\right)^2 + \sqrt{2}\left(3.078\frac{1-z^{-1}}{1+z^{-1}}\right) + 1}$$

and after simple rearrangements:

$$H(z) = 0.067\frac{1 + 2z^{-1} + z^{-2}}{1 - 1.143z^{-1} + 0.412z^{-2}}$$

Therefore the filter difference equation has the form:

$$y(n) = 0.067x(n) + 0.134x(n-1) + 0.067x(n-2) + 1.143y(n-1) - 0.412y(n-2)$$

The impulse response and frequency characteristics of designed filter are shown in Fig. 5.2.

Example 5.2 Design a bandpass IIR filter, having cutoff frequencies 50 and 150 Hz. Assume sampling rate 1000 Hz and analog prototype approximation of Butterworth II order.

Solution According to (5.16–5.18) one should substitute for s in the filter transfer function

$$H(s) = \frac{1}{s^2 + \sqrt{2}s + 1}$$

appropriate form, with the constant coefficients as follows:

$$B = ctg[0.5(\omega_{dCH} - \omega_{dCL})T_S) = ctg(0.5 * (150 - 50) * 2\pi * 0.001) = 3.08,$$

$$a = \frac{\cos[0.5(\omega_{dCH} + \omega_{dCL})T_S)}{\cos[0.5(\omega_{dCH} - \omega_{dCL})T_S)} = \frac{\cos(0.5 * (150 + 50) * 2\pi * 0.001)}{\cos(0.5 * (150 - 50) * 2\pi * 0.001)} = 0.85.$$

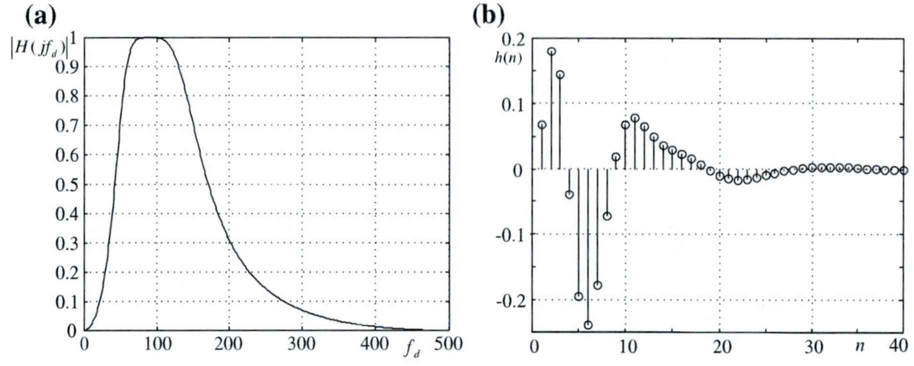

Fig. 5.3 Impulse response **a** and frequency response **b** of digital filter from Example 5.2

Thus the substitution for Laplace operator is:

$$s = B\frac{1 - 2az^{-1} + z^{-2}}{1 - z^{-2}} = 3.08\frac{1 - 1.7z^{-1} + z^{-2}}{1 - z^{-2}}$$

Then, after simple transformations, one obtains the following transfer function of sought IIR filter:

$$H(z) = \frac{0.0674(1 - 2z^{-2} + z^{-4})}{1 + b(1)z^{-1} + b(2)z^{-2} + b(3)z^{-3} + b(4)z^{-4}},$$

where

$$b(1) = -2.672, \quad b(2) = 2.99, \quad b(3) = -1.674, \quad b(4) = 0.413.$$

The filter difference equation is then:

$$\begin{aligned} y(n) = {} & 0.0674[x(n) - 2 * x(n-2) + x(n-4)] + 2.672y(n-1) - 2.99y(n-2) \\ & + 1.674y(n-3) - 0.413y(n-4). \end{aligned}$$

The impulse and frequency responses of the designed filter are shown in Fig. 5.3. One can see that the impulse response lasts for approximately 1.5 cycle of 50 Hz (some 30 samples). The transition band from pass to rejection regions is quite steep. In case if this slope is not sufficient, one can increase the filter order (both analog prototype and resulting digital filter), but this usually results also in longer duration of filter transient response in time domain.

5.2.2 Application of Impulse Response Invariance Method

The impulse response invariance method relies on design of digital filter, which has impulse response obtained by sampling of impulse response of the analog origin:

$$h(n) = h_a(t)|_{t=nT_s}, \tag{5.21}$$

where $h_a(t)$ is an impulse response of analog filter with transfer function $H_a(s)$.

Frequency response of digital filter having impulse response as above is Fourier transform of equation:

$$h^*(t) = \sum_{n=0}^{\infty} h_a(nT_S)\delta(t - nT_S). \tag{5.22}$$

and is equal to:

$$H^*(j\omega) = \frac{1}{T_S}\sum_{k=-\infty}^{\infty} H_a\left[j\left(\omega - k\frac{2\pi}{T_S}\right)\right]. \tag{5.23}$$

Though the impulse response of analog filter is in a way "stored" in digital system, it may happen that frequency responses of analog and digital filters differ substantially. The reason of that is aliasing known from sampling theorem. It is very important then to match parameters of analog origin and the sampling frequency so that the replicas of analog filter frequency responses do not (or just a little) overlap.

Described method is rarely applied in above-presented way. The most frequently the difference equation of sought digital filter is obtained from analog origin transfer function presented in the form of partial expansion (transfer function poles must be known):

$$L\{h_a(t)\} = H_a(s) = \sum_{k=1}^{N} \frac{A_k}{s - s_k}. \tag{5.24}$$

Then:

$$h_a(t) = \sum_{k=1}^{N} A_k \exp(s_k t) \tag{5.25}$$

and

$$h(n) = \sum_{k=1}^{N} A_k \exp(s_k n T_S), \tag{5.26}$$

which further yields:

$$H(z) = \sum_{k=1}^{N} \frac{A_k}{1 - \exp(s_k T_s) z^{-1}}. \tag{5.27}$$

The resulting discrete transfer function (5.27) allows getting the sought difference equation needed for digital filter realization.

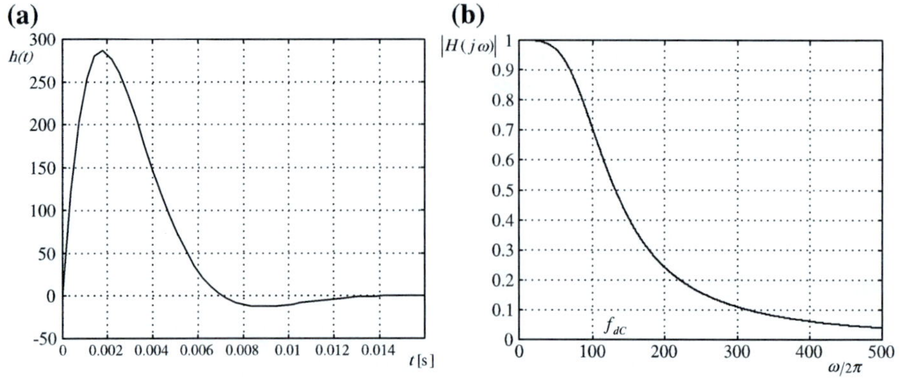

Fig. 5.4 Impulse response **a** and frequency response **b** of analog filter from Example 5.3; cut-off frequency $\omega_{dC} = 628$.

Concluding one can say that procedure of digital IIR filter design using invariance impulse response method is as follows:

- design low-pass normalized analog filter having frequency response according to requirements (choose type and order of approximation),
- transfer this filter to required cut off frequency,
- obtained transfer function present using partial expansion,
- transfer above to discrete transfer function,
- write difference equation and realize the filter.

Example 5.3 Applying the method of impulse response invariance design a low-pass filters of the parameters as in Example 5.1 (same analog approximation, cut-off and sampling frequencies).

Solution According to the design procedure outlined above, first the analog filter transfer function should be "shifted" to the new (required) cut-off frequency $\omega_{dC} = 628$.

Using the substitution:

$$s \Rightarrow s/\omega_{dC}$$

one obtains:

$$H(s) = \frac{\omega_{dC}^2}{s^2 + \sqrt{2}\omega_{dC}s + \omega_{dC}^2}.$$

The impulse and frequency responses of the resulting analog filter are shown in Fig. 5.4. Now one can calculate the poles of filter transfer function:

$$s_{1,2} = \frac{1}{\sqrt{2}}\omega_{dC}(-1 \pm j),$$

therefore sought partial expansion of the filter transfer function is:

$$H(s) = \frac{\omega_{dC}}{j\sqrt{2}}\left(\frac{1}{s - s_1} - \frac{1}{s - s_2}\right).$$

Thus the transfer function of designed digital filter amounts (according to (5.27)):

$$H(z) = \frac{\omega_{dC}}{j\sqrt{2}}\left(\frac{1}{1 - z^{-1}\exp(s_1 T_S)} - \frac{1}{1 - z^{-1}\exp(s_2 T_S)}\right).$$

Introducing the values of poles, one obtains after simple rearrangements:

$$H(z) = \frac{\sqrt{2}\omega_{dC}\exp(-\beta)\sin(\beta)z^{-1}}{1 - 2z^{-1}\exp(-\beta)\cos(\beta) + z^{-2}\exp(-2\beta)},$$

where

$$\beta = \frac{\omega_{dC} T_S}{\sqrt{2}}.$$

Finally, for assumed cut-off and sampling frequencies, one gets:

$$H(z) = \frac{244.92z^{-1}}{1 - 1.158z^{-1} + 0.412z^{-2}}.$$

From the above results that filter gains for low frequencies are approximately equal to the value of sampling frequency. In order to get filter gain equal to unity, one can rescale the filter, multiplying its transfer function by the sampling period, which gives:

$$H(z)T_S = H_1(z) = \frac{0.244z^{-1}}{1 - 1.158z^{-1} + 0.412z^{-2}}.$$

From the above one can derive sought difference equation of digital filter:

$$y(n) = 0.244x(n-1) + 1.158y(n-1) - 0.412y(n-2).$$

The impulse and frequency responses of designed filter are presented in Fig. 5.5. It is clear that the assumed value of sampling frequency is too low, having significant effects in the rejection region (overlapping of replicas of analog filter frequency responses). One can reduce this effect selecting higher value of sampling frequency. The results for $f_S = 4000$ Hz can be seen in Fig. 5.6. It becomes evident that with higher sampling rate the resulting frequency response of designed digital filter are better and better, closer to the analog prototype characteristics.

The above examples show clearly that bilinear transformation gives better results than impulse invariance method. The first one is simpler and more accurate as far as analog origin and digital realization are concerned.

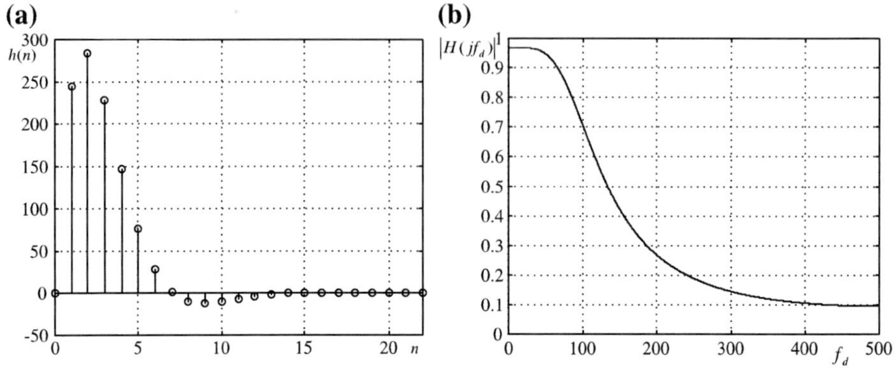

Fig. 5.5 Impulse response **a** and frequency response **b** of digital filter from Example 5.3; sampling frequency $f_S = 1000$ Hz

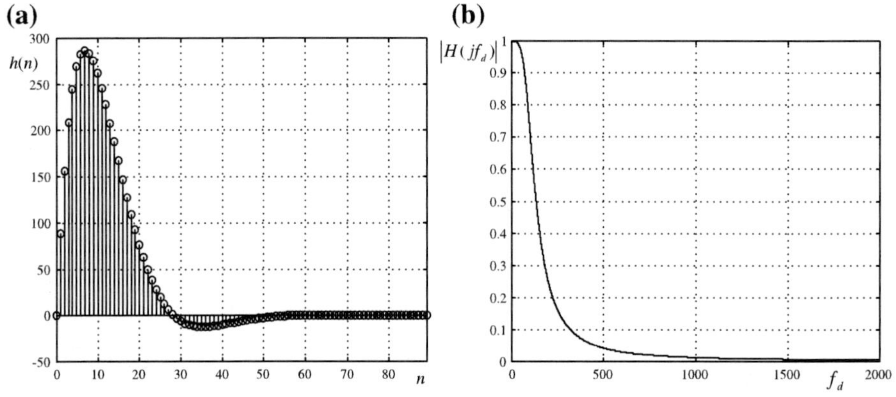

Fig. 5.6 Impulse response **a** and frequency response **b** of digital filter from Example 5.3; sampling frequency $f_S = 4000$ Hz

References

1. Jackson LB (1996) Digital filters and signal processing. Kluwer Academic Publishers, Boston
2. Lam HY-F (1979) Analog and digital filters, design and realization. Prentice-Hall, Englewood Cliffs
3. Oppenheim AV, Schafer RW (1975) Digital signal processing. Prentice-Hall, Englewood Cliffs
4. Rabiner LR, Gold B (1975) Theory and application of digital signal processing. Prentice-Hall, Englewood Cliffs
5. Szafran J, Wiszniewski A (2001) Measurement and decision algorithms of digital protection and control. WNT, Warszawa (in Polish)
6. Vegte de JV (2002) Fundamentals of digital signal processing. Prentice-Hall, Englewood Cliffs

Chapter 6
Finite Impulse Response Filters

6.1 Finite Impulse Response Filter Fundamentals

Digital finite impulse response (FIR) filters produce output signal realizing weighting sum of actual and preceding input signal samples. This relationship is also called discrete convolution, being expressed in the form, [2, 4, 7, 8]:

$$y(n) = \sum_{k=0}^{N-1} a(k)x(n-k), \tag{6.1}$$

where x, y are input and output signal samples, respectively, $a(k)$ are filter coefficients, N is a number of coefficients creating so called filter window.

The FIR filters may be considered as a special case of IIR filters described by Eq. 5.1, with coefficients $b(k)$ equal to zero. The FIR filters have many advantages and are realized in digital technique only. For such filters one introduces the so called "window" of coefficients having a length equivalent to N samples used in weighting sum (6.1). This window length in terms of time is equal to:

$$T_w = NT_S, \tag{6.2}$$

where T_S is a sampling period.

The FIR filters have two very important features. The first one is expressed by the filter name—FIR. Finite impulse response is very important for applications in power system control and protection because it allows to foresee the length of transients after a step change of input signal parameters. The transient time is here equal to a product of the number of samples of filter window N and the sampling period. It means that the filter response length is equal to its window.

The transfer function of a FIR filter can be derived either from Eq. 6.1 by means of Z transform or by substituting zero coefficients $b(k)$ in Eq. 5.2. One obtains then:

W. Rebizant et al., *Digital Signal Processing in Power System Protection and Control,* Signals and Communication Technology,
DOI: 10.1007/978-0-85729-802-7_6,

$$H(z) = \sum_{k=0}^{N-1} a(k) z^{-k}. \tag{6.3}$$

The frequency response of the FIR filter is obtained by substituting $e^{(j\omega T_S)}$ for z in Eq. 6.3, which for FIR filter yields:

$$H^*(j\omega) = \sum_{k=0}^{N-1} a(k) e^{-jk\omega T_S}. \tag{6.4}$$

The other advantage of FIR filters is a possibility to design the filters having linear phase (versus frequency), allowing to obtain a pair of orthogonal filters easily.

It can be proved that FIR filter has linear phase when its impulse response is either even or odd function of k, which may be written in the form:

$$a(k) = a(N - 1 - k). \tag{6.5a}$$

$$a(k) = -a(N - 1 - k). \tag{6.5b}$$

It may also be proved that each of the filters fulfilling the conditions (6.5) has linear phase and, moreover, that the phase difference between their phase equals $\pi/2$ for any frequency.

The frequency response of a FIR filter can also be given in the form:

$$H(j\Omega) = \sum_{k=0}^{N-1} a(k) \exp(-jk\Omega), \tag{6.6}$$

where $\Omega = \omega T_S = 2\pi f / f_S$ is a ratio of angular frequency to sampling frequency.

The frequency transfer function (6.6) can be analyzed in order to prove the filter phase linearity as well as orthogonality for a pair of filters fulfilling the symmetry conditions. Rearranging (6.6) one gets:

$$\begin{aligned} H(j\Omega) &= \sum_{k=0}^{N-1} a(k) \exp(-jk\Omega) \\ &= \sum_{k=0}^{N/2-1} \{a(k) \exp[-jk\Omega] + a(N-1-k) \exp[-j(N-1-k)\Omega]\} \end{aligned}$$

and then:

$$\begin{aligned} H(j\Omega) = \exp\left[-j\left(\frac{N-1}{2}\right)\Omega\right] \sum_{k=0}^{N/2-1} a(k) \Big\{ \exp\left[j\left(\frac{N-1}{2} - k\right)\Omega\right] \\ \pm \exp\left[-j\left(\frac{N-1}{2} - k\right)\Omega\right] \Big\} \end{aligned} \tag{6.8}$$

The plus sign in (6.8) appears for even symmetry of filter coefficients and minus—for odd symmetry. In the first case one obtains:

$$H(j\Omega) = 2\exp\left[-j\left(\frac{N-1}{2}\right)\Omega\right]\sum_{k=0}^{N/2-1} a(k)\cos\left[\left(\frac{N-1}{2}-k\right)\Omega\right] = H_e(j\Omega) \tag{6.9}$$

and in the second case:

$$H(j\Omega) = 2j\exp\left[-j\left(\frac{N-1}{2}\right)\Omega\right]\sum_{k=0}^{N/2-1} a(k)\sin\left[\left(\frac{N-1}{2}-k\right)\Omega\right] = H_o(j\Omega). \tag{6.10}$$

Since in both cases the sums are real, the filter arguments are given by:

$$\arg[H_e(j\Omega)] = -\frac{N-1}{2}\Omega, \tag{6.11}$$

$$\arg[H_o(j\Omega)] = -\frac{N-1}{2}\Omega + \frac{\pi}{2}. \tag{6.12}$$

Concluding, one can stress that if the FIR filters have even and odd impulse responses, respectively, their transfers function have linear phase and are orthogonal, i.e. the difference of their phases equals $\pi/2$. The filters applied in protection systems should have that features, which may be very helpful in simplifying substantially development of algorithms for criterion values measurement.

6.2 Analysis of Standard FIR Filters

6.2.1 Filters with Walsh Windows

The first five so called Walsh windows are shown in Fig. 6.1. Their impulse responses have coefficients being equal either plus or minus one [1, 3, 7]. This feature gives the simplest possible equations of filter realization, which is important in fast protection systems.

Digital FIR filter realized applying Walsh function of zero order, called also rectangular window, produces output signal being the sum of N most recent samples of the input signal, which is described by very simple equation:

$$y(n) = \sum_{k=0}^{N-1} x(n-k). \tag{6.13}$$

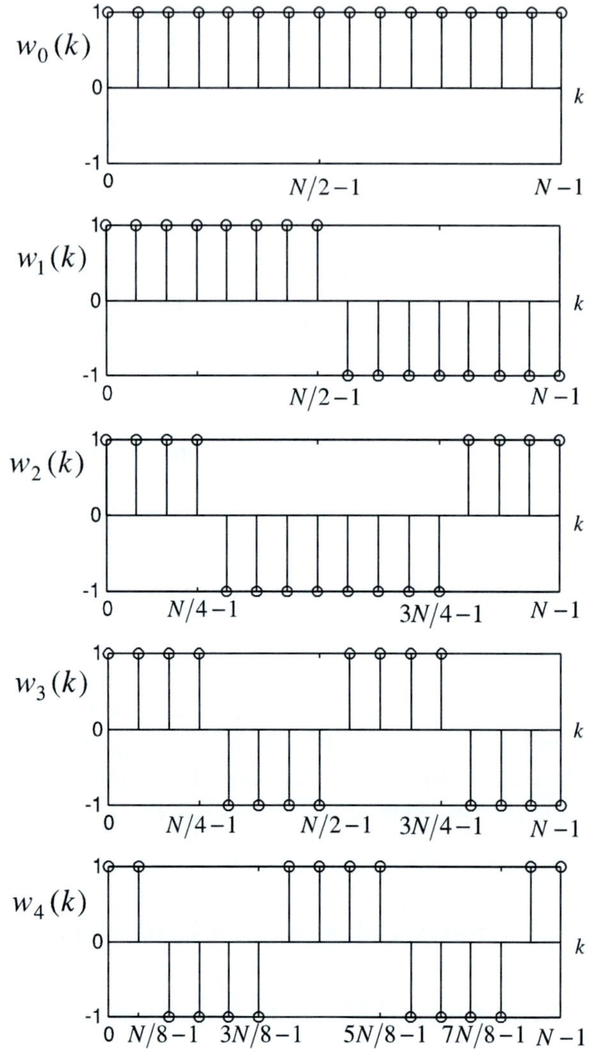

Fig. 6.1 First five Walsh impulse functions

Taking Z transform of (6.10) one obtains transfer function of the filter:

$$W_0(z) = \sum_{k=0}^{N-1} z^{-k} = \frac{1 - z^{-N}}{1 - z^{-1}} \tag{6.14}$$

and substituting $\exp(j\Omega)$ for z one can get the frequency response of the filter:

$$W_0(j\Omega) = \exp\left[-j\frac{N-1}{2}\Omega\right]\frac{\sin\left(N\frac{\Omega}{2}\right)}{\sin\left(\frac{\Omega}{2}\right)}. \tag{6.15}$$

As expected, the filter has linear phase shift versus frequency. It is also seen that filter gain is equal to N for angular frequency Ω equal to zero, while the gain equals zero for frequencies fulfilling equation $N\Omega/2 = m\pi$, and then

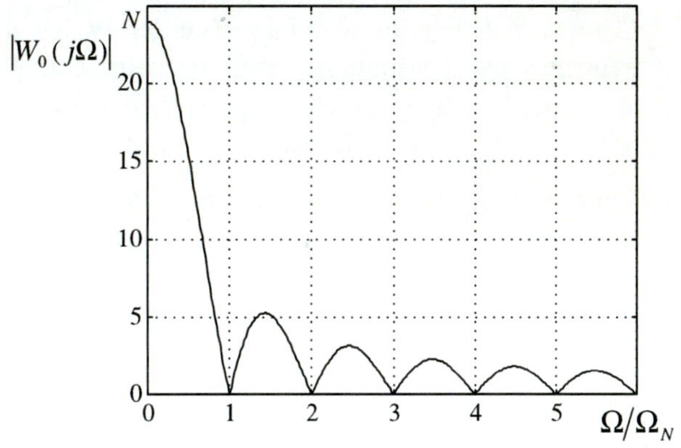

Fig. 6.2 Normalized frequency response of the 0-order Walsh filter

$$|W_0(j\Omega)| = 0 \quad \text{for} \quad \frac{\Omega}{\Omega_N} = m, \tag{6.16}$$

where $\Omega_N = 2\pi/N$.

It is seen from the last two equations that filter frequency response is fully determined by its window length N (number of samples) for given sampling frequency. Equation 6.15 can be written in the form expressing stronger conclusions:

$$W_0(j\Omega) = \exp\left(-j\pi\frac{N-1}{N}\frac{\Omega}{\Omega_N}\right)\frac{\sin\left(\pi\frac{\Omega}{\Omega_N}\right)}{\sin\left(\frac{\pi}{N}\frac{\Omega}{\Omega_N}\right)}. \tag{6.17}$$

The equation is graphically presented versus ratio Ω/Ω_N in Fig. 6.2. It is seen that the 0-order Walsh window is a "weak" low-pass filter that, however, allows for rejecting frequency Ω_N and all its harmonics. For the other frequencies dumping of noise is really poor.

The significance of this filter results from its very easy algorithm that can be realized with minimum burden of processor and memory. The other important conclusion is that any FIR filter has impulse response being a product of impulse responses of an ideal filter and a rectangular window. Thus, knowing the magnitude and phase shift functions of rectangular window versus frequency one can simplify development of any arbitrary FIR filter with the desired frequency characteristics.

It is obvious that rectangular window filter can not be used when useful information for protection operation is included in the signal component with relative angular velocity Ω_N since for that frequency the filter gain equals zero. On the other hand, one can use that to evaluate frequency deviations, for instance.

The filters having rectangular window of different window lengths can be applied. An example could be a filter having the window equivalent to half period of the fundamental component and then it is called half cycle zero order Walsh filter (or window).

Example 6.1 For the averaging 0-order Walsh filter select either the sampling frequency by given number of filter coefficients $N = 20$, or select the number of filter coefficients for given sampling frequency $f_S = 1000$, so that the component of frequency $f = 125$ Hz was suppressed to zero.

Solution From Eq. 6.17 describing the filter frequency response one can take the condition for zeroing the transfer function magnitude:

$$\sin\left(N\frac{\pi f}{f_S}\right) = 0, \quad \text{if } \frac{Nf}{f_S} = k,$$

where k is any integer number.

Therefore, the sought sampling frequency amounts to:

$$f_S = \frac{Nf}{k} = \frac{2500}{k}.$$

For the consecutive integer numbers k one gets: $f_S = 2500, 1250, 2500/3, \ldots,$ $500\,\text{Hz} > 250\,\text{Hz}$.

On the other hand, the number of filter coefficients for the sampling frequency 1000 Hz results from the equation:

$$N = \frac{kf_S}{f} = 8k$$

and thus $N = 8, 16, 24, \ldots$

It is worth to observe that not for every value of frequency such an easy solution is reached. For instance, if the signal component of frequency 311 Hz is to be rejected, then:

$$f_S = \frac{6220}{k},$$

which yields quite strange sampling frequencies and non-integer filter window lengths:

$$N = \frac{1000k}{311}.$$

Example 6.2 For the 0-order Walsh filter select the number of filter coefficients N so that the frequencies $f_1 = 100\,\text{Hz}$ and $f_2 = 125\,\text{Hz}$ are both suppressed to zero.

Assume sampling rates: (a) $f_S = 1000\,\text{Hz}$ and (b) $f_S = 600\,\text{Hz}$.

Solution Referring to the preceding example and Eq. 6.20 one obtains:

$$\frac{Nf_1}{f_S} = k_1 \quad \text{and} \quad \frac{Nf_2}{f_S} = k_2,$$

where k_1, k_2 are integer constants. Dividing the above equations side by side yields:

$$\frac{k_1}{k_2} = \frac{f_1}{f_2} = \frac{4}{5},$$

representing ratio of the smallest possible integer factors k_1, k_2. Therefore, considering the first equation one gets (for the cases (a) and (b)) the following window lengths:

$$N_a = \frac{f_S k_1}{f_1} = \frac{1000 * 4}{100} = 40; \quad N_b = \frac{f_S k_2}{f_2} = \frac{600 * 4}{100} = 24$$

The output signal of the FIR filter applying first order Walsh function is obtained using the following (also very simple) difference equation:

$$y(n) = \sum_{k=0}^{N/2-1} x(n-k) - \sum_{k=N/2}^{N-1} x(n-k). \tag{6.18}$$

In this case one also calculates a sum (with proper sign) of actual and preceding input samples. The filter algorithm is here again as easy as before and requires addition of a half of input samples and subtraction of remaining half of the samples observed within the filter window (all coefficients are equal to either one or minus one). Transfer function of the filter is calculated applying Z transform to both sides of Eq. 6.18:

$$W_1(z) = \sum_{k=0}^{N/2-1} z^{-k} - \sum_{k=N/2}^{N-1} z^{-k} = \frac{\left(1 - z^{-N/2}\right)^2}{1 - z^{-1}}. \tag{6.19}$$

Substituting $\exp(j\Omega)$ for operator z and rearranging one gets the frequency response of the first order Walsh function:

$$W_1(j\Omega) = 2j \exp\left[-j\frac{N-1}{2}\Omega\right] \frac{\sin^2\left(N\frac{\Omega}{4}\right)}{\sin\left(\frac{\Omega}{2}\right)}. \tag{6.20}$$

The formula (6.20) represents again a linear phase filter, however, the resulting phase shift is by $\pi/2$ greater than before, giving the filter orthogonal to that of zero order. Both magnitude and phase of frequency responses depend on window length N as before. Doing the same calculations as before one can present these responses graphically (Fig. 6.3). It is seen that this time a band-pass filter is obtained, not of high quality, but having very simple algorithm and easy to be applied with low processor and memory burden.

The FIR filter applying second order Walsh function produces output signal according to equation:

$$y(n) = -\sum_{k=0}^{N/4-1} x(n-k) + \sum_{k=N/4}^{3N/4-1} x(n-k) - \sum_{k=3N/4}^{N-1} x(n-k) \tag{6.21}$$

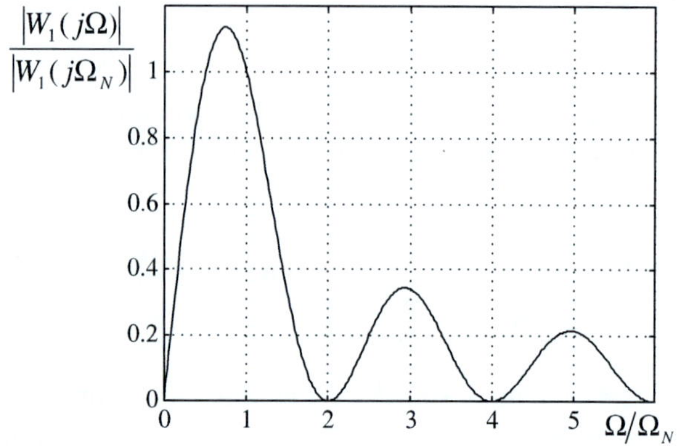

Fig. 6.3 Normalized frequency response of the first order Walsh filter

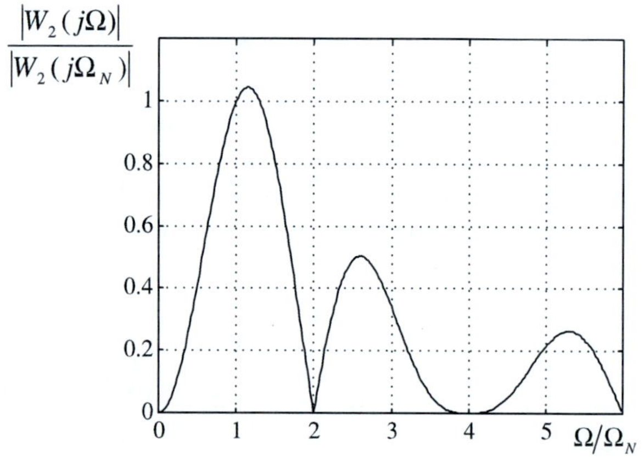

Fig. 6.4 Normalized frequency response of the second order Walsh filter

The resulting filter transfer function is:

$$W_2(z) = -\sum_{k=0}^{N/4-1} z^{-k} + \sum_{k=N/4}^{3N/4-1} z^{-k} - \sum_{k=3N/4}^{N-1} z^{-k} = \frac{\left(1 - z^{-N/4}\right)^2 \left(1 - z^{-N/2}\right)}{1 - z^{-1}} \tag{6.22}$$

From (6.25a, b, c), after known substitution, one obtains the filter frequency response:

$$W_2(j\Omega) = 2\exp\left[-j\frac{N-1}{2}\Omega\right] \frac{\sin\left(N\frac{\Omega}{4}\right)\left[1 - \cos\left[\left(N\frac{\Omega}{4}\right)\right]\right]}{\sin\left(\frac{\Omega}{2}\right)} \tag{6.23}$$

From (6.26) results that the 2nd order Walsh filter has also linear phase, it is furthermore orthogonal to the first order Walsh function. As before the frequency responses depend on window length N only (for given sampling frequency). The responses are presented graphically in Fig. 6.4. It is seen that in this case one also gets a band-pass filter of not very high quality, but very simple to be realized.

Frequency responses of the filters using last two Walsh functions from Fig. 6.1 are shown in Fig. 6.5. Certain improvement of filter features can be seen, first of all

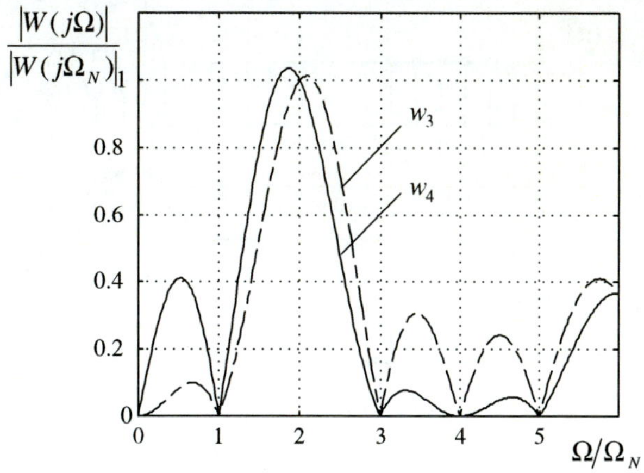

Fig. 6.5 Normalized frequency responses of the remaining two Walsh functions from Fig. 6.1

selectivity and transient region, however the filters have the highest gain for the frequencies two times higher than the filters analyzed before. They could be considered as orthogonal filters of the second harmonic of the power system nominal frequency.

Application of Walsh functions to orthogonal FIR filters can be concluded in the following way. The most important seem to be band-pass filters applying 1st and 2nd order Walsh functions. In such a case the filter window lengths N are equal to number of samples N_1 in one period of fundamental component and that is why the filters are called full cycle filters. Taking into account frequency responses obtained before, one can get for such filters:

$$|W_1(j\Omega_1)| = |W_2(j\Omega_1)| = \frac{2}{\sin(\pi/N_1)}, \tag{6.24a}$$

$$\arg[W_1(j\Omega_1)] = \frac{\pi}{2} - \frac{N_1 - 1}{2}\Omega_1 = -\frac{\pi}{2} + \frac{\pi}{N_1}, \tag{6.24b}$$

$$\arg[W_2(j\Omega_1)] = -\frac{N_1 - 1}{2}\Omega_1 = -\pi + \frac{\pi}{N_1}. \tag{6.24c}$$

Another important case is application of Walsh functions in the filters having "half period window". This is the case when $N = N_1/2$, i.e. filter window length is equal to number of samples within half period of fundamental frequency component. Similarly as before one can calculate filter magnitudes and phase shifts, remembering that now Walsh functions of the zero and first order are applied (as a result of shortening of the window). Calculated values are following:

$$|W_0(j\Omega_1)| = |W_1(j\Omega_1)| = \frac{1}{\sin(\pi/N_1)}, \tag{6.25a}$$

$$\arg[W_0(j\Omega_1)] = -\frac{N_1/2 - 1}{2}\Omega = -\frac{\pi}{2} + \frac{\pi}{N_1}, \tag{6.25b}$$

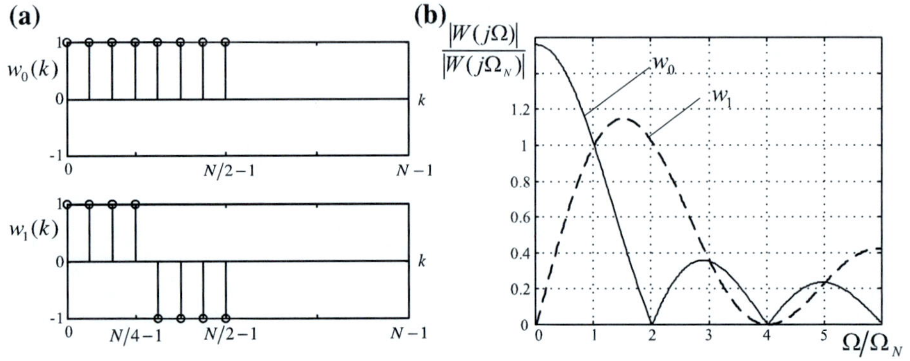

Fig. 6.6 Normalized impulse (**a**) and frequency responses (**b**) of half cycle Walsh filters w_1, w_2

$$\arg[W_1(j\Omega_1)] = \frac{\pi}{2} - \frac{N_1/2 - 1}{2}\Omega_1 = +\frac{\pi}{N_1}. \tag{6.25c}$$

These two cases of pairs of full and a half cycle Walsh orthogonal functions are very important for practical applications in power system protection devices. Identical filter gains and full orthogonality allow to get very simple algorithms of protection criterion values. The normalized frequency responses of the filters are shown in Fig. 6.6.

Example 6.3 Calculate relative gains of the first and second order Walsh filters (with respect to the gains for the values for the fundamental frequency 50 Hz) for the third and fifth harmonic components. Assume sampling frequency 1000 Hz.

Solution The number of samples in one cycle of fundamental frequency equals $N_1 = f_S/f_1 = 20$. Thus, from Eqs. 6.20, 6.23 and 6.24a, b, c one gets:

$$\Omega_1 = 0.1\pi, \quad \Omega_3 = 0.3\pi, \quad \Omega_5 = 0.5\pi,$$

$$|W_1(j\Omega_1)| = |W_2(j\Omega_1)| = \frac{2}{\sin(\Omega_1/2)} = 12.785,$$

$$\frac{|W_1(j\Omega_3)|}{|W_1(j\Omega_1)|} = \frac{2\sin^2(0.25N_1\Omega_3)}{|W_1(j\Omega_1)|\sin(0.5\Omega_3)} = \frac{2.2}{12.785} = 0.17,$$

$$\frac{|W_1(j\Omega_5)|}{|W_1(j\Omega_1)|} = 0.11,$$

$$\frac{|W_2(j\Omega_3)|}{|W_2(j\Omega_1)|} = \frac{2\sin(0.25N_1\Omega_3)[1 - \cos(0.25N_1\Omega_3)]}{|W_1(j\Omega_1)|\sin(0.5\Omega_3)} = \frac{5.28}{12.785} = 0.41,$$

$$\frac{|W_2(j\Omega_5)|}{|W_2(j\Omega_1)|} = 0.266.$$

As one can see (also in Figs. 6.3, 6.4), the second order Walsh function (w_2) particularly poorly suppresses the odd harmonic components.

6.2.2 Filters with Sine and Cosine Windows

Though digital filters applying Walsh functions give the simplest filter algorithms, the most popular are FIR filters having sine and cosine windows. The reason is very simple—they give very effective suppression or rejection of noise, having also simple algorithms, especially in recursive version [1, 3, 7, 8].

Output signals of a pair of orthogonal filters having sine, cosine windows and keeping the virtue of orthogonality for any window length N are given by the equations:

$$y_c(n) = \sum_{k=0}^{N-1} x(n-k)\cos\left[\left(\frac{N-1}{2} - k\right)\Omega_0\right], \tag{6.26}$$

$$y_s(n) = \sum_{k=0}^{N-1} x(n-k)\sin\left[\left(\frac{N-1}{2} - k\right)\Omega_0\right]. \tag{6.27}$$

Thus the filter coefficients, retaining the above discussed even and odd symmetry giving orthogonality, are described by equations:

$$a_c(k) = \cos\left[\left(\frac{N-1}{2} - k\right)\Omega_0\right], \tag{6.28}$$

$$a_s(k) = \sin\left[\left(\frac{N-1}{2} - k\right)\Omega_0\right], \tag{6.29}$$

where $0 \le k \le N-1$ and window length N is arbitrary.

There are many ways of calculation of the filter frequency responses. The simplest one is perhaps the method using representation of sine, cosine window as exponential functions and then the FIR is obtained multiplying that window by rectangular window (Walsh function w_0). Such approach allows for easy calculation of Z transforms and further transfer functions of the filters. From that we can calculate frequency responses using known substitution $z = \exp(j\Omega)$:

$$A_c(j\Omega) = 0.5\exp\left[-j\frac{N-1}{2}\Omega\right]\left[\frac{\sin\left(N\frac{\Omega-\Omega_0}{2}\right)}{\sin\left(\frac{\Omega-\Omega_0}{2}\right)} + \frac{\sin\left(N\frac{\Omega+\Omega_0}{2}\right)}{\sin\left(\frac{\Omega+\Omega_0}{2}\right)}\right], \tag{6.30}$$

$$A_s(j\Omega) = 0.5\exp\left[-j\frac{N-1}{2}\Omega\right]\left[\frac{\sin\left(N\frac{\Omega-\Omega_0}{2}\right)}{\sin\left(\frac{\Omega-\Omega_0}{2}\right)} - \frac{\sin\left(N\frac{\Omega+\Omega_0}{2}\right)}{\sin\left(\frac{\Omega+\Omega_0}{2}\right)}\right]. \tag{6.31}$$

The filters' frequency responses are determined by two parameters: natural relative angular frequency Ω_0 and filter window length N. For any arbitrary values of the

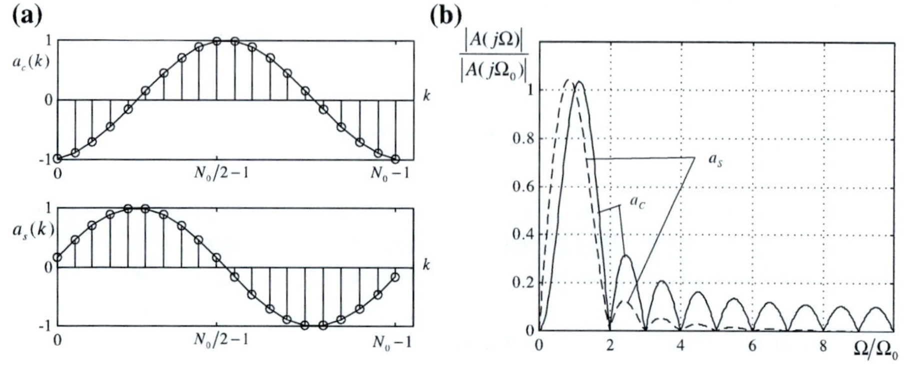

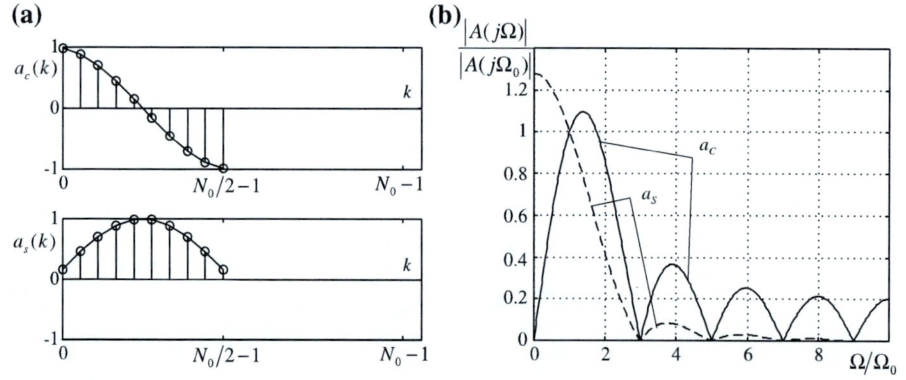

Fig. 6.8 Characteristics of half cycle sine/cosine filters: **a** impulse responses, **b** frequency spectra

parameters the difference between their arguments equal always 90°, however, for specific parameter relationships only the filters have identical gains for chosen frequencies. These identical gains are especially important for frequency equal to frequency of fundamental component of current and voltage, allowing to get simplified algorithms of measurement of criterion values. Assuming that the filters are designed to get identical gains for this frequency one arrives at the conclusion that it is obtainable when the second terms of Eqs. 6.30 and 6.31 equal zero. It is reached when: $N\Omega_0 = l\pi$, where $l = 1, 2. \ldots$, and then:

$$N = lN_0/2. \tag{6.32}$$

It means that the gain of the filters are identical for frequency equal to natural filters frequency when the filter window length is equal to a multiple of half period of their impulse response. One can also notice that in such cases the filter gains are equal to a half of filter window length, i.e. $N/2$ for frequency $\Omega = \Omega_0$.

Two cases are important in practice and are widely applied: filters having full cycle window ($N = N_0$) and having half cycle window ($N = N_0/2$). Normalized magnitudes of frequency responses are shown in Figs. 6.7 and 6.8, respectively. It is seen that especially advantageous are filters with full cycle windows. They are band-pass filters that reject all harmonics of fundamental frequency. The characteristics of half cycle filters are worse, especially in respect of passing DC component, however, they are two times faster. For that filters, very popular in protection applications, we can collect their coefficients, gains and arguments for fundamental frequency:

- for full cycle window, i.e. for $N = N_0$:

$$a_c(k) = -\cos[(k+0.5)\Omega_0], \tag{6.33a}$$

$$a_s(k) = \sin[(k+0.5)\Omega_0], \tag{6.33b}$$

$$|A_c(j\Omega_0)| = |A_s(j\Omega_0)| = N_0/2, \tag{6.33c}$$

$$\arg[A_c(j\Omega_0] = -\pi + \pi/N_0, \tag{6.33d}$$

$$\arg[A_s(j\Omega_0] = -\pi/2 + \pi/N_0, \tag{6.33e}$$

- and for half cycle window, i.e. for $N = N_0/2$:

$$a_c(k) = \sin[(k+0.5)\Omega_0], \tag{6.34a}$$

$$a_s(k) = \cos[(k+0.5)\Omega_0], \tag{6.34b}$$

$$|A_c(j\Omega_0)| = |A_s(j\Omega_0)| = N_0/4, \tag{6.34c}$$

$$\arg[A_s(j\Omega_0)] = -\pi/2 + \pi/N_0, \tag{6.34d}$$

$$\arg[A_s(j\Omega_0)] = \pi/N_0, \tag{6.34e}$$

where $\Omega_0 = 2\pi/N_0$.

Next figures show frequency responses of the filters modified in a different way, i.e. having different natural relative angular frequency Ω_0 and different window length N. The case when the filter natural frequency is the same as before but the filter window is two times longer $(N = 2N_0)$ is shown in Fig. 6.9. This pass band filter has even better ability to reject or dump signal noise. When such window length is allowed taking into account protection dynamics such filters might be recommended. The case with filters of half cycle window $(N = N_0/2)$ but two times greater natural frequency $(\Omega_0 = 4\pi/N_0)$ is presented in Fig. 6.10. Such frequency was chosen to get filter frequency response rejecting DC for both orthogonal filters, which may be important in many applications. The effect of such natural frequency matched filter window is improving the filter frequency response in region of small frequencies and worsening in the region of high frequencies. Such filters may be applied in case of low-level of expected high frequency noise and for high content of DC or decaying DC component, with additional requirements of fast response. One can see that increasing this filter

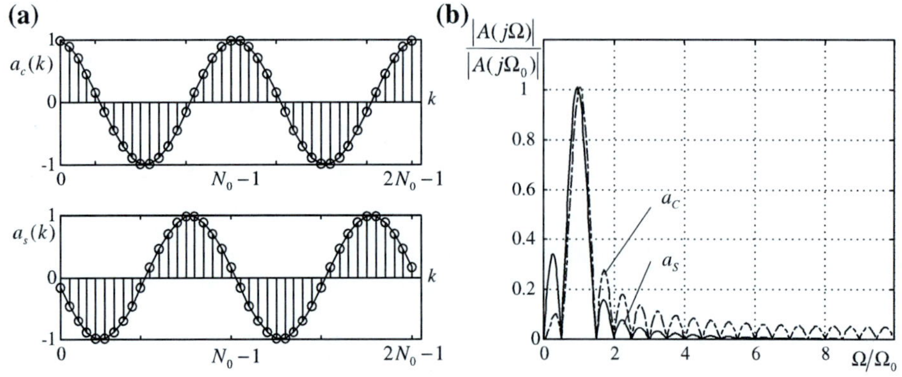

Fig. 6.9 Characteristics of double-cycle sine/cosine filters: **a** impulse responses **b** frequency spectra

window to full cycle (by the same natural frequency) one gets excellent band-pass filter of second harmonic.

Example 6.4 Calculate the level of suppression of the frequencies $f_a = 125$ and $f_b = 175$ Hz by the full cycle sine, cosine filters (in respect to filter gains for the fundamental frequency $f_1 = 50$ Hz). Compare results for two sampling frequencies $f_s = 1000$ and 600 Hz.

Solution Equations 6.30, 6.31 and 6.33a, b, c, d, e are used for the calculations

(1) Sampling frequency 1000 Hz

Number of samples per cycle is $N_1 = f_S/f_1 = 20$ and:

$$\Omega_a = 2\pi f_a/f_S = 2\pi * 125/1000 = 0.25\pi,\ \Omega_b = 0.35\pi,\ \Omega_1 = 0.1\pi.$$

From (6.30) and (6.31) one obtains:

$$|A_c(j\Omega)| = \left|0.5\frac{\sin[0.5N_1(\Omega-\Omega_1)]}{\sin[0.5(\Omega-\Omega_1)]} + 0.5\frac{\sin[0.5N_1(\Omega+\Omega_1)]}{\sin[0.5(\Omega+\Omega_1)]}\right|,$$

$$|A_s(j\Omega)| = \left|0.5\frac{\sin[0.5N_1(\Omega-\Omega_1)]}{\sin[0.5(\Omega-\Omega_1)]} - 0.5\frac{\sin[0.5N_1(\Omega+\Omega_1)]}{\sin[0.5(\Omega+\Omega_1)]}\right|$$

and from (6.33a, b, c, d, e): $|A_c(j\Omega_1)| = |A_s(j\Omega_1)| = N_1/2 = 10$.
Substituting the calculated values yields:

$$\begin{aligned}\frac{|A_c(j\Omega_a)|}{|A_c(j\Omega_1)|} &= 0.1\left|0.5\frac{\sin[5(0.25\pi-0.1\pi)]}{\sin[0.5(0.25\pi-0.1\pi)]} + 0.5\frac{\sin[5(0.25\pi+0.1\pi)]}{\sin[0.5(0.25\pi+0.1\pi)]}\right| \\ &= 0.05(4.27+1.91) = 0.309,\end{aligned}$$

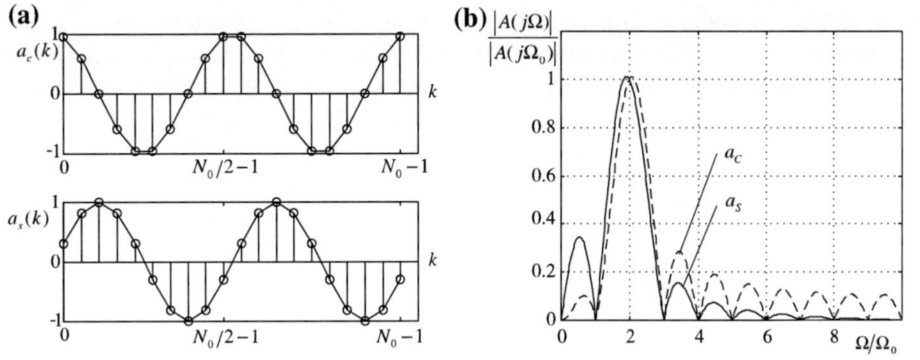

Fig. 6.10 Characteristics of full cycle sine/cosine 2nd harmonic filters: **a** impulse responses **b** frequency spectra

$$\begin{aligned}\frac{|A_s(j\Omega_a)|}{|A_s(j\Omega_1)|} &= 0.1\left|0.5\frac{\sin[5(0.25\pi - 0.1\pi)]}{\sin[0.5(0.25\pi - 0.1\pi)]} - 0.5\frac{\sin[5(0.25\pi + 0.1\pi)]}{\sin[0.5(0.25\pi + 0.1\pi)]}\right| \\ &= 0.05(4.27 - 1.91) = 0.118,\end{aligned}$$

$$\frac{|A_c(j\Omega_b)|}{|A_c(j\Omega_1)|} = 0.207 \quad \text{and} \quad \frac{|A_s(j\Omega_b)|}{|A_s(j\Omega_1)|} = 0.0535.$$

(2) Sampling frequency 600 Hz

Number of samples per cycle is $N_1 = f_S/f_1 = 12$ and:

$$\Omega_a = 2\pi f_a/f_S = 2\pi 125/600 = 0.417\pi,\ \Omega_b = 0.583\pi,\ \Omega_1 = 0.167\pi.$$

In this case one gets:

$$\frac{|A_c(j\Omega_a)|}{|A_c(j\Omega_1)|} = 0.323, \quad \frac{|A_s(j\Omega_a)|}{|A_s(j\Omega_1)|} = 0.113,$$

$$\frac{|A_c(j\Omega_b)|}{|A_c(j\Omega_1)|} = 0.227, \quad \frac{|A_s(j\Omega_b)|}{|A_s(j\Omega_1)|} = 0.047.$$

One can see that the cosine and sine filters have different abilities of suppression of oscillating components (it results also from the filters spectra shown), the latter ones are more effective in this respect. The sampling frequency has some (limited) effects on the filters characteristics.

- Recursive algorithm of a pair of orthogonal filters

Algorithms of filters having sine, cosine windows are more complex and require more calculations and more computational burden than those applying Walsh functions. This is due to variable coefficients, however, it can be reduced using recursive procedures. It is possible to find many such procedures but perhaps the most effective and the simplest result can be obtained using complex notation of a pair of orthogonal sine, cosine filters. Such representation for the filters given by Eqs. 6.26 and 6.27 is the following $(\cos(x) + j\sin(x) = \exp(jx))$:

$$y(n) = y_c(n) + jy_s(n) = \sum_{k=0}^{N-1} x(n-k)\exp\left[j\left(\frac{N-1}{2} - k\right)\Omega_0\right]. \tag{6.35}$$

Rearranging (6.35) yields:

$$y(n) = \exp\left(j\frac{N-1}{2}\Omega_0\right)\sum_{k=0}^{N-1} x(n-k)\exp(-jk\Omega_0) \tag{6.36}$$

and writing each component of the filters, output at instant n we obtain:

$$y(n) = \exp\left(j\frac{N-1}{2}\Omega_0\right)\{x(n) + x(n-1)\exp(-j\Omega_0) + \cdots + x(n-N+1)$$
$$\exp[-j(N-1)\Omega_0]\}. \tag{6.37}$$

Doing the same at instant $n - 1$ one gets:

$$y(n-1) = \exp\left(j\frac{N-1}{2}\Omega_0\right)\{x(n-1) + x(n-2)\exp(-j\Omega_0) + \cdots + x(n-N)$$
$$\exp[-j(N-1)\Omega_0]\}. \tag{6.38}$$

Multiplying the last equation by adequately chosen complex function and subtracting it from Eq. 6.37 one can reduce many components simplifying the final equation to the form:

$$y(n) - y(n-1)\exp(-j\Omega_0) = \exp\left(j\frac{N-1}{2}\Omega_0\right)[x(n) - x(n-N)\exp(-j\Omega_0)]. \tag{6.39}$$

Comparing real and imaginary parts of the equation one can get finally the recursive algorithm of a pair of orthogonal sine, cosine FIR filters.

$$Y(n) - CY(n-1) + DX, \tag{6.40}$$

where

$$Y(n) = \begin{bmatrix} y_c(n) \\ y_s(n) \end{bmatrix};\ X = \begin{bmatrix} x(n) \\ x(n-N) \end{bmatrix};\ C = \begin{bmatrix} \cos(\Omega_0) & \sin(\Omega_0) \\ -\sin(\Omega_0) & \cos(\Omega_0) \end{bmatrix};$$
$$D = \begin{bmatrix} \cos(\alpha) & -\cos(\beta) \\ \sin(\alpha) & \sin(\beta) \end{bmatrix}$$

$$\alpha = 0.5(N-1)\Omega_0;\quad \beta = 0.5(N+1)\Omega_0.$$

The recursive procedure can be applied to the filters having different parameters. They influence matrices and constants of Eq. 6.40 and can be calculated substituting:

- for full cycle filters: $N = N_0$; $\Omega_0 = 2\pi/N_0$; $\alpha = \pi - \Omega_0/2$; $\beta = \pi + \Omega_0/2$;
- for half cycle filters: $N = N_0/2$; $\Omega_0 = 2\pi/N_0$; $\alpha = \pi/2 - \Omega_0/2$; $\beta = \pi/2 + \Omega_0/2$;
- for two cycle filters: $N = 2N_0$; $\Omega_0 = 2\pi/N_0$; $\alpha = -\Omega_0/2$; $\beta = +\Omega_0/2$;
- for half cycle filter having two times greater natural frequency: $N = N_0/2$; $\Omega_0 = 4\pi/N_0$; $\alpha = \pi - \Omega_0/2$; $\beta = \pi + \Omega_0/2$.

Example 6.5 Derive recursive equations of the full cycle sine, cosine filters. Assume sampling frequency 1000 Hz and fundamental frequency 50 Hz.

Solution Number of samples per cycle $N_1 = f_S/f_1 = 20$ and $\Omega_0 = \Omega_1 = 0.1\pi$

The matrices C and D of Eq. 6.40 are then equal:

$$C = \begin{bmatrix} \cos(0.1\pi) & \sin(0.1\pi) \\ -\sin(0.1\pi) & \cos(0.1\pi) \end{bmatrix} = \begin{bmatrix} 0.951 & 0.309 \\ -0.309 & 0.951 \end{bmatrix},$$

$$D = \begin{bmatrix} \cos(\alpha) & -\cos(\beta) \\ \sin(\alpha) & \cos(\beta) \end{bmatrix} = \begin{bmatrix} -\cos(0.05\pi) & \cos(0.05\pi) \\ \sin(0.05\pi), & -\sin(0.05\pi) \end{bmatrix}$$

$$= \begin{bmatrix} -0.988 & 0.988 \\ 0.156 & -0.156 \end{bmatrix},$$

where $\alpha = \pi - \Omega_0/2 = 2 = \pi - 0.05\pi$; $\beta = \pi + \Omega_0/2 = \pi + 0.05\pi$.

Substituting them to Eq. 6.40, after rearrangement, the following recursive equations are reached:

$$y_c(n) = 0.951y_c(n-1) + 0.309y_s(n-1) + 0.988[x(n-20) - x(n)],$$

$$y_s(n) = -0.309y_c(n-1) + 0.951y_s(n-1) - 0.156[x(n-20) - x(n)].$$

The above formulae can be directly used for practical implementation and realization of filtration. Analogously one can also derive recursive procedures for other sampling frequencies and/or other window lengths.

6.3 Synthesis of FIR Filters

FIR filter analysis was presented in the preceding section. Time and frequency domain characteristics of the orthogonal filters applied in power system protection were discussed. Some general conditions allowing to get orthogonal filters were also given. Among many filters basing on sine, cosine and Walsh windows it is possible to choose in almost all cases an adequate one for given application. However, sometimes special requirements concerning for instance the shape or parameters of frequency response of FIR filters may appear, which are not satisfied with use of typical known filters. In such cases it is necessary to realize the full synthesis of FIR filters, which means in general calculating their impulse response (i.e. coefficients) from assumed frequency response.

The are many methods allowing to realize this task. Two of them are presented here, [6, 7]:

- the method applying complex Fourier series, together with special smoothing filter windows, and
- the method applying Fast Fourier Transform (FFT), which can be used for any arbitrary frequency response of the filter.

6.3.1 Application of Complex Fourier Series

It is well known from general theory that sampling process results in periodic frequency response with a period equal to sampling angular frequency ω_S. This periodic function (frequency response of sampled signal) can be described using infinite complex Fourier series given by a pair of equations:

$$H_i(\Omega) = \sum_{n=-\infty}^{\infty} h_i(n)\exp(-jn\Omega), \tag{6.41}$$

$$h_i(n) = \frac{1}{2\pi}\int_{-\pi}^{\pi} H_i(\Omega)\exp(jn\Omega)\mathrm{d}\Omega, \tag{6.42}$$

where $H_i(\Omega)$ is a filter frequency response, $h_i(n)$is a filter impulse response and its coefficients, simultaneously.

This pair of equations and the second one especially can be used for direct synthesis of FIR filters. However, some additional operations are needed, since the Fourier series is infinite, but the filter designed is to be finite, having the number of coefficients $h_i(n)$ limited to N. With the method considered it is realized by adequate cutting of infinite series of coefficients $h_i(n)$ and leaving only N of them. The operation can be considered as a product of this series of infinite coefficients and rectangular window $w(n)$:

$$\begin{gathered} h(n) = h_i(n)w(n), \\ \text{where } w(n) = \begin{cases} 1 \text{ for } 0 \le n \le N-1. \\ 0 \text{ for remaining } n \end{cases} \end{gathered} \tag{6.43}$$

The result of this cutting is frequency response of the real filter $H(\Omega)$ (having N coefficients), different from an ideal filter $H_i(\Omega)$ (6.42). This real frequency response can be calculated using equation:

$$H(\Omega) = \sum_{n=0}^{N-1} h_i(n) \exp(-jn\Omega). \tag{6.44}$$

From (6.44) one cannot find the relation between frequency responses of ideal and real filter. To get that one should use general Laplace and Fourier transform relationships. A product of two functions in time domain results in the convolution of their transforms in frequency domain (and the other way round):

$$H(\Omega) = H_i(\Omega) * W(\Omega) = \frac{1}{2\pi} \int_{-\pi}^{\pi} H_i(\Omega')W(\Omega - \Omega')\mathrm{d}\Omega', \tag{6.45}$$

where $H_i(\Omega), W(\Omega)$ are frequency responses of ideal filter and rectangular window, respectively.

It may be concluded that frequency responses of real filters obtained in such a way will have oscillatory bands (ripples) in both pass and rejection regions. It results from the known frequency response of rectangular window (zero order Walsh function presented before). This oscillation has the biggest peak values in the region of transition period of the filter characteristic. The phenomenon is known as the Gibbs effect. To limit the effect we can use different than rectangular cutting windows, designed to get small values of coefficients at both window ends. Known examples of such windows are: Blackman, Bartlett, Hamming, Hanning, Kaiser, etc. Their impulse responses of some of them are following [5, 7]:

- Hanning window

$$h(n) = 0.5\left[1 - \cos\left(\frac{2\pi n}{N-1}\right)\right], \quad \text{for } 0 \leq n \leq N-1,$$

- Hamming window

$$h(n) = 0.54 - 0.46 \cos\left(\frac{2\pi n}{N-1}\right), \quad \text{for } 0 \leq n \leq N-1,$$

- Blackman window

$$h(n) = 0.42 - 0.5 \cos\left(\frac{2\pi n}{N-1}\right) + 0.08 \cos\left(\frac{4\pi n}{N-1}\right), \quad \text{for } 0 \leq n \leq N-1.$$

All these windows decrease oscillations of frequency responses of the filters both at pass and stop bands, however, they increase the transition band. These effects are presented in Fig. 6.11 for an example of low-pass filter and application of rectangular window and the other windows as above. It is seen that the noise at stop band is decreased more effectively (in the case of Blackman window even by 80 dB), which is an evident advantage, however, the transition period becomes longer. Sometimes it may lead to bad results in power system protection and that is why application of given widows should be considered carefully.

Example 6.6 Design a low-pass digital FIR filter with N coefficients and the cut-off frequency Ω_0, having unity transfer function magnitude in the pass region and zero in the rejection band. The filter should have linear phase versus frequency.

Solution According to the requirements specified, the frequency response of the filter can be defined as:

$$H_i(\Omega) = \begin{cases} \exp(-j\Omega\lambda) & \text{for } -\Omega_0 \le \Omega \le \Omega_0 \\ 0 & \text{for other frequencies} \end{cases},$$

where λ is a group delay, to be selected for given value of the filter coefficients N.

The filter coefficients are calculated with use of Eq. 6.42:

$$h_i(n) = \frac{1}{2\pi}\int_{-\pi}^{\pi} H_i(\Omega)\exp(jn\Omega)\mathrm{d}\Omega = \frac{1}{2\pi}\int_{-\Omega_0}^{\Omega_0} \exp[j\Omega(n-\lambda)]\mathrm{d}\Omega = \frac{\sin[\Omega_0(n-\lambda)]}{\pi(n-\lambda)}.$$

From the above infinite number of coefficients one should take only N, cutting off the rest. Additionally one should assure that the remaining coefficients would have even or odd symmetry with respect to the window center. The latter requirement can be fulfilled by selecting appropriate group delay value equal $(N-1)/2$. With this in mind the FIR filter coefficients are given by:

$$h(n) = \frac{\sin\left[\Omega_0\left(n - \frac{N-1}{2}\right)\right]}{\pi\left(n - \frac{N-1}{2}\right)} \quad \text{for } 0 \le n \le N-1.$$

The designed filter coefficients as well as the frequency responses of the ideal (IIR) and real (FIR) filters are shown in Fig. 6.11. One can see that the frequency characteristic of the real filter exhibits significant oscillations, which can be reduced if an appropriate smoothing window is applied. As an effect of Hanning filter the ripples are greatly limited (curve 3), however, the resulting filter has much broader transition band, which may not always be acceptable in power system protection applications. Utilization of smoothing windows with the standard sine, cosine window filters can lead to significant change of the band-pass filter characteristics and unacceptable errors for signal components lower frequencies, which in turn may result in improper operation of protection devices. Therefore application of smoothing windows is to be done with care.

Example 6.7 Determine the impulse response (coefficients) of the band-pass FIR filter, having unity gain for frequencies between $f_1 = 50$ and $f_2 = 150$ Hz, and zero gain outside this band. Apply design method with complex Fourier series. Draw this impulse response and resulting frequency response of the filter for $N = 20$ and sampling frequency $f_S = 1000$ Hz.

Fig. 6.11 Impulse **a** and frequency **b** responses of low-pass FIR filter designed with use of complex Fourier series method; (*1*) ideal filter, (*2*) real filter, (*3*) response after application of smoothing Hanning window (Example 6.6)

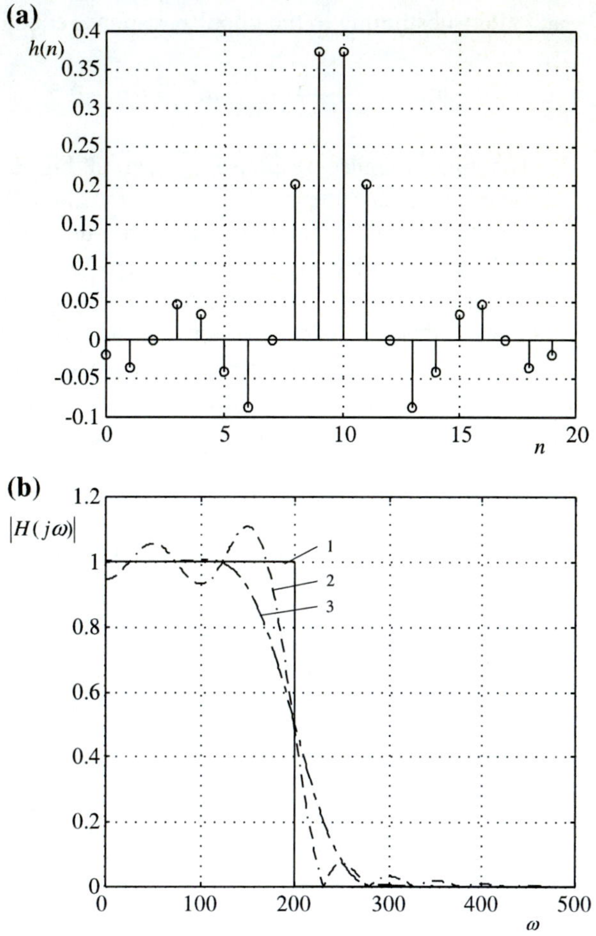

Solution According to (6.42) one can write:

$$h(n) = \frac{1}{2\pi}\int_{-\Omega_2}^{-\Omega_1} \exp[j\Omega(n-\lambda)]\mathrm{d}\Omega + \frac{1}{2\pi}\int_{\Omega_1}^{\Omega_2} \exp[j\Omega(n-\lambda)]\mathrm{d}\Omega$$

and after calculation of integrals and rearrangements, taking as before the group delay equal $0.5(N-1)$ (symmetry conditions), one obtains:

$$h(n) = \frac{1}{\pi[n-0.5(N-1)]}\{\sin[\Omega_2(n-0.5(N-1))] - \sin[\Omega(n-0.5(N-1))]\}.$$

For the specified filter parameters one can calculate:

$$0.5(N-1) = 9.5;\ \Omega_1 = 2\pi f_1/f_S = 2\pi 50/1000 = 0.1\pi;\ \Omega_2 = 0.3\pi$$

and after substituting to the impulse response equation $h(n)$ one gets the final form:

$$h(n) = \frac{1}{\pi(n-9.5)}\{\sin[0.3\pi(n-9.5)] - \sin[0.1\pi(n-9.5)]\}.$$

The filter impulse response, shown in Fig. 6.12a, is symmetrical, thus the designed filter has linear phase. From the filter frequency response (Fig. 6.12b) calculated with use of FFT it is seen that designed filter is quite good band-pass filter. Narrowing its pass band to the range 75–125 Hz (separate design performed) results in increase of ripples of the frequency characteristics (Fig. 6.12c). One can counteract this by increasing the number of filter coefficients (longer impulse response). In Fig. 6.12d the new characteristic is shown for $N = 100$, representing very good band-pass filter with steep transition bands. Almost ideal results are obtained after superimposing the smoothing Hanning window (Fig. 6.12e).

Example 6.8 Design high-pass FIR filter with the cut-off frequency 150 Hz, applying the complex Fourier series method. Assume sampling frequency 1000 Hz. Determine filter frequency response for the number of coefficients $N = 20$ and 100.

Solution The high-pass digital filter can be treated as a band-pass one with the higher cut-off frequency equal to half of sampling rate. According to Eq. 6.42 and preceding example designed with the same approach, one can write now:

$$h(n) = \frac{1}{2\pi}\int_{-\Omega_p/2}^{-\Omega} \exp[j\Omega(n-\lambda a)]\mathrm{d}\Omega + \frac{1}{2\pi}\int_{\Omega_0}^{\Omega_p/2} \exp[j\Omega(n-\lambda)]\mathrm{d}\Omega,$$

where

$$\lambda = \frac{N-1}{N};\ \Omega_0 = 2\pi\frac{f_0}{f_S};\ \frac{\Omega_S}{2} = 2\pi\frac{0.5f_S}{f_S} = \pi.$$

Finally, one obtains:

$$h(n) = \frac{1}{\pi[n-0.5(n-1)]}\{\sin[\pi(n-0.5(N-1)] - \sin[\Omega_0(n-0.5(N-1)]\}.$$

The filter impulse response for $N = 20$ is shown in Fig. 6.13a, while its frequency response is presented in Fig. 6.13b. Superimposing the smoothing window one gets the resulting filter spectrum as seen in Fig. 6.13c. The steep slope of the transition band can be improved (increased) taking more filter coefficients (Fig. 6.13d). The final filter characteristic with $N = 100$ and additional smoothing window is shown in Fig. 6.13e. One should understand, however, that taking higher number of N results in longer transient response of the filter in time domain. One can shorten it (by the same number of coefficients) when higher sampling frequency is selected, if possible.

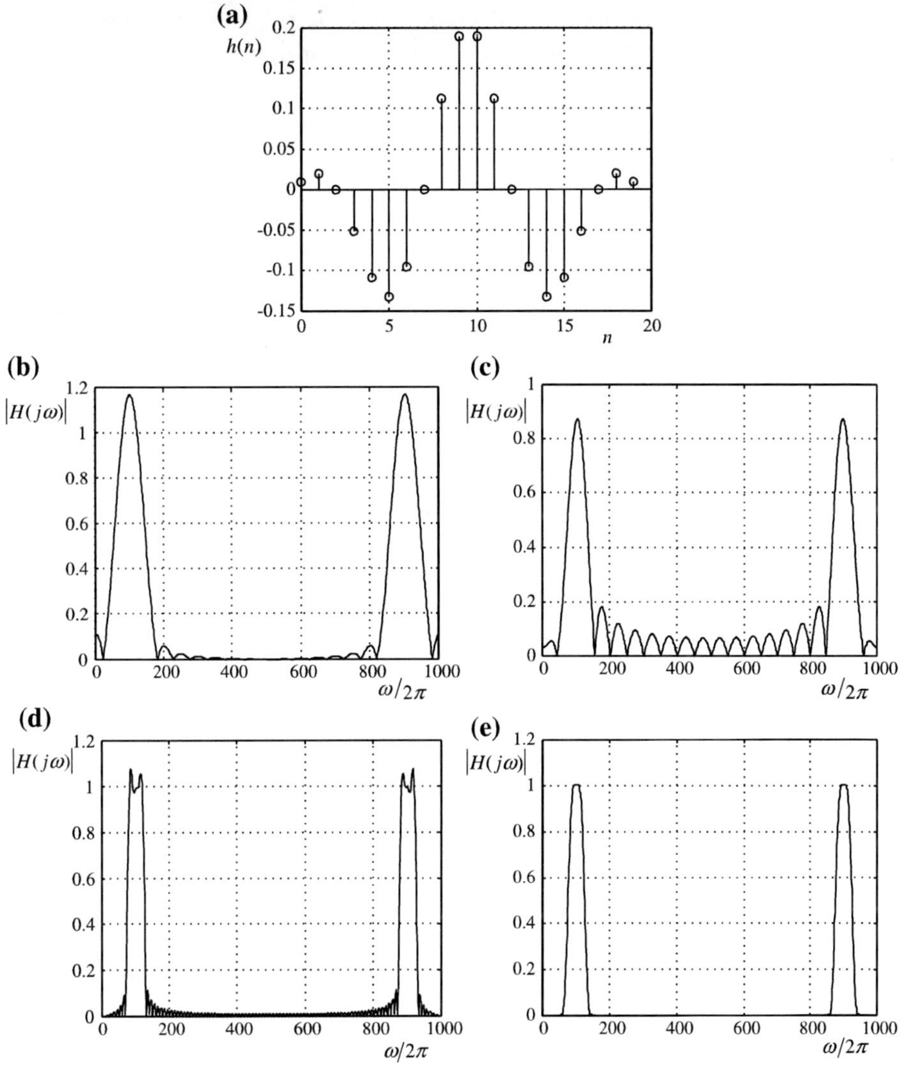

Fig. 6.12 Impulse response (*a*) and filter frequency characteristics for: **b** $N = 20, f_1 = 50$ Hz, $f_2 = 150$ Hz, **c** $N = 20, f_1 = 75$ Hz, $f_2 = 125$ Hz, **d** $N = 100, f_1 = 75$ Hz, $f_2 = 125$ Hz, **e** as **d**, with application of Hamming window

6.3.2 Application of Fast Fourier Transform

The method presented here is based on sampling in frequency domain that relies on substituting for continuous frequency response its samples taken for identical frequency distance and application to these samples inverse discrete Fourier transform (DFT) to calculate filter coefficients. To follow the relationship between the filter coefficients and its spectrum let us assume that the filter coefficients (impulse response) are known. In order to find the frequency response one can calculate Z transform first, which for N elements of the sum yields:

$$H(z) = \sum_{n=0}^{N-1} h(n) z^{-n}. \tag{6.46}$$

From analysis of discrete systems it results that the points on unity circle on z plane (counterclockwise) have fixed values of frequency. One turn along this circle is equivalent to sampling frequency. When one divides this circle into N equal sections, the distance between two consecutive points equals $\Delta f = f_S/N$. The values of frequency at any point of the circle are given by the equation:

$$z_k = \exp\left(j\frac{2\pi}{N}k\right), \quad \text{for } k = 0, 1, \ldots, N-1, \tag{6.47}$$

where $\frac{k}{N} = \frac{f_k}{f_S}$.

Substitution of (6.47) into (6.46) allows finding frequency response at these discrete points. This response is now a function of k:

$$H(k) = H\left[\exp\left(j\frac{2\pi}{N}k\right)\right], \quad \text{for } k = 0, 1, \ldots, N-1. \tag{6.48}$$

Substituting the above to (6.46) yields:

$$H(k) = \sum_{n=0}^{N-1} h(n) \exp\left(-j\frac{2\pi}{N}kn\right), \quad \text{for } k = 0, 1, \ldots, N-1. \tag{6.49}$$

Equation 6.49 describes the relationship between N-element discrete series in time and frequency domains. Equation 6.49 is one of the pair of equations describing DFT, which is numerically implemented as FFT. The second equation of this DFT, allowing to calculate N-element series in time domain having N elements in frequency domain is:

$$h(n) = \frac{1}{N} \sum_{k=0}^{N-1} H(k) \exp\left(j\frac{2\pi}{N}kn\right), \quad \text{for } n = 0, 1, \ldots, N-1. \tag{6.50}$$

The pair of Eqs. 6.49 and 6.50 are discrete versions of Fourier transform allowing to jump from time to frequency domain and the other way round. The latter of the equations, i.e. (6.50) is very important for synthesis of FIR filters, allowing for direct calculation of filter coefficients using samples in frequency domain. One must remember, however, that on one hand the magnitude of frequency response should be received as assumed, but on the other hand—the values at other frequency points may show substantial differences. It can be estimated analytically substituting (6.50) to (6.46) and rearranging. In this way one gets continuous frequency response of the designed filter:

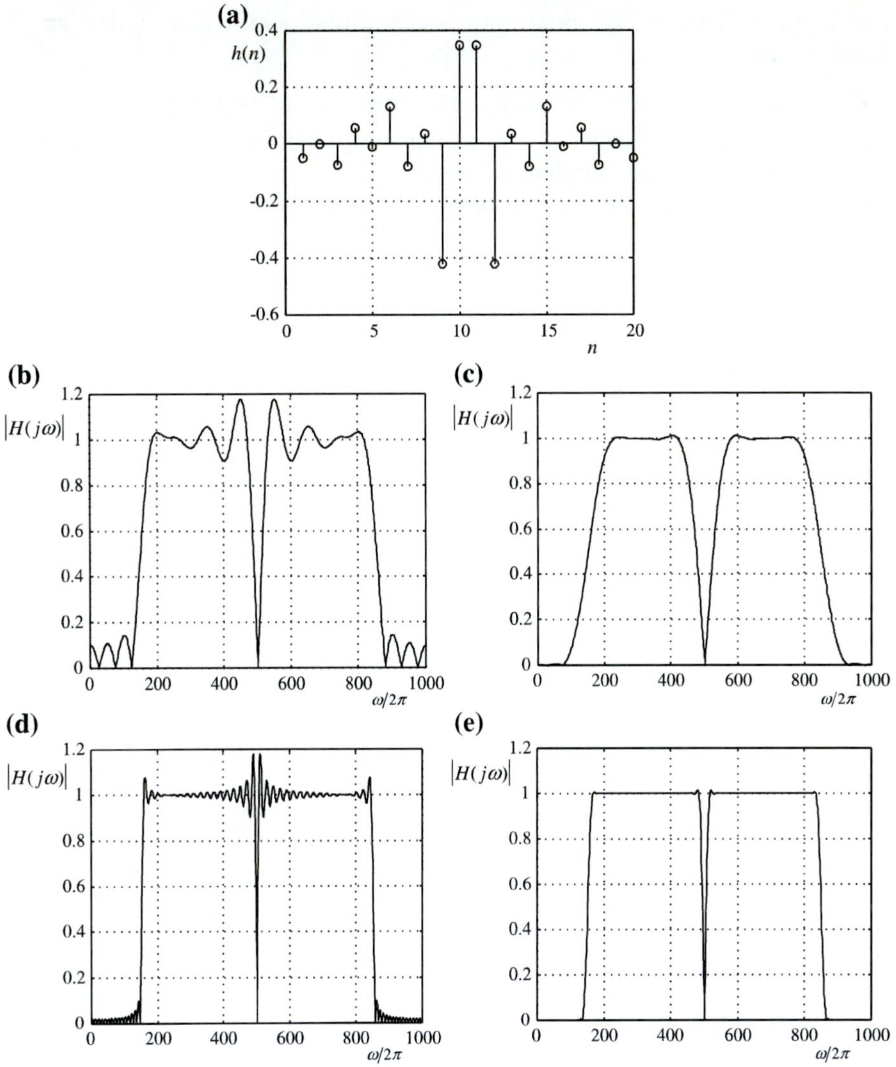

Fig. 6.13 Impulse response (**a**) and filter frequency characteristics for: **b** $N = 20$, **c** as **b**, with application of Hamming window, **d** $N = 100$, **e** as **d**, with application of Hamming window

$$H(\Omega) = \sum_{k=0}^{N-1} H(k) \frac{\sin\left(N\frac{\Omega}{2} - \pi k\right)}{N\,\sin\left(\frac{\Omega}{2} - \pi\frac{k}{N}\right)} \exp\left[-j\frac{N-1}{2}\left(\Omega - 2\pi\frac{k}{N}\right)\right], \tag{6.51}$$

where $\Omega = 2\pi\frac{f}{f_S}$.

It is seen that for fixed discrete frequency values ($f_k = kf_S/N$) the assumed and obtained values are the same. The spectra at other frequencies are given by (6.51).

If the differences between assumed continuous frequency and obtained one are for any frequency too big, one can apply smoothing windows as described above.

The procedure of FIR filter design using sampling in frequency domain is the following:

- assume given continuous frequency response of the digital filter,
- assume sampling frequency f_S and number of filter coefficients N (according to general rules of the sampling theorem),
- present the filter frequency response in the range of frequency from zero to sampling frequency and realize the sampling in frequency domain ($\Delta f = f_S/N$),
- use the values to calculate discrete inverse Fourier transform in order to receive filter coefficients (impulse response),
- check the resulting continuous frequency response and in case of high oscillations either in pass or in stop band choose and introduce special smoothing window.

Example 6.9 This course of design is illustrated with an example of low-pass filter synthesis. It is assumed that the filter should have unity magnitude of frequency response from zero to 200 Hz and zero gain for other frequencies. Further parameters are: $N = 20$ and $f_S = 1000\,\text{Hz}$. Thus the assumed digitalized frequency response of the filter is as shown in Fig. 6.14, which can also be expressed by:

$$|H(k)| = \begin{cases} 1 & \text{for} \quad k = 0, 1, 2, 3, 4 \text{ and } 16, 17, 18, 19, \\ 0 & \qquad\ \ \text{other values of } k, \end{cases}$$

$$\arg[H(k)] =$$

$$\begin{cases} \exp[-jk\pi(N-1)/N] = \exp(-j\pi k 19/20) & \text{for} \quad k = 0, 1, 2, 3, 4 \\ \exp[j(N-k)\pi(N-1)/N] = \exp[j(20-k)\pi 19/20] & \qquad\ \ k = 16, 17, 18, 19. \end{cases}$$

In the next step inverse digital Fourier transform is performed, which yields coefficients of designed FIR filter $h(n)$, shown in Fig. 6.15a. The filter can now be realized and its frequency response calculated. The resulting frequency characteristic is presented in Fig. 6.15b, compared with the one obtained after application of a smoothing window that helps to remove the ripples in original filter characteristic.

Example 6.10 Applying the method of sampling in frequency domain design the low-pass linear phase FIR filter with unity gain between 50 and 150 Hz, zero outside of this range. Assume sampling frequency 1000 Hz and number of filter coefficients $N = 20$.

Solution From the specified parameters it results that the distance between consecutive discrete frequency points of the filter frequency response equals:

$$\Delta f_k = f_S/N = 1000/20 = 50\,\text{Hz}.$$

The filter spectrum is symmetrical with respect to the Nyquist frequency and in the range between 0 and 950 Hz there exist 20 points of this spectrum, given as follows:

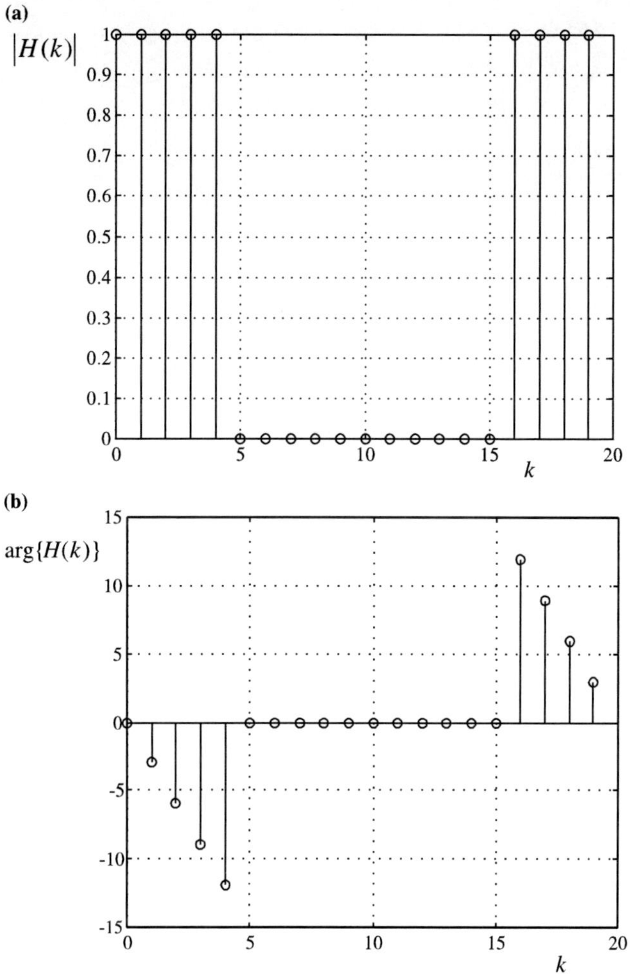

Fig. 6.14 Magnitude **a** and phase **b** of the filter to be designed (Example 6.9)

$$|H(k)| = \begin{cases} 1 & \text{for}\, k = 1, 2, 3, \text{ i.e. for} f = 50, 100, 150\,\text{Hz} \\ 1 & \text{for}\, k = 17, 18, 19, \text{ i.e. for } = 1000 - 150, 1000 - 100, 1000 - 50\,\text{Hz} \\ 0 & \text{for other frequencies.} \end{cases}$$

From the requirement of linear phase it follows:

$$\arg[H(k)] = \begin{cases} -k\pi\frac{N-1}{N} = -k\pi 19/20 & \text{for } k = 1, 2, 3 \\ (N-k)\pi\frac{N-1}{N} = (20-k)\pi 19/20 & \text{for } k = 17, 18, 19. \end{cases}$$

Thus the complex filter spectrum (including both magnitude and phase) is described as:

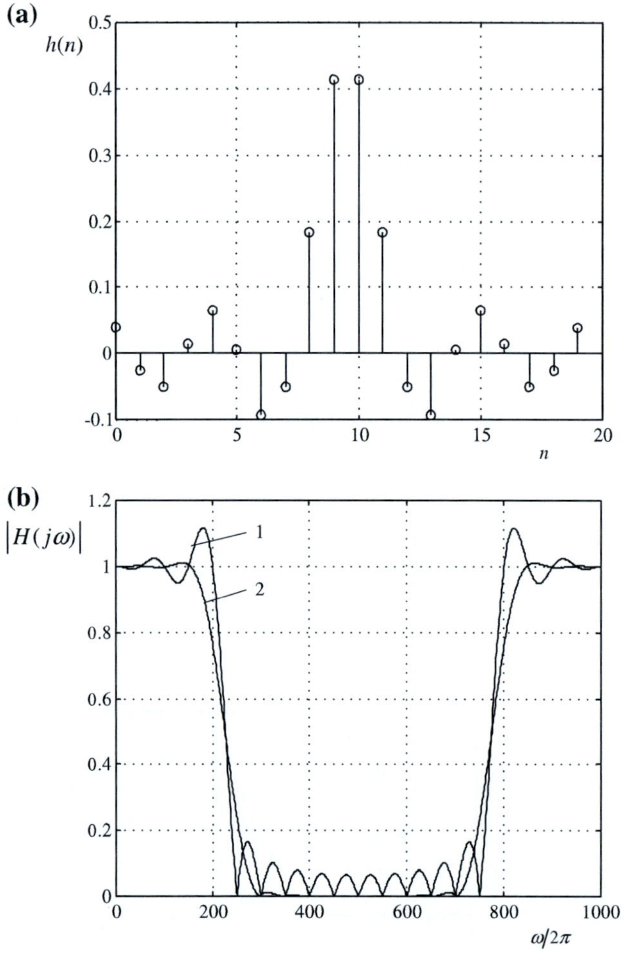

Fig. 6.15 Impulse **a** and frequency **b** responses of the designed filter (Example 6.9); *1*—original filter spectrum, *2*—spectrum modified by application of a smoothing window

$$H(k) = \begin{cases} \exp[-jk\pi(N-1)/N] = \exp(-jk\pi 19/20) & \text{for } k = 1,2,3 \\ \exp[j(N-k)\pi(N-1)/N = \exp[j(20-k)\pi N - 1/N] & \text{for } k = 17,18,19. \end{cases}$$

The above filter spectrum (Fig. 6.16a) has been subjected to inverse Fourier transformation, giving the following impulse response (filter coefficients):

$h(0) = h(19) = -0.0928$; $h(1) = h(18) = -0.046$; $h(2) = h(17) = 0$;
$h(3) = h(16) = -0.0054$;
$h(4) = h(15) = -0.0654$; $h(5) = h(14) = -0.1249$; $h(6) = h(13) = -0.1121$;
$h(7) = h(12) = 0$;
$h(8) = h(11) = 0.1635$; $h(9) = h(10) = 0.2830$;

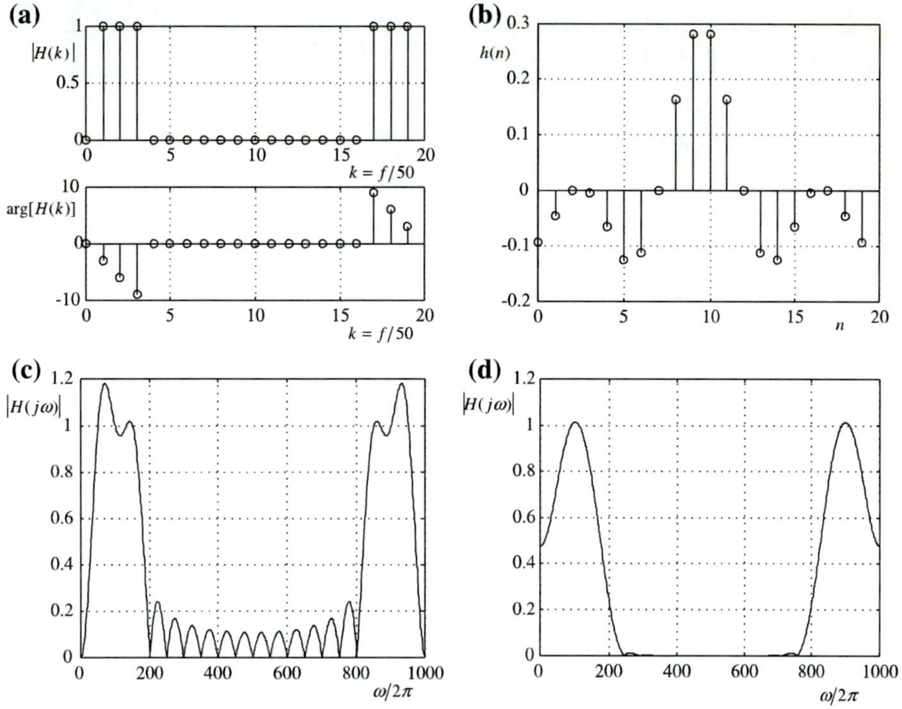

Fig. 6.16 Characteristics of the band-pass filter designed with frequency response sampling method (Example 6.10): **a** assumed filter frequency response, **b** filter impulse response, **c** effective filter frequency response, **d** frequency response after application of Hamming window

The filter impulse response is presented in Fig. 6.16b, whereas the corresponding filter frequency spectrum is seen in Fig. 6.16c. As expected (refer to previous examples) the frequency response of designed filter reveals some oscillations in both pass and stop bands that can be reduced with use of a smoothing window (Fig. 6.16d). One can see that the oscillations disappeared, however, it is at the cost of worse steepness of the transition band. This can be again improved taking more filter coefficients (compare example of band-pass filter design with previous method).

Example 6.11 Design a high-pass FIR filter with cut-off frequency 150 Hz, applying the method of sampling in frequency domain. Assume $N = 20$ and sampling frequency 1000 Hz.

Solution The filter spectrum (Fig. 6.17a) can now be expressed as:

$$H(k) = \begin{cases} 0 & \text{for} \begin{cases} k = 0, 1, 2 \text{ and } k = 18, 19 \\ k = 10 \text{ from the condition of symmetry} \end{cases} \\ \exp[-jk(N-1)/N] & \text{for } k = 3\ldots9 \\ \exp[j(N-k)\pi(N-1)/N] & \text{for } k = 11\ldots17. \end{cases}$$

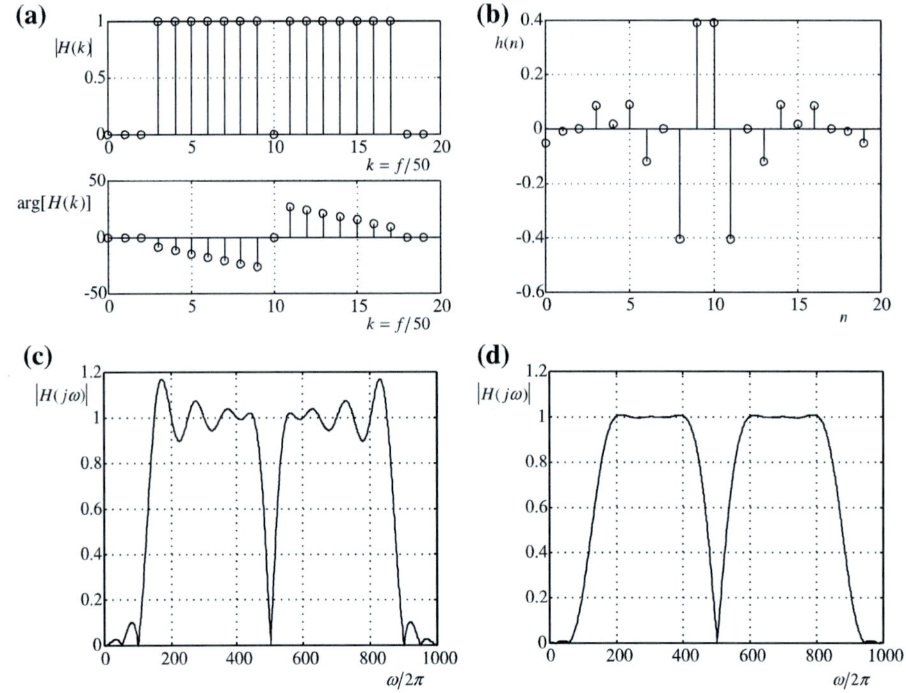

Fig. 6.17 Characteristics of the high-pass filter designed with frequency response sampling method (Example 6.11): **a** assumed filter frequency response **b** filter impulse response **c** effective filter frequency response **d** frequency response after application of Hamming window

Applying inverse Fourier transformation results in the following filter coefficients:

$$h(0) = h(19) = -0.0503;\ h(1) = h(18) = -0.0077;\ h(2) = h(17) = 0;$$
$$h(3) = h(16) = 0.0848;$$
$$h(4) = h(15) = 0.018;\ h(5) = h(14) = 0.088;\ h(6) = h(13) = -0.1162;$$
$$h(7) = h(12) = 0;$$
$$h(8) = h(11) = -0.4061;\ h(9) = h(10) = 0.3914;$$

In Fig. 6.17b–d the filter impulse response along with corresponding filter spectrum and spectrum after application of Hamming window are presented. All remarks from preceding examples apply also to this case (spectrum oscillations, effects of smoothing window, effects of higher number of filter coefficients), thus they will not be repeated here again.

References

1. Altuve CM, Diaz I, Vasquez E (1996) Fourier and Walsh digital filtering algorithms for distance protection. IEEE Trans Power Syst 11:457–462
2. Azizi SA (1990) Entwurf und Realisierung digitaler filter. R. Oldenburg, Verlag, München
3. Blackman NM (1974) Sinusoids versus Walsh functions. Proc IEEE 62:346–354
4. Bozic M (1979) Digital and Kalman filtering. Edward Arnold Ltd., London
5. Hamming RW (1989) Digital filters. General Publishing Company, Toronto
6. Jackson LB (1996) Digital filters and signal processing. Kluwer Academic Publisher, Boston
7. Szafran J, Wiszniewski A (2001) Measurement and decision algorithms of digital protection and control. WNT, Warszawa (in Polish)
8. Ungrad H, Winkler W, Wiszniewski A (1995) Protection techniques in electrical energy systems. Marcel Dekker Inc., New York

Chapter 7
Correction of Errors Introduced by Instrument Transformers

Whatever sophisticated algorithms of signal processing are further used, the overall performance of the protection relay is also a function of quality of the analog signal pre-processing path including current and voltage transformers, analog antialiasing filters and A/D converters. In this chapter the effects of VT and CT errors on protection operation are discussed and some approaches and algorithms for compensation of the errors are presented.

7.1 Correction of Voltage Transformers Performance

Inductive voltage transformers, if selected in a proper way, do not introduce errors which could affect operation of protective devices. It is so in both steady state and during transients. Therefore there is no need to make special arrangements which would correct the errors.

However, in case of capacitive voltage transformers it is otherwise. Since the CVTs are equipped with the compensating inductance and the ferroresonance suppressing circuits, they may introduce substantial errors if there is a sudden change of the primary voltage. The errors may cause maloperation of protective devices, particularly if there is a sudden drop of primary voltage. Therefore correction of expected errors may be advantageous.

The equivalent circuit of a capacitive voltage transformer is presented in Fig. 7.1. The ferroresonance suppressing circuit is represented by the elements R_f, L_f, C_f.

The entire circuit may be considered linear, therefore one may write the following Laplace transform relation:

$$\frac{v_2}{v_1} = N_u F(s), \tag{7.1}$$

W. Rebizant et al., *Digital Signal Processing in Power System Protection and Control,* Signals and Communication Technology,
DOI: 10.1007/978-0-85729-802-7_7,

Fig. 7.1 Equivalent circuit of a capacitive voltage transformer

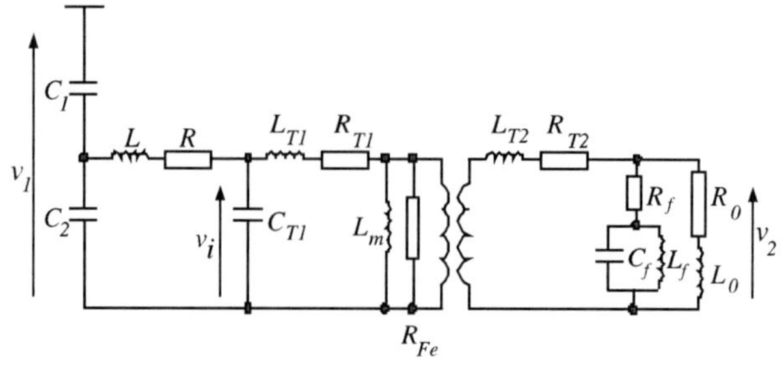

where N_u is a transformation ratio of the CVT, $F(s)$ is a Laplace transfer function of the circuit, and v_1, v_2 are primary and secondary CVT voltages.

The correction ought to process the secondary voltage by a transfer function which is equal to inverse of the CVT transfer function. Therefore, the corrected secondary voltage v_{2c} becomes:

$$v_{2c} = v_2 F_1(s) \approx v_1 N_u, \tag{7.2}$$

where $F_1(s) = 1/(F(s))$.

Digital representation of the transfer function $F_1(s)$ may be obtained by introducing discrete operators of integration, e.g. the Euler's operator:

$$s^{-1} \Rightarrow \frac{T_S z^{-1}}{1 - z^{-1}}, \tag{7.3}$$

remembering, that multiplication by z^{-1} in the Z-transform domain represents delay by one sampling period T_S in time domain.

7.2 Correction of Current Transformer Errors

7.2.1 Formulation of the Problem

The problems with errors of current transformers are much greater than the ones of voltage transformers. It is so because:

- errors of the CTs are much more damaging for proper operation of digital devices,
- current transformers are strongly non-linear, because of the non-linear magnetizing characteristic of the cores, therefore their errors are much more difficult to calculate and correct,
- due to the hysteresis loop of the magnetization characteristic of the core it is hardly possible to establish the starting value of the core flux,
- range of levels of primary currents and the time constants of their DC components are great.

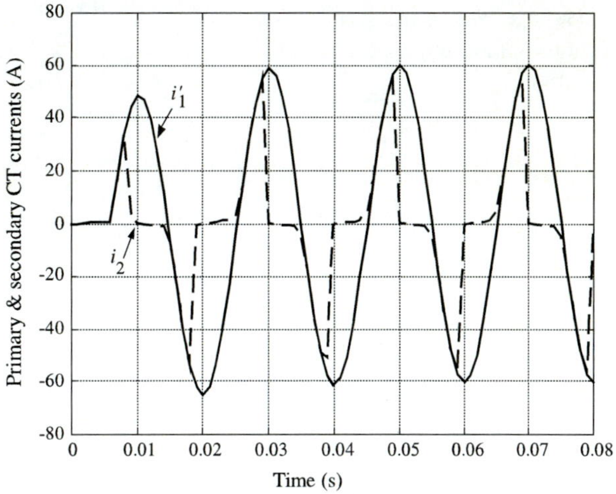

Fig. 7.2 Steady state errors of the CT due to large input signal magnitude

If during the steady state operation the primary current levels of the transformer are within the accuracy range (they are below the rated level multiplied by the accuracy limit factor) the errors of transformation do not affect operation of protective relays. However, if the primary currents are larger—it is otherwise. Typical waveshapes of primary and secondary currents in case of a CT with purely resistive burden are presented in Fig. 7.2.

It is obvious that protection criterion values calculated on the basis of saturated CT secondary signal may fall quite distantly from their correct values, which might have been determined if the CT primary unsaturated signal was available. Erroneous measurement may in consequence lead to false decisions (e.g. underreaching of overcurrent relays, overestimation of fault loop impedance in distance relays) and protection maloperation. All those problems are well described e.g. in [6]. Thus, it can be stated that CT saturation phenomenon may impair protection system reliability if appropriate algorithms for saturation detection and/or correction are not applied to eliminate the problem.

There are three ways of overcoming the problem and make correct operation of the protective devices possible. They are:

- application of the CTs with larger accuracy limit factor, i.e., CTs that do not saturate during steady state operation,
- application of the relay algorithms, which are insensitive to the errors (for example they make measurements only in the fraction of periods, when the transformation is correct, or they introduce large restraint),
- calculation of the primary current on the ground of the secondary, corrupted waveshape.

The first way is obvious, but in some cases it requires the CTs with the cores of very large cross-sections. The second and third ways do not require very large cross-sections, but they demand identification of the fraction of periods, when the CT is not saturated.

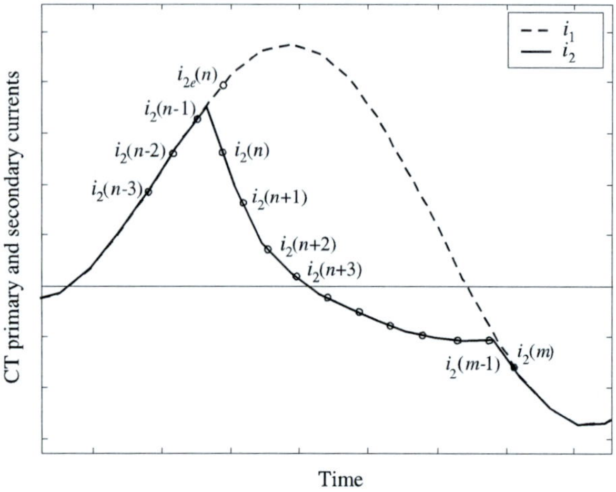

Fig. 7.3 Transient waveform of the saturated CT secondary current

7.2.2 Detection of the Unsaturated Fragment of the Waveshape

An exemplary detailed waveshape of the CT saturated secondary current is shown in Fig. 7.3, that can be used as illustration and basis for the developed saturation detection methods.

One may see that the CT saturates between the samples $(n - 1)$ and (n). It means that the core flux reached the saturation level. The saturation ends between the samples $(m - 1)$ and (m). Therefore, the transformation is correct, which means that the secondary current is proportional to the primary current $(i_2 = N_i i_1)$, until the sample (n), and after the sample (m).

Determination of the sample (n) is based on the comparison of the measured secondary current $i_2(n)$ and the estimated secondary current $i_{2e}(n)$:

$$|i_{2e}(n) - i_2(n)| \geq \Delta |i_{2e}(n)| \tag{7.4}$$

and

$$|i_{2e}(n)| - |i_2(n)| \geq 0 \tag{7.5}$$

If the difference between the two exceeds a part Δ of the estimated level, it means that there is a sudden change of the waveshape and saturation happened between the samples.

If the condition (7.4) is satisfied, but

$$|i_{2e}(n)| - |i_2(n)| \leq 0 \tag{7.6}$$

like between the samples (m) and $(m - 1)$ in Fig. 7.3, it marks the end of saturation.

Estimation of $i_{2e}(n)$ may be done with the assumption that the derivatives of the secondary current between the samples $(n - 1)$ and $(n - 2)$ are the same, as between the samples (n) and $(n-1)$. The simplest way is to assume that the first order derivatives are equal:

$$\frac{i_{2e}(n) - i_2(n-1)}{T_S} \approx \frac{i_2(n-1) - i_2(n-2)}{T_S}, \tag{7.7}$$

which leads to the formula:

$$i_{2e}(n) \approx 2i_2(n-1) - i_2(n-2). \tag{7.8}$$

And now one may apply the condition (7.4) to find out whether there was a rapid change in the waveshape.

Second and third order derivatives can be calculated with the well known formulae:

$$i''(n) \approx \frac{i'(n) - i'(n-1)}{T_S}, \tag{7.9}$$

$$i'''(n) \approx \frac{i''(n) - i''(n-1)}{T_S}. \tag{7.10}$$

If one assumes that the second order derivative is the same in nth sampling period, then the value of $i_{2e}(n)$ may be estimated by the formula:

$$i_{2e}(n) \approx 3i_2(n-1) - 3i_2(n-2) + i_2(n-3), \tag{7.11}$$

which is far more accurate than (7.8). However, if one wants to increase accuracy even more, one may assume that the third derivative is to be the same between the samples, what gives the condition:

$$i_{2e}(n) \approx 4i_2(n-1) - 6i_2(n-2) + 4i_2(n-3) - i_2(n-4), \tag{7.12}$$

The same procedure may be repeated to estimate the sample $i_{2e}(m)$ what marks the end of saturation.

Saturations of CTs caused by the primary currents with large DC components are much more likely than in case of transformation of the steady state symmetrical currents. It is because the DC components, particularly the ones with long time constants, increase the magnetic flux in the core more intensely than the sinusoidal currents. Therefore, correction of the errors caused by saturation of CTs in transient states are more needed than the correction in steady states.

Determination of the moments, which mark the beginning of core saturation and the end of it, may be done by the same formulae, which were developed for symmetrical currents (Eqs. 7.4–7.12). This means that between the samples (n) and (m) the transformation is corrupted by the saturation of the magnetic core.

The other method of marking the CT saturation ending point can be based on the fact (for CTs with resistive burden) that the integral of the secondary current within the time span of saturation must be zero:

$$\int_{t_n}^{t_m} i_2(t)\mathrm{d}t = 0, \tag{7.13}$$

where t_n is a beginning of saturation, and t_m is an end of saturation.

Applying the rectangle integrating formula (Euler's procedure) the continuous integral (7.13) can also be approximated numerically as follows:

$$S(m) = T_S \sum_{k=n}^{m} i_2(k). \tag{7.14}$$

Now, the detection of the first sample in the new unsaturated period $m + 1$ may be based on the formula:

$$\mathrm{Sign}[S(m)] \neq \mathrm{Sign}[S(m+1)]. \tag{7.15}$$

7.2.3 Correction of the Secondary Current

Despite the transient errors caused by the CT core saturation, operation of the protective devices may be correct, however, it demands careful processing of the signals, what may be done in a number of ways.

If the aim of processing is to make proper operation of differential relays, the best way is to use current signals only in the fraction of periods, when the CTs at both sides are not saturated. Therefore, the differential current is detected taking into consideration non-corrupted current samples of the input currents: $i_2(n-1), i_2(n-2), i_2(n-3)$, etc., and $i_2(m), i_2(m+1), i_2(m+2)$, etc., until the beginning of saturation in the next period. In fact, this unsaturated fragment may be artificially expanded particularly in the cases, when the secondary current samples $i_{2e}(n), i_{2e}(n+1)$ and $i_{2e}(n+2)$ are estimated by the formula (7.12), which returns very accurate results.

If the aim of the processing is the proper operation of protective devices that are based on the calculation of the fundamental frequency components (overcurrent, distance, etc.), the amplitude I_1 of the current ought to be determined on the ground of the distorted current signal. The following simple secondary current signal model valid in the fraction of period during which the CT core is not saturated and the transformation of the current is correct, is represented by the formula:

$$i_2 = N_i i_1 \cong I_a \mathrm{e}^{-\frac{t}{T_a}} - I_{1m} \cos(\omega t - \alpha), \tag{7.16}$$

where N_i is a current transformer ratio, i_1 is a primary current, I_a is a DC decaying component initial value, T_a is a time constant of the decay, I_{1m} is an amplitude of the AC component, and α is a phase angle of the AC component.

7.2.3.1 Calculation of the Amplitude I_1 Based on the Unsaturated Fractions of the Waveshape

There are several methods of calculation of the amplitude, considering only the samples of the secondary current taken in the fraction of each period during which the CT core is not saturated. Perhaps the easiest one is to consider the value of the secondary current i_2, and the first derivative of i_2', correlating both of them with the sinusoidal function:

$$b_1(n) = \left(\frac{2T_S}{T_w - \frac{T_1 \sin(\omega T_w)}{2\pi}}\right) \sum_{m=n-p}^{n} i_2(m) \sin\left[\omega_1 (m-n) T_S + \frac{T_w}{2}\right] = f_-, \quad (7.17)$$

$$a_1(n) = \frac{T_1}{2\pi}\left(\frac{2T_S}{T_w - \frac{T_1 \sin(\omega T_w)}{2\pi}}\right) \sum_{m=n-p}^{n} i_2'(m) \sin\left[\omega_1 (m-n) T_S + \frac{T_w}{2}\right] = f_+, \quad (7.18)$$

where

$$i_2(m) \approx \frac{i_2(m) + i_2(m-1)}{2}$$

$$i_2'(m) \approx \frac{i_2(m) - i_2(m-1)}{T_S}$$

Eventually:

$$I_1 \approx \sqrt{a_1^2 + b_1^2}. \quad (7.19)$$

7.2.3.2 Reproduction of the Distorted Waveshape

Equivalent circuit of the CT reduced to the secondary side is presented in Fig. 7.4 and the magnetizing characteristic of the core in Fig. 7.5.

Estimation of the true values of the secondary current by means of the formula (7.12) is sufficiently accurate while calculating a maximum of two samples of the current after saturation. Therefore, one may assume that:

$$i_{2e}(n) \approx i_1(n),$$

$$i_{2e}(n+1) \approx i_1(n+1).$$

Thus, the samples of the magnetizing current become:

$$i_m(n) \approx i_{2e}(n) - i_2(n), \quad (7.20)$$

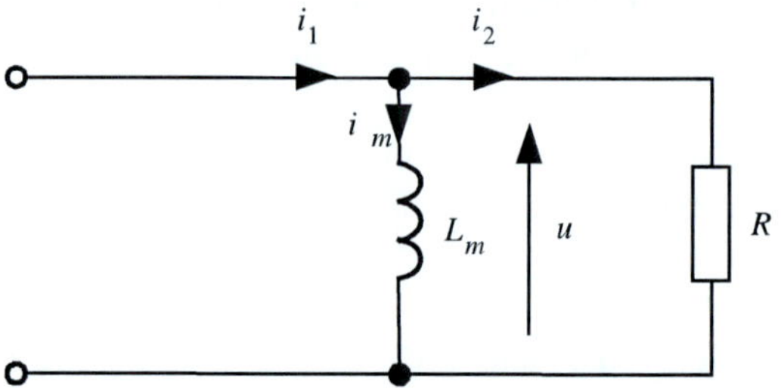

Fig. 7.4 Simplified equivalent circuit of the saturated CT

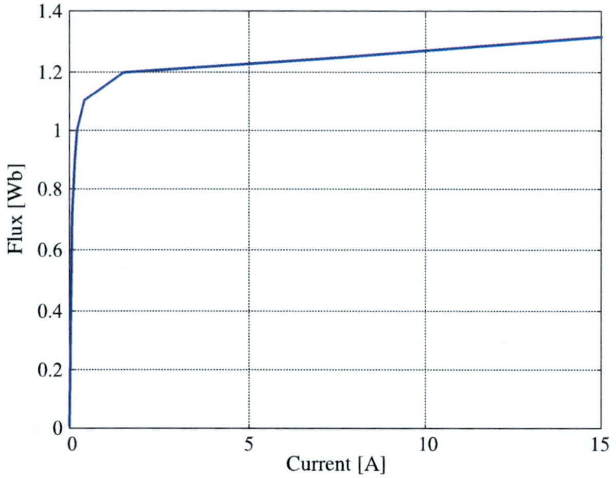

Fig. 7.5 Magnetizing characteristic of the CT core

$$i_m(n+1) \approx i_{2e}(n+1) - i_2(n+1). \tag{7.21}$$

Then the change of the magnetizing current between the samples (n) and $(n+1)$ becomes:

$$\Delta i_m(n+1) \approx i_m(n+1) - i_m(n). \tag{7.22}$$

The mean value of the CT secondary voltage between the samples (n) and $(n+1)$ equals:

$$u(n+1) \approx R_2 \frac{i_2(n+1) + i_2(n)}{2} + L_2 \frac{i_2(n+1) - i_2(n)}{T_S}. \tag{7.23}$$

Therefore, the increase of the flux linkage Ψ in one sampling period becomes:

$$\Delta\psi(n+1) \approx T_S u(n+1). \tag{7.24}$$

The value of the magnetizing inductance between the samples is given by the formula:

$$L_m(n+1) \approx \frac{\Delta\psi(n+1)}{\Delta i_m(n+1)}. \tag{7.25}$$

Now, assuming that between the samples $(n + 1)$ and $(n + 2)$ the magnetizing inductance L_m has the same value, as during the previous sample, one may write:

$$\Delta i_m(n+2) \approx \frac{\Delta\psi(n+2)}{L_m(n+1)} \approx \Delta i_m(n+1)\frac{\Delta\psi(n+2)}{\Delta\psi(n+1)}, \tag{7.26}$$

where $\Delta\psi(n+2) \approx T_S u(n+2)$.

Further, for the next time instant one obtains:

$$u(n+2) \approx R_2\frac{i_2(n+2)+i_2(n+1)}{2} + L_2\frac{i_2(n+2)-i_2(n+1)}{T_S}, \tag{7.27}$$

therefore:

$$i_m(n+2) \approx i_m(n+1) + \Delta i_m(n+2) \tag{7.28}$$

and in consequence:

$$i_{2e}(n+2) \approx i_2(n+2) + i_m(n+2). \tag{7.29}$$

This algorithm may be repeated, bearing in mind that:

$$L_m(n+2) \approx \frac{\Delta\psi(n+2)}{\Delta i_m(n+2)}, \tag{7.30}$$

$$\Delta i_m(n+3) \approx \Delta i_m(n+2)\frac{\Delta\psi(n+3)}{\Delta(n+2)}, \tag{7.31}$$

and consequently:

$$i_{2e}(n+3) \approx i_2(n+3) + i_m(n+3). \tag{7.32}$$

The process of calculation of the estimated values of the secondary current samples continues, i.e.,

$$\Delta i_m(n+k) \approx \Delta i_m(n+k-1)\frac{\Delta\psi(n+k)}{\Delta\psi(n+k-1)}, \tag{7.33}$$

$$i_{2e}(n+k) \approx i_2(n+k) + i_m(n+k), \tag{7.34}$$

until the end of the saturated fraction of the period (see block scheme of the procedure shown in Fig. 7.6).

If the waveshape of the secondary current is reproduced, the phasor may be calculated using the data window duration T_w equal to the period of the

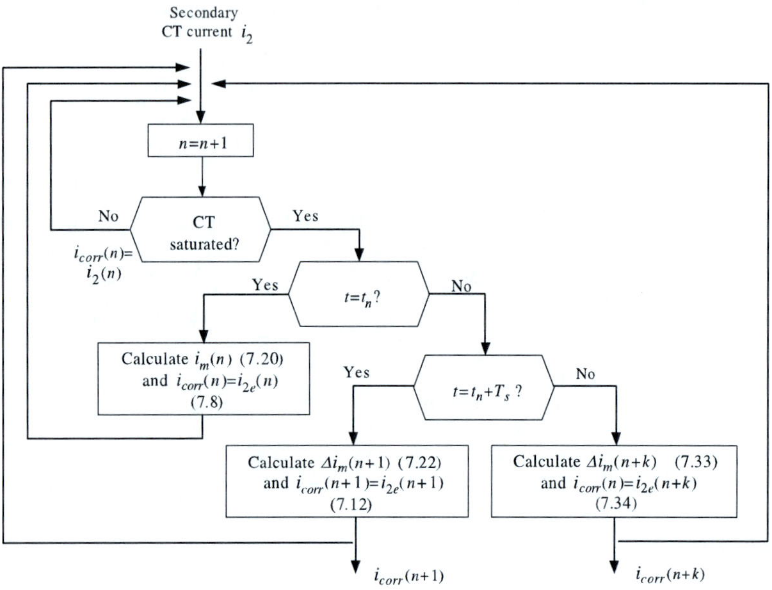

Fig. 7.6 Block scheme of the CT saturation correction procedure

fundamental frequency component T_1, either calculating the non-rotating or rotating phasor components.

The above considerations show how digital representation of the secondary current may be efficiently used to determine fractions of the period when the CT magnetic core is not saturated, thus transformation is correct, and how distorted waveshape may be processed to determine phasors. The presented methods are not unique, but perhaps represent the best compromise between the accuracy and efficiency. More details as well as simulative examples of the CT saturation detection/correction procedures operation can be found in [13]. Slightly modified version of the procedure for CT saturation is presented in [11], where an approach based on numerical solution of the differential equation describing the saturated current transformer is adopted.

7.2.4 Other Methods of CT Saturation Detection and Secondary Current Reconstruction

A large number of journal and conference papers, which dealt with the CT saturation detection and correction issue, may be divided into three groups:

- no CT current correction is performed, but the information on CT saturation is used for other purpose, e.g. for adaptation of protection settings [9] or performing calculations during CT unsaturated periods only [1],

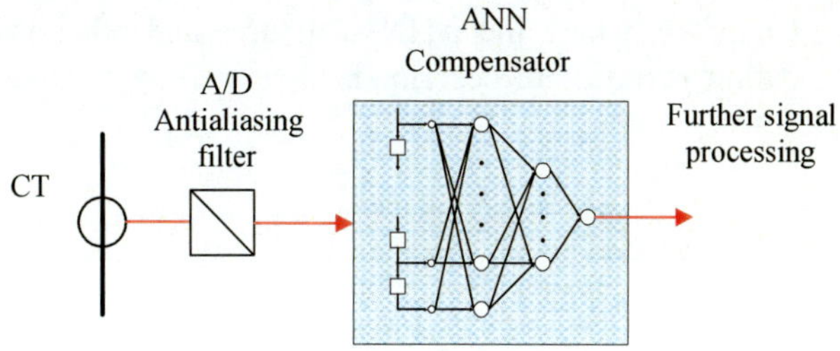

Fig. 7.7 Basic scheme of CT saturation detection/ compensation with ANN

- correction performed is based on the information extracted from the secondary current during the saturation interval, which reproduces the primary current waveshape [3, 4],
- correction is based on information extracted from the secondary current during the non-saturation interval, which aims at determination of the fundamental and DC component of the primary current [5, 11].

The first step of all the correction procedures is to determine the beginning and end of each saturation span. It is not a simple task, however a few approaches may be applied to solve it:

- analysis of wavelet details of the secondary current [7],
- comparison of the fundamental component and its scaled 2nd derivative [2],
- calculation of the difference between outputs of mean and median filters [2],
- application of neural networks for the purpose [8, 10, 12, 14].

Among the algorithms for CT correction the ones based on algorithmic approach as well as novel Artificial Intelligence schemes are proposed.

The method of reconstruction of the secondary CT current on the ground of the current data during the non-saturation period presented in [5] requires that the CT magnetizing characteristic and its load are a priori known, which is not always the case. On the contrary, the algorithms developed by the authors of this book and given in [11] are free of this limitation.

Promising results are also reported for application of Artificial Neural Networks. The basic idea of CT saturation compensation with neural networks is depicted in Fig. 7.7. The CT secondary current, after low-pass filtering and converting to digital form, is passed to an ANN, at the output of which the reconstructed primary CT current should appear. The ANN output is further digitally processed to get certain relay criterion values, the ones needed for decision making in the particular relay under consideration.

Independent of the ANN version (non-recursive or recursive) the neural network is trained to resemble the non-linear inverse "transfer function" of the CT. As final result the corrected secondary current samples are expected to be produced at the output of ANN that should be as close as possible to the non-measurable samples of the unsaturated primary CT current.

The performance of ANN-based correction methods depend to a large degree on the cases used to train the networks. More details on that one can be found in

Chap. 12, where the ANN structures and other issues important for their application purpose are described.

References

1. Bunyagul T, Crossley P, Galac P (2001) Overcurrent protection using signals derived from saturated measurement CTs. In: Proceedings of PES summer meeting, Vancouver, vol 1, pp 103–108
2. Chen KW, Glad ST (1991) Estimation of the primary current in a saturated transformer. In: Proceedings of 30th conference on decision and control, Brighton, England, vol 3, pp 2363–2365
3. Kang YC, Park JK, Kang SH, Johns AT, Aggarwal RK (1996) Development and hardware implementation of a compensating algorithm for the secondary current of current transformers. IEE Proc Electr Power Appl 243:41–49
4. Kang YC, Park JK, Kang SH, Johns AT, Aggarwal RK (1997) An algorithm for compensating secondary current of current transformer. IEEE Trans Power Deliv 12:116–124
5. Kang YC, Lim UJ, Kang SH (2004) Compensating algorithm for the secondary current for use with measuring type current transformers. In: Proceedings of the international conference advanced power system automation and protection, Jeju, Korea, pp 3–8
6. Kasztenny B, Mazereeuw J, Jones K (2001) CT saturation in industrial applications—analysis and application guidelines. Publ GET-8501 GE Multilin, Canada
7. Li F, Li Y, Aggarwal RK (2002) Combined wavelet transform and regression technique for secondary current compensation of current transformer. IEE Proc Gener Transm Distrib 149:497–503
8. Rebizant W, Bejmert D (2007) Current-transformer saturation detection with genetically optimized neural networks. IEEE Trans Power Deliv 22:820–827
9. Rebizant W, Hayder T, Schiel L (2004) Prediction of CT saturation period for differential relay adaptation purposes. In: Proceedings of the international conference on advanced power system automation and protection, Jeju, Korea, pp 17–22
10. Rebizant W, Bejmert D, Staszewski J, Schiel L (2007) CT saturation detection and correction with artificial neural networks. In: Proceedings of the 2nd international conference on advanced power system automation and protection, Jeju, Korea, paper 504
11. Rebizant W, Wiszniewski A, Schiel L (2008) CT saturation correction based on the estimated CT saturation time constant. In: Proceedings of the 8th international IET conference developments in power system protection, Glasgow, UK, pp 174–179
12. Saha MM, Iżykowski J, Łukowicz M, Rosołowski E (2001) Application of ANN method for instrument transformer correction in transmission line protection. In: Proceedings of the IEE development in power system protection, Conf Publ No 479, pp 303–306
13. Wiszniewski A, Rebizant W, Schiel L (2008) Correction of current transformers transient performance. IEEE Trans Power Deliv 23:624–632
14. Yu DC, Cummins JC, Wang Z, Yoon HJ, Kojovic LA (2001) Correction of current transformer distorted secondary current due to saturation using artificial neural networks. IEEE Trans Power Deliv 16:189–194

Chapter 8
Measurement Algorithms for Digital Protection

8.1 Fundamentals of Digital Measurements

In case of protection devices of pre-digital era that were made in analogue technique the processes of measurement and decision-making were integrated. Magnitude or phase comparators were applied and thus the measurement of criterion values could not be realized directly. From the digital signal processing block schemes presented in Fig. 3.12 it becomes evident that with the advent of digital technology the principles of design of protection relays have been changed substantially. In digital protection devices the measurement of criterion values and decision–making were separated in principle from the very beginning of digital era, thus allowing for optimization of each of the processes separately. The measured criterion values could now be used for other purposes, moreover, development of multi-criteria decision-making as well as application artificial intelligence techniques was also possible.

Great number digital measurement algorithms of criterion values have appeared in the past [1, 3, 4, 6, 8–10]. At the very beginning they were applying signals averaging and some other methods of analogue technique origin. Very soon, however, digital technique with its advantages, especially FIR digital filtering and orthogonal signal components became a fundamental and exceptional solution.

Criterion values of digital protections are magnitudes of fundamental components of currents and voltages, powers, impedance or admittance and their components, frequency, phase shift, the values for symmetrical components and the others. Very frequently many of them are used for given protection simultaneously (multi-criteria solutions). Measurement algorithms for all of them are presented in this chapter, whereas some special features of the algorithms including their dynamics and frequency responses are described in Chap. 9.

Before the actual measurement of protection criteria can be performed the input signals have to be pre-processed, which is referred to as digital signal conditioning, Fig. 8.1. Various methods allowing for extraction of noise-free information-rich

W. Rebizant et al., *Digital Signal Processing in Power System Protection and Control,* Signals and Communication Technology,
DOI: 10.1007/978-0-85729-802-7_8,

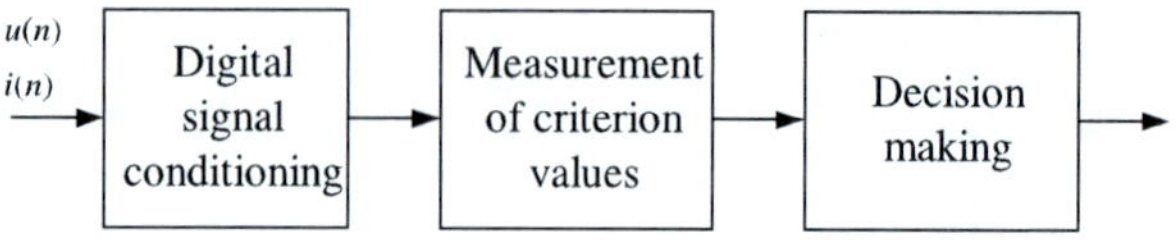

Fig. 8.1 Basic scheme of signal processing in a digital protection relay

protection signals have been developed, some of them are described in next subsection.

Then a number of measurement algorithms belonging to several families are described. The averaging methods are presented first, then, more deeply the methods of measurement of criterion values emerging from digital technique are described, with special attention paid to the ones applying orthogonal components of input signals.

8.1.1 Digital Signal Conditioning

Digital signal conditioning is aimed at realizing several important functions. The first, perhaps the most important one, is extraction of useful information from the signals on disposal. This is done by either dumping or rejection of noise. The second task is producing such signal components which make calculation of protection criterion values possible in the simplest way. It should be mentioned that all the above is to be realized within the shortest possible period of time. It is easy to notice that the tasks are in most cases contradictory according to the known speed–accuracy dilemma and thus proposed solutions are always a compromise. Since particular requirements are different depending on protected or controlled element of the power system, then sometimes the most important is the operation time and sometimes measurement accuracy or noise reduction. According to given requirements many methods are applied in signal conditioning and further signal processing. They can be divided for those designed and optimized in either frequency or time domain. An example of the first one could be digital filter synthesis starting from required frequency response (Sect 6.3). An example of the second group could be synthesis of Kalman filter on basis of mean square error and the fastest possible time response.

The most important methods of digital signal conditioning are following:

- Fast Fourier transform (Chap. 4),
- digital filters (described in Chaps. 5 and 6),
- separate methods of orthogonalization,
- digital methods basing on LSE,
- correlation, etc.

The processes of filtering/extracting of useful signal components and their orthogonalization can be joint (using appropriate FIR filter pairs) or separate (i.e. without filtering that should be implemented separately, if needed). Since the first approach has already been presented in Chap. 6, the second approach with its

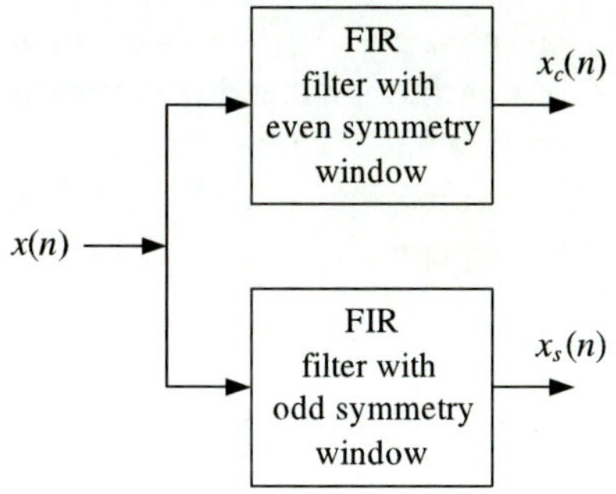

Fig. 8.2 Joint filtering/signal orthogonalization with use of FIR filters

advantages and drawbacks is described in detail below. Least squares estimation (LSE) technique, allowing to get the sought values of signal parameters within comparatively short period of time, is also outlined. This section ends with a brief description of the correlation technique.

8.1.1.1 Joint Filtering and Signal Orthogonalization

The FIR orthogonal filters presented in Chap. 6 realize two basic tasks: they eliminate or suppress noise and produce orthogonal signal components that can be applied in almost all measurement algorithms of protection criterion values. According to Fig. 8.2 the orthogonal components are produced at the outputs of two filters having even and odd symmetry window functions, respectively (e.g. Walsh or sine/cosine filters). The characteristics of filters both in time and frequency domains have already been discussed in Chap. 6.

8.1.1.2 Separate Methods of Signal Orthogonalization

As mentioned above, the processes of filtration and orthogonalization can be separated, i.e. noise filtration is performed with use of one filter only (or for low noise level there is no filter at all) and required orthogonal components are obtained using special orthogonalization methods [6, 8]. The methods concerned apply certain linear combinations of delayed input signal samples and therefore they are sometimes called as methods of orthogonalization by single time delay or double time delay. All of them, with exception for particular cases, can also be treated as certain types of digital filters, usually with quite poor frequency responses.

Orthogonalization by Single Time Delay

Orthogonalization by single time delay is one of the simplest methods for obtaining orthogonal components. The method has several variants, two of them are presented below. One can state in general that the methods are similar to the use of the signal and its derivative, though in digital version the finite difference is calculated.

Let the signal,in which orthogonal components are to be calculated, be given by the equation:

$$x(n) = X_{1m}\cos(n\Omega_1 + \varphi) \tag{8.1}$$

The signal delayed by k samples is given by:

$$\begin{aligned} x(n-k) &= X_{1m}\cos[(n-k)\Omega_1 + \varphi)] \\ &= X_{1m}\cos(k\Omega_1)\cos(n\Omega_1 + \varphi) + X_{1m}\sin(k\Omega_1)\sin(n\Omega_1 + \varphi) \\ &= x_c(n)\cos(k\Omega_1) + x_s(n)\sin(k\Omega_1) \end{aligned} \tag{8.2}$$

From both equations one gets:

$$x_c(n) = x(n), \tag{8.3a}$$

$$x_s(n) = \frac{x(n-k) - x(n)\cos(k\Omega_1)}{\sin(k\Omega_1)} \tag{8.3b}$$

The values of number of delay samples can be different: from one sample of delay giving the fastest algorithm up to a number of samples equivalent to a quarter of fundamental frequency component. The simplest digital orthogonalization method is thus given by the equations:

$$x_c(n) = x(n), \tag{8.4a}$$

$$x_s(n) = x(n - N_1/4). \tag{8.4b}$$

The latter case, though "time consuming", is very convenient since orthogonalization being in this case simple signal delaying is not a digital filter and does not change the signal spectrum. In general, however, in all remaining cases for k between one and $N_1/4$ samples the orthogonalization means realizing certain digital filter, whose settling time and frequency response depend on k. To get the frequency response of such a filter one has to calculate its discrete transfer function, then substitute $\exp(j\Omega)$ for operator z, and calculate magnitude and phase shift. One obtains then:

$$\frac{X_s(j\Omega)}{X(j\Omega)} = G_{1s}(j\Omega) = \frac{\exp(-jk\Omega) - \cos(k\Omega_1)}{\sin(k\Omega_1)} \tag{8.5}$$

and from that the magnitude and phase shift are following:

$$|G_{1s}(j\Omega_1)| = \frac{\sqrt{1 - 2\cos(k\Omega)\cos(k\Omega_1) + \cos^2(k\Omega_1)}}{\sin(k\Omega_1)}, \tag{8.6a}$$

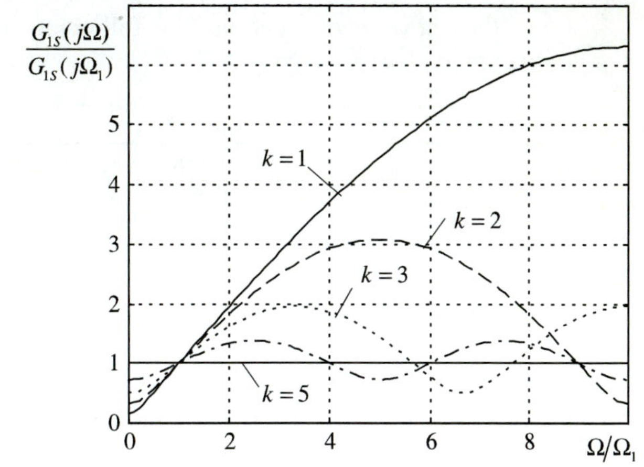

Fig. 8.3 Frequency response of orthogonalization by single time delay

$$\arg[G_{1s}(j\Omega)] = -\text{arctg}\frac{\sin(k\Omega)}{\cos(k\Omega) - \cos(k\Omega_1)}. \tag{8.6b}$$

It is easy to notice that the filter gain equals unity for Ω equal to Ω_1 while the argument is equal to $-\pi/2$ independent of parameter k. However, when Ω is different from Ω_1 the argument is not equal to $-\pi/2$. It means that during frequency deviations from nominal value the product of the method is not orthogonal to the signal.

Frequency response of the method depends on assumed value of parameter k. The minimum value of this parameter equals one and its maximum value results from equation $k\Omega_1 = \pi/2$, which depends on sampling frequency. The frequency responses for sampling frequency equal to 1000 Hz and k varying from 1 to 5 are shown in Fig. 8.3. They are a very good illustration of the well-known dilemma: speed–accuracy.

Example 8.1 Calculate magnitudes of the frequency response of orthogonalization by time delay for two selected frequencies: zero (DC component) and second harmonic. Assume sampling at 1000 Hz and fundamental frequency 50 Hz.

Solution From the assumed values it results:

$$N_1 = 20; \quad \Omega_1 = 0.1\pi.$$

Use of (8.6a) leads to (for the frequency equal zero):

$$|G_{1s}(j0)| = tg(0.5k\Omega_1) = tg(0.05k\pi)$$

and for the second harmonic:

$$|G_{1s}(j2\Omega_1)| = \frac{\sqrt{1 + \cos^2(0.1k\pi) - 2\cos(0.1k\pi)\cos(0.2k\pi)}}{\sin(0.1k\pi)}$$

From the above two equations the following table can be obtained for the assumed range of k:

k	$\lvert G_{1s}(j0)\rvert$	$\lvert G_{1s}(j2\Omega_1)\rvert$
1	0.158	1.957
2	0.325	1.83
3	0.51	1.46
4	0.726	1.35
5	1	1

As one can see, lower delay values induce stronger suppression of the DC component, but at the same time—evoke stronger amplification of the second harmonic (effect of differentiation).

Another variant of orthogonalization by single delay applies addition and subtraction of a signal and a signal delayed by k samples (Eqs. 8.1 and 8.2). It is easy to notice that the orthogonal components are then described by equations:

$$x_c\left(n-\frac{k}{2}\right)=\frac{x(n-k)+x(n)}{2\cos(k\Omega_1/2)}, \tag{8.7a}$$

$$x_s\left(n-\frac{k}{2}\right)=\frac{x(n-k)-x(n)}{2\sin(k\Omega_1/2)}, \tag{8.7b}$$

Similar to the method of orthogonalization by single delay one can obtain the filter transfer functions and equivalent frequency responses (magnitudes and phase shifts) resulting from (8.7a, b). The two filters obtained (one of them is low pass filter and the other high pass) have the following transfer functions:

$$G_c(j\Omega)=\frac{\exp(-jk\Omega)+1}{2\cos(k\Omega_1/2)}, \tag{8.8a}$$

$$G_s(j\Omega)=\frac{\exp(-jk\Omega)-1}{2\sin(k\Omega_1/2)}. \tag{8.8b}$$

Their magnitudes and phase shifts versus frequency are:

$$|G_c(j\Omega)|=\frac{\cos(k\Omega/2)}{\cos(k\Omega_1/2)}, \tag{8.9a}$$

$$|G_s(j\Omega)|=\frac{\sin(k\Omega/2)}{\sin(k\Omega_1/2)}, \tag{8.9b}$$

$$tg(\psi_c-\psi_s)=\frac{tg(\psi_c)-tg(\psi_s)}{1+tg(\psi_c)tg(\psi_s)}=\frac{-tg(k\Omega/2)-ctg(k\Omega/2)}{1+[-tg(k\Omega/2)]ctg(k\Omega/2)}=\pm\infty, \tag{8.9c}$$

where $\psi_c=\arg(G_c)$ and $\psi_s=\arg(G_s)$.

It can be noted that for the smallest possible delay value ($k = 1$) Eqs. 8.7a, b represents signal value and its derivative. Using them for magnitude measurement (e.g. by substituting to (8.50)) results in one of the simplest and fastest estimation algorithms.

Orthogonalization by Double Time Delay

The method can be derived easily taking the signal samples at given instant and delayed by $2k$ sampling periods:

$$\begin{aligned} x(n-2k) &= x(n-k-k) \\ &= X\cos[(n-k)\Omega_1 + \varphi]\cos(k\Omega_1) + X\sin[(n-k)\Omega_1 + \varphi]\sin(k\Omega_1) \end{aligned}$$

$$\begin{aligned} x(n) &= x(n-k+k) \\ &= X\cos[(n-k)\Omega_1 + \varphi]\cos(k\Omega_1) - X\sin[(n-k)\Omega_1 + \varphi]\sin(k\Omega_1) \end{aligned}$$

It is seen at once that orthogonal components are given by equations:

$$x_c(n) = x(n-k) \tag{8.10a}$$

$$x_s = \frac{x(n-2k) - x(n)}{2\sin(k\Omega_1)} \tag{8.10b}$$

"Pure" signal delay, required to get the first of orthogonal components, is not a digital filter, however, it introduces certain phase shift depending on frequency. The second of orthogonal components is obtained using a transformation as certain digital filter. Its frequency response calculated in the same way as before is given by:

$$G_{2s}(j\Omega) = \frac{\exp(-j2k\Omega) - 1}{2\sin(k\Omega_1)} = \frac{\cos(2k\Omega) - 1 - j\sin(2k\Omega)}{2\sin(k\Omega_1)} \tag{8.11}$$

The magnitude of frequency response has especially simple form:

$$|G_{2s}(j\Omega)| = \frac{\sin(k\Omega)}{\sin(k\Omega_1)} \tag{8.12}$$

Figure 8.4 shows the curves of the magnitude (8.12) versus frequency depending on the parameter k (number of samples of delay) for the range from 1 to 5, assuming sampling frequency of 1 kHz. It is seen that the frequency response of the method is better than in the case of orthogonalization by single delay, especially in the range of high frequencies. However, one must remember that for the same value of a parameter k, the orthogonalization time is two times longer in the case of double delay.

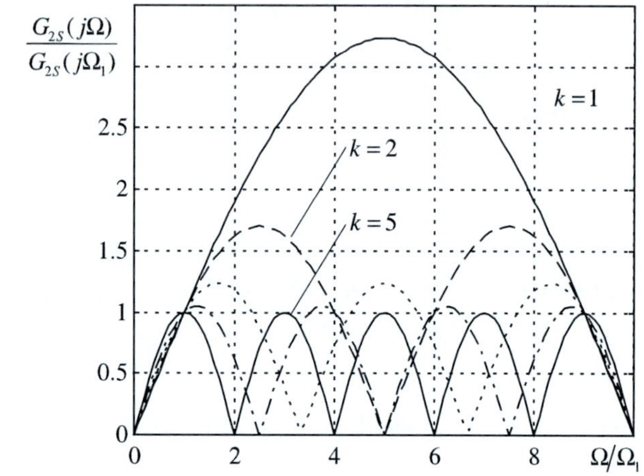

Fig. 8.4 Frequency response of orthogonalization by double time delay

To calculate phase difference resulting from (8.10a, b) one can use frequency response (8.11) and the frequency response of the transformation required to get direct signal component $x_c(n)$ given by:

$$G_{2c}(j\Omega) = \exp(-jk\Omega). \tag{8.13}$$

The tangent of this phase difference is equal to:

$$tg(\psi_{2c} - \psi_{2s}) = \frac{tg(\psi_{2c}) - tg(\psi_{2s})}{1 + tg(\psi_{2c})tg(\psi_{2s})} = \frac{-tg(k\Omega) - ctg(k\Omega)}{1 + [-tg(k\Omega)]ctg(k\Omega)} = \pm\infty, \tag{8.14}$$

where $\psi_{2c} = \arg(G_{2c})$ and $\psi_{2s} = \arg(G_{2s})$, which proves that the phase difference is equal $\pi/2$ for all frequencies.

8.1.1.3 Least Squares Estimation Technique

The FIR orthogonal filters presented in Chap. 6 are aimed at extracting particular (mostly—fundamental) frequency component and at producing the signal orthogonal components useful for further processing in the measurement unit. The transient period of FIR filters lasts as long as their window length and cannot be shortened without significant loss of quality in frequency domain. In order to minimize the transient state after sudden change of signal parameters one can apply the LSE techniques.

The LSE method is a commonly known approach to signal parameters estimation with a minimum of least square error. The technique is applied in many areas of engineering,also including electrical power engineering [2, 5, 8]. In many problems of power system protection and control parameters of input signal (phasor magnitude and phase) are often to be determined whereas the signal model structure is usually clearly defined. For a single component discrete fundamental frequency signal such a model consists of two terms plus noise, according to:

$$x(n) = X_{1m}\cos(n\Omega_1 + \varphi) = X_{1C}\cos(n\Omega_1) + X_{1S}\sin(n\Omega_1) + e(n), \tag{8.15}$$

where Ω_1 is a relative fundamental angular frequency, $X_{1C} = X_{1m}\cos(\varphi)$, $X_{1S} = X_{1m}\sin(\varphi)$ are unknown parameters, $e(n)$ is an error signal representing all other frequency components and noise.

The signal magnitude and phase can be determined provided the signal model parameters are estimated with use of a proper algorithm. The parameters X_{1C}, and X_{1S} can be used for the purpose, being themselves close equivalents of the signal orthogonal components obtainable from FIR filter outputs.

The signal model (8.15) can be extended by inclusion of one (or more) additional frequency components which, on the one hand, enables direct estimation of parameters of that component and on the other hand—improvement of the fundament frequency component measurement accuracy. In case when one harmonic component is taken into account the signal model becomes:

$$x(n) = X_{1C}\cos(n\Omega_1) + X_{1S}\sin(n\Omega_1) + X_{kC}\cos(nk\Omega_1) + X_{kS}\sin(nk\Omega_1) + e(n), \tag{8.16}$$

with X_{kC} and X_{kS} representing the parameters of the kth harmonic.

In general, depending on the needs the signal model (8.15) can be extended theoretically without limits, however, this entails higher computational burden during practical implementation of the method for signal parameters estimation. Equations 8.15 and 8.16 can be generalized to the form:

$$x(n) = [\cos(n\Omega_1)\ \sin(n\Omega_1)\ \cos(nk\Omega_1)\ \sin(nk\Omega_1)\ldots][X_{1C}\ X_{1S}\ X_{kC}\ X_{kS}\ldots]^{\mathrm{T}} + e(n). \tag{8.17}$$

For a vector containing K components there are $2K$ parameters to be determined, which is possible if at least $N \geq 2K$ signal samples are available, which leads to the following matrix equation:

$$\boldsymbol{x}_N = \boldsymbol{h}_N \boldsymbol{X} + \boldsymbol{e}_N, \tag{8.18}$$

where $\boldsymbol{x}_N = [x(0), x(1), \ldots, x(N-2), x(N-1)]^{\mathrm{T}}$ is a vector of N signal samples, $\boldsymbol{e}_N = [e(0), e(1), \ldots, e(N-2), e(N-1)]^{\mathrm{T}}$ is a vector of N error samples, $\boldsymbol{X} = [X_{1C}, X_{1S}, \ldots, X_{KC}, X_{KC}]^{\mathrm{T}}$ is a vector of unknown $2K$ signal parameters and

$$\boldsymbol{h}_N = \begin{bmatrix} \cos(0\Omega_1) & \sin(0\Omega_1) & \ldots & \cos(0K\Omega_1) & \sin(0K\Omega_1) \\ \cos(1\Omega_1) & \sin(1\Omega_1) & \ldots & \cos(1K\Omega_1) & \sin(1K\Omega_1) \\ & & & & \\ \cos((N-1)\Omega_1) & \sin((N-1)\Omega_1) & & \cos((N-1)K\Omega_1) & \sin((N-1)K\Omega_1) \end{bmatrix}$$

is a matrix of known model elements for different sampling instants.

The sought parameters $\boldsymbol{X}$ can be estimated with high accuracy if only the mean square error for all the N samples taken is minimized, i.e. if:

$$\boldsymbol{e}_N^{\mathrm{T}} \boldsymbol{e}_N = \sum_{n=0}^{N-1} e_n^2 \rightarrow \min. \tag{8.19}$$

It can easily be derived and the parameters of signal model are then calculated from:

$$\hat{\boldsymbol{X}} = \left\{ \boldsymbol{h}_N^{\mathrm{T}} \boldsymbol{h}_N \right\}^{-1} \boldsymbol{h}_N^{\mathrm{T}} \boldsymbol{x}_N = \boldsymbol{P}_N \boldsymbol{x}_N, \tag{8.20}$$

where $\hat{\boldsymbol{X}}$ is a approximate vector of sought parameters determined with minimum least square error and $\boldsymbol{P}_N$ is a LSE transformation matrix.

It should be noted that the matrix $\boldsymbol{P}_N$ in (8.20) for given numbers of K and N is fully determined before the calculations start and thus it can be prepared well in advance and stored in look-up tables in the protection memory. Application of the estimation procedure in on-line mode requires execution of (8.20) at consecutive time instants, with new sets of signal samples captured within fixed—data window.

Example 8.2 For the signal model (8.15) containing the fundamental frequency component only determine the model parameters and calculate the signal magnitude assuming availability of $N = 2$, 6, 12 signal samples. Assume sampling frequency $f_S = 600$ Hz, unity level of the signal and its initial phase equal to $\pi/6$. Draw the course of measured signal magnitude for the case of pure fundamental frequency signal and for the case of signal distorted by the second harmonic component (10% of the fundamental).

Solution One should note that the minimum window length for estimation of signal parameters is here equal to two sampling periods (two parameters of the model are sought). Extending-data window length makes the problem overdetermined which should result in improved accuracy, especially when the signal differs from assumed model (distortions, noise). The window length $N = N_1 = 12$ covers full cycle of the signal, thus the expected results should be close to the ones obtained when full-cycle Fourier filters are applied.

For the assumed sampling rate the relative fundamental angular frequency is $\Omega_1 = 2\pi f_1/f_S = 2\pi 50/600 = \pi/6$. The signal model can be expressed in the form:

$$\begin{aligned} x(n) &= X_{1C} \cos(n\Omega_1) + X_{1S} \sin(n\Omega_1) + e(n) \\ &= X_{1C} \cos(n\pi/6) + X_{1S} \sin(n\pi/6) + e(n). \end{aligned}$$

For the case when there are no distortions the signal samples are equal:

$$\begin{aligned} x(n) &= X_{1m} \cos(n\Omega_1 + \varphi) = \cos(n\pi/6 + \pi/4) \\ &= 0.707 \cos(n\pi/6) - 0.707 \sin(n\pi/6). \end{aligned}$$

For the window lengths $N = 2$, 6, 12 the matrix $\boldsymbol{h}_N^{\mathrm{T}}$ is equal:

$$\boldsymbol{h}_2^{\mathrm{T}} = \begin{bmatrix} \cos(0) & \cos(\pi/6) \\ \sin(0) & \sin(\pi/6) \end{bmatrix} = \begin{bmatrix} 1 & 0.866 \\ 0 & 0.5 \end{bmatrix},$$

$$\begin{aligned} \boldsymbol{h}_6^{\mathrm{T}} &= \begin{bmatrix} \cos(0) & \cos(\pi/6) & \ldots & \cos(4\pi/6) & \cos(5\pi/6) \\ \sin(0) & \sin(\pi/6) & \ldots & \sin(4\pi/6) & \sin(5\pi/6) \end{bmatrix} \\ &= \begin{bmatrix} 1 & 0.866 & \ldots & -0.5 & -0.866 \\ 0 & 0.5 & \ldots & 0.866 & 0.5 \end{bmatrix}, \end{aligned}$$

$$\begin{aligned} \boldsymbol{h}_{12}^{\mathrm{T}} &= \begin{bmatrix} \cos(0) & \cos(\pi/6) & \ldots & \cos(10\pi/6) & \cos(11\pi/6) \\ \sin(0) & \sin(\pi/6) & \ldots & \sin(10\pi/6) & \sin(11\pi/6) \end{bmatrix} \\ &= \begin{bmatrix} 1 & 0.866 & \ldots & 0.5 & 0.866 \\ 0 & 0.5 & \ldots & -0.866 & -0.5 \end{bmatrix} \end{aligned}$$

thus the matrix $\boldsymbol{P}_N$ and the estimation result $\hat{\boldsymbol{X}}$ become:

$$\boldsymbol{P}_2 = \begin{bmatrix} 1 & 0 \\ -1.732 & 2 \end{bmatrix}, \quad \hat{\boldsymbol{X}} = \boldsymbol{P}_2 \boldsymbol{x}_2 = \begin{bmatrix} 0.707 \\ -0.707 \end{bmatrix},$$

$$\boldsymbol{P}_6 = \begin{bmatrix} 0.333 & 0.289 & \ldots & -0.167 & -0.289 \\ 0 & 0.167 & \ldots & 0.289 & 0.167 \end{bmatrix}, \quad \hat{\boldsymbol{X}} = \boldsymbol{P}_6 \boldsymbol{x}_6 = \begin{bmatrix} 0.707 \\ -0.707 \end{bmatrix},$$

$$\boldsymbol{P}_{12} = \begin{bmatrix} 0.333 & 0.144 & \ldots & -0.083 & -0.144 \\ 0 & 0.083 & \ldots & 0.144 & 0.083 \end{bmatrix}, \quad \hat{\boldsymbol{X}} = \boldsymbol{P}_{12} \boldsymbol{x}_{12} = \begin{bmatrix} 0.707 \\ -0.707 \end{bmatrix}.$$

From the above one can see that the estimates of signal parameters are identical and equal to correct values irrespective of the applied window length. The situation changes, however, when the signal contains some distortion, here—10% of the second harmonic. The results of signal magnitude calculation ($X_{1m} = \sqrt{X_{1C}^2 + X_{1S}^2}$) for such a case are illustrated in Fig. 8.5b (for pure cosine wave—Fig. 8.5a). One can see that the impact of distortion component decreases with the increase in window length (time from estimation start) and the results are accurate when N reaches full−cycle length N_1.

Some peculiar advantages of the LSE algorithm are obtained when one assumes additional probabilistic features of the error or when the number of samples N is adequately adjusted to the signal period, what leads to the DFT or the correlation algorithms, respectively.

Considering the special case when $N = N_1$ the transformation matrix P_N becomes:

$$\boldsymbol{P}_N = \frac{2}{N_1} \boldsymbol{h}_N^{\mathrm{T}}. \tag{8.21}$$

Thus, for the $\boldsymbol{h}_{\mathrm{N}}$ matrix form as above the expressions for particular signal model parameters are:

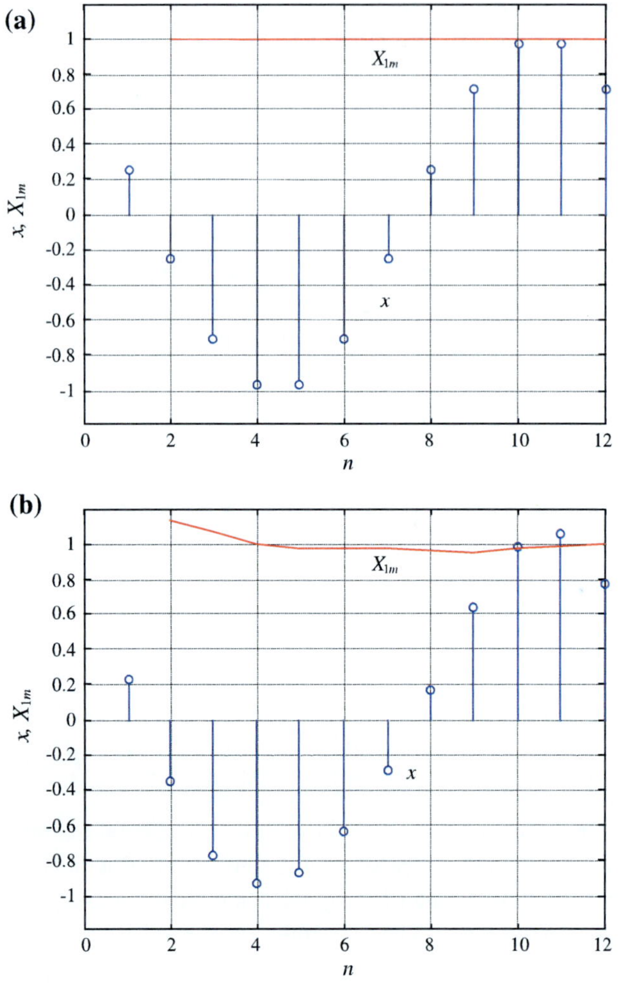

Fig. 8.5 Signal magnitude estimation results with LSE approach for: **a** pure sine wave and **b** signal containing 10% of the 2nd harmonic

$$X_{kC} = \frac{2}{N_1}\sum_{n=0}^{N_1-1} x(n)\cos(nk\Omega_1) = \frac{2}{N_1}\sum_{n=0}^{N_1-1} x(n)\cos\left(nk\frac{2\pi}{N_1}\right), \tag{8.22}$$

$$X_{kS} = \frac{2}{N_1}\sum_{n=0}^{N_1-1} x(n)\sin(nk\Omega_1) = \frac{2}{N_1}\sum_{n=0}^{N_1-1} x(n)\sin\left(nk\frac{2\pi}{N_1}\right) \tag{8.23}$$

and

$$X(k) = X_C(k) - jX_S(k) = \frac{2}{N_1}\sum_{n=0}^{N_1-1} x(n)\exp\left(-jnk\frac{2\pi}{N_1}\right), \tag{8.24}$$

which is another form of Discrete Fourier Transform (compare (4.26)) enabling for transformation of the N_1-element series in time domain into an equivalent N_1-element series in frequency domain (exact to the constant coefficient $2/N_1$).

8.1.1.4 Correlation Method

The digital correlation consists in representing the considered real signal with a series of mutually orthogonal functions [8]. For power–system protection applications such representation is mostly limited to the components of fundamental frequency only. The expansion components can be obtained from (8.22) and (8.23) assuming calculation for the discrete frequency $k = 1$, which yields:

$$X_{1C} = \frac{2}{N_1} \sum_{n=0}^{N_1-1} x(n) \cos\left(n \frac{2\pi}{N_1}\right), \tag{8.25}$$

$$X_{1S} = \frac{2}{N_1} \sum_{n=0}^{N_1-1} x(n) \sin\left(n \frac{2\pi}{N_1}\right). \tag{8.26}$$

For a pure fundamental frequency signal the values resulting from (8.25) and (8.26) are:

$$X_{1C} = X_{1m} \cos(\varphi), \tag{8.27}$$

$$X_{1S} = X_{1m} \sin(\varphi), \tag{8.28}$$

being constant, in contrast to the outputs of digital Fourier filters that are time–depending functions. Nevertheless, also here the harmonic distortions (if any) are completely rejected by the correlation algorithm and the obtained values X_{1C} and X_{1S} are orthogonal, which enables calculation of signal magnitude, e.g. with application of (8.50).

If the correlation algorithm is to be applied on-line, where time instant n changes, it is more convenient to use the algorithms in the form:

$$X_{1C}(n) = \frac{2}{N_1} \sum_{m=0}^{N_1-1} x(n-m) \cos\left((n-m) \frac{2\pi}{N_1}\right), \tag{8.29}$$

$$X_{1S}(n) = \frac{2}{N_1} \sum_{m=0}^{N_1-1} x(n-m) \sin\left((n-m) \frac{2\pi}{N_1}\right). \tag{8.30}$$

The advantage of correlation approach is a possibility of obtaining simple recursive algorithms. Observing that for the samples n and $n-1$ the sums (8.29) and (8.30) differ only by two elements, one can derive the following recursive relationships:

$$X_{1C}(n) = X_{1C}(n-1) + \frac{2}{N_1}[x(n) - x(n-N_1)]\cos\left(n\frac{2\pi}{N_1}\right), \tag{8.31}$$

$$X_{1S}(n) = X_{1S}(n-1) + \frac{2}{N_1}[x(n) - x(n-N_1)]\sin\left(n\frac{2\pi}{N_1}\right). \tag{8.32}$$

One can prove that the components X_{1C} and X_{1S} remain orthogonal also when the correlation window is an integer multiple of the half of measured signal period, i.e., when $N = cN_1/2$. Suitable recursive algorithms have then the form:

$$X_{1C}(n) = X_{1C}(n-1) + \frac{2}{N}\left[x(n)\cos\left(n\frac{2\pi}{N_1}\right) - x(n-N)\cos\left((n-N)\frac{2\pi}{N_1}\right)\right], \tag{8.33}$$

$$X_{1S}(n) = X_{1S}(n-1) + \frac{2}{N}\left[x(n)\sin\left(n\frac{2\pi}{N_1}\right) - x(n-N)\sin\left((n-N)\frac{2\pi}{N_1}\right)\right], \tag{8.34}$$

which enables calculation of the orthogonal components X_{1C} and X_{1S} for any conceivable window length. One should of course understand that choosing the number c of half cycles has adequate consequences in the frequency domain, which will not be discussed further here. The readers are encouraged to refer to [8] for more detailed information.

8.1.2 Averaging Measurement Methods

8.1.2.1 Measurement of the Magnitude of Voltage or Current Signal (Sinusoidal Variables)

Let samples of fundamental component of voltage or current be given by the equation:

$$x_1(n) = X_{1m}\cos(n\Omega_1 + \varphi_1) \tag{8.35}$$

where

$$\Omega_1 = \omega_1 T_S = 2\pi\frac{T_S}{T_1} = \frac{2\pi}{N_1}$$

and N_1 is a number of samples in one signal period.

Averaging of absolute value of this signal over the time being multiple of half period is one of the earliest methods of magnitude measurement. In this case, of course, finite sum of signal samples instead of continuous integral is calculated. It is evident that the sum of samples depends not only on magnitude of the signal but on the instant of zero crossing as well. The error depends on sampling frequency and decreases when this frequency increases. The error analysis is given in Chap. 9, where it is shown that it is possible to find the coefficients minimizing the error for given sampling frequency. In the case of half period one can obtain:

$$\sum_{n=0}^{N_1/2-1} |x_1(n)| = S_{av} X_{1m} \tag{8.36}$$

where

$$S_{av} = 0.5ctg\left(\frac{\pi}{2N_1}\right)$$

and then the magnitude can be measured according to the equation:

$$X_{1m} = 2tg\left(\frac{\pi}{2N_1}\right) \sum_{n=0}^{N_1/2-1} |x_1(n)|. \tag{8.37}$$

The equation can be simplified for high sampling frequencies:

$$X_{1m} \cong \frac{\pi}{2N_1} \sum_{n=0}^{N_1/2-1} |x_1(n)| \tag{8.38}$$

The equation can be written for time instant n as follows:

$$X_{1m}(n) = 2tg\left(\frac{\pi}{2N_1}\right) \sum_{k=0}^{N_1/2-1} |x_1(n-k)| \tag{8.39}$$

and for m half periods:

$$X_{1m,m} = m2tg\left(\frac{\pi}{2N_1}\right) \sum_{k=0}^{mN_1/2-1} |x_1(n-k)|. \tag{8.40}$$

For high number of samples it is also possible to use recursive procedure:

$$X_{1m}(n) = X_{1m}(n-1) + 2tg\left(\frac{\pi}{2N_1}\right)[|x_1(n)| - |x_1(n - N_1/2)|], \tag{8.41}$$

where x are signal samples and X is magnitude value.

One can avoid the error described above by averaging signal samples squared:

$$x_1^2(n) = 0.5X_{1m}^2\{1 + \cos(2n\Omega_1 + 2\varphi_1)\}$$

which, as it is seen, consist of DC component proportional to magnitude squared and the second harmonic of the signal. It can be noticed easily that averaging it over the time being multiple of half period gives the result:

$$\sum_{n=0}^{N-1} x_1^2(n) = 0.5X_{1m}^2 N \tag{8.42}$$

where

$$N = m * \frac{N_1}{2}$$

and then the algorithm of magnitude measurement is given in the form:

$$X_{1m}(n) = \sqrt{\frac{2}{N}}\sqrt{\sum_{k=0}^{N-1} x_1^2(n-k)}. \tag{8.43}$$

It should be underlined that the result is independent on zero crossing instant.

8.1.2.2 Measurement of Active and Reactive Power

Similar methods of averaging can be applied to measure the powers. Let current and voltage samples be given by the equations:

$$u_1(n) = U_{1m}\cos(n\Omega_1 + \varphi_{1U}), \tag{8.44a}$$

$$i_1(n) = I_{1m}\cos(n\Omega_1 + \varphi_{1I}) \tag{8.44b}$$

and let voltage samples delayed by a quarter of period be also on disposal:

$$u_1(n - N_1/4) = U_{1m}\sin(n\Omega_1 + \varphi_{1U}) \tag{8.44c}$$

where

$$\Omega_1 = \frac{2\pi}{N_1}.$$

Adequate products of voltage and current samples are given in the form:

$$u_1(n)i_1(n) = 0.5U_{1m}I_{1m}\{\cos(\varphi_{1U} - \varphi_{1I}) + \cos(2n\Omega_1 + \varphi_{1U} - \varphi_{1I})\},$$

$$u_1(n - N_1/4)i_1(n) = 0.5U_{1m}I_{1m}\{\sin(\varphi_{1U} - \varphi_{1I}) + \sin(2n\Omega_1 + \varphi_{1U} - \varphi_{1I})\}.$$

It is seen that each product consists of DC component proportional to power and the second harmonic. Calculating the sum of these products over multiple of half period results in the following algorithms of power calculation:

$$P_1(n) = \frac{1}{N}\sum_{k=0}^{N-1} u_1(n-k)i_1(n-k), \tag{8.45}$$

$$Q_1(n) = \frac{1}{N}\sum_{k=0}^{N-1} u_1(n-k-N_1/4)i_1(n-k), \tag{8.46a}$$

and similarly, when the current is delayed:

$$Q_1(n) = -\frac{1}{N}\sum_{k=0}^{N-1} u_1(n-k)i_1(n-k-N_1/4), \tag{8.46b}$$

where $N = mN_1/2$.

It can be noticed that it is possible to get very simple algorithms using averaging, which results in Eqs. 8.39, 8.43, 8.45 and 8.46a, b. In the first case only the measurement error depends on sampling frequency. On the other hand, the algorithms are sensitive to noise of either high or low frequency. The resulting errors can be removed by prefiltering.

Example 8.3 With the averaging approach provide the equations of half- and full-cycle algorithms for signal magnitude and power components measurement for the sampling rate equal to 1000 Hz. Assess influence of sampling frequency on the measurement errors. Specify sampling frequency and/or measurement period to get errors less than 0.1%.

Solution For the assumed sampling frequency the number of samples per full cycle of fundamental frequency component equals $N_1 = f_S/f_1 = 1000/50 = 20$. Thus, for half-cycle algorithms one obtains $N = N_1/2 = 10$ and from Eqs. 8.39, 8.43, 8.45 and 8.46a, b one gets:

$$X_{1m}(n) = 0.157\sum_{k=0}^{9} |x_1(n-k)|,$$

$$X_{1m}(N) = 0.447\sqrt{\sum_{k=0}^{9} x_1^2(n-k)},$$

$$P_1 = 0.1\sum_{k=0}^{9} u_1(n-k)i_1(n-k),$$

$$Q_1 = 0.1\sum_{k=0}^{9} u_1(n-k)i_1(n-k-5).$$

Full-cycle algorithms are obtained by taking twice longer averaging window, i.e. for $N = N_1 = 20$. Comparing to the above the scaling coefficients and summing limits are changed, yielding:

$$X_{1m}(n) = 7.85 \times 10^{-3}\sum_{k=0}^{19} |x_1(n-k)|,$$

$$X_{1m}(n) = 0.316\sqrt{\sum_{k=0}^{19} x_1^2(n-k)},$$

$$P_1 = 0.05 \sum_{k=0}^{19} u_1(n-k) i_1(n-k),$$

$$Q_1 = 0.05 \sum_{k=0}^{19} u_1(n-k) i_1(n-k-5).$$

Among the four algorithms, both in half- and full-cycle versions, the accuracy of measurement does not depend on sampling frequency for averaging of samples squared. In the case of summing of absolute values the results depend on f_s value, and for full-cycle algorithm ($N = N_1 = 20$) the following error value can be derived:

$$\Delta S = \frac{S_{\max} - S_{av}}{S_{\max}} = 0.5[1 - \cos(\pi/N_1)] = 6 \times 10^{-3}$$

From the above one can arrive at a formula for determination of the required number of samples (per cycle) that assures given level of error. Here, for the assumed maximum error at the level of 0.1% one gets:

$$N > \frac{\pi}{\arccos(1 - 2\Delta S)} = 49.6 \cong 50.$$

To achieve required accuracy level 50 samples of signal are needed. It means that either the measurement should last for five half-cycles (by sampling at 1000 Hz) or the sampling rate should be increased to 50×50 Hz = 2500 Hz.

8.1.3 Measurement with Use of Orthogonal Components

The most important and the most frequently used algorithms of measurement of criterion values use orthogonal signal components [6, 8–10]. The components can be obtained in various ways, as it was described above. They can be divided in general for such obtained from rotating phasor, as in the case when we use digital filters, and the ones resulting from standing phasor, as in case of correlation. The first case is more general since AC is more adequate for further processing. That is why mainly the first one is considered here.

8.1.3.1 Measurement of Sinusoidal Signal Magnitude

It is assumed that having on disposal current or voltage samples a pair of orthogonal components given by the equations can be produced:

$$x_{1C}(n) = X_{1m} \cos(n\Omega_1 + \varphi_1), \tag{8.47a}$$

$$x_{1S}(n) = X_{1m}\sin(n\Omega_1 + \varphi_1). \tag{8.47b}$$

Rotating, time-dependent phasor, is a complex function of these components:

$$\underline{x}_1(n) = x_{1C}(n) + jx_{1S}(n) = X_{1m}\exp[j(n\Omega_1 + \varphi_1)] \tag{8.48}$$

One can easily notice that the product of this phasor and its conjugate is equal to magnitude squared, being also equal to sum of squared orthogonal components:

$$\underline{x}_1(n)\underline{x}_1^*(n) = X_{1m}^2 = x_{1C}^2(n) + x_{1S}^2(n). \tag{8.49}$$

Then the simplest and the most typical algorithm of magnitude measurement is as follows:

$$X_{1m} = \sqrt{x_{1C}^2(n) + x_{1S}^2(n)}. \tag{8.50}$$

More general version can be obtained using the product of the phasor and its delayed conjugate. Then one obtains:

$$\underline{x}_1(n)\underline{x}_1^*(n-k) = X_{1m}^2\exp(jk\Omega_1) \tag{8.51}$$

Calculating the same, but applying orthogonal components, one gets:

$$\begin{aligned}\underline{x}_1(n)\underline{x}_1^*(n-k) = x_{1C}(n)x_{1C}(n-k) + x_{1S}(n)x_{1S}(n-k) + j[x_{1S}(n)x_{1C}(n-k)\\ - x_{1S}(n-k)x_{1C}(n)]\end{aligned} \tag{8.52}$$

Comparing two above equations, i.e. their real and imaginary components, yields:

$$X_{1m} = \sqrt{\frac{x_{1C}(n)x_{1C}(n-k) + x_{1S}(n)x_{1S}(n-k)}{\cos(k\Omega_1)}}, \tag{8.53}$$

$$X_{1m} = \sqrt{\frac{x_{1S}(n)x_{1C}(n-k) - x_{1S}(n-k)x_{1C}(n)}{\sin(k\Omega_1)}}. \tag{8.54}$$

The first of these equations is more general version of (8.50) for delay value equal to zero. The second is a new measurement possibility, which requires delays different from zero.

The three presented algorithms required orthogonal components. There is also a possibility to design the algorithm for signal samples only, given in the form:

$$X_{1m} = \sqrt{\frac{x_1(n-k)x_1(n-m) - x_1(n)x_1(n-k-m)}{\sin(k\Omega_1)\sin(m\Omega_1)}}. \tag{8.55}$$

Example 8.4 Specify the algorithms for current or voltage magnitude measurement (other than averaging) that do not employ orthogonal components. Sampling rate is assumed at 1000 Hz.

Solution Similar to the previous example, the number of samples per fundamental frequency cycle N_1 equals to 20 and the relative angular frequency $\Omega_1 = 2\pi/N_1 = \pi/10$. There is only one algorithm fulfilling the requirements, given by Eq. 8.55. Its parameters to be selected are the values of delay, which allows getting either very fast algorithms or the ones with very simple equation. Examples:

(a) both delays equal one sample

$$X_{1m} = k_1\sqrt{x^2(n-1) - x(n)x(n-2)},$$

$$\text{where } k_1 = \frac{1}{\sin(\Omega_1)} = \frac{1}{\sin(\pi/10)} = 3.236,$$

(b) both delays equal to quarter of cycle, i.e. five samples

$$X_{1m} = \sqrt{x^2(n-5) - x(n)x(n-10)},$$

(c) $m = 1$, $k = 5$ samples

$$X_{1m} = k_3\sqrt{x(n-1)x(n-5) - x(n)x(n-6)},$$

where

$$k_2 = \frac{1}{\sqrt{\sin(\Omega_1)}} = 1.799.$$

8.1.3.2 Measurement of Power

Digital algorithms of power measurement can be found in the same way as the ones of magnitude. To calculate them it is assumed that orthogonal components of current and voltage are on disposal:

$$u_1(n) = u_{1C}(n) = U_{1m}\cos(n\Omega_1 + \varphi_{1U}), \tag{8.56a}$$

$$i_1(n) = i_{1C}(n) = I_{1m}\cos(n\Omega_1 + \varphi_{1I}), \tag{8.56b}$$

$$u_{1S}(n) = U_{1m}\sin(n\Omega_1 + \varphi_{1U}), \tag{8.56c}$$

$$i_{1S}(n) = I_{1m}\sin(n\Omega_1 + \varphi_{1I}), \tag{8.56d}$$

The components can be combined to give phasors:

$$\underline{u}_1(n) = u_{1C}(n) + ju_{1S}(n) = U_{1m}\exp[j(n\Omega_1 + \varphi_{1U})], \tag{8.57a}$$

$$\underline{i}_1(n) = i_{1C}(n) + ji_{1S}(n) = I_{1m}\exp[j(n\Omega_1 + \varphi_{1I})]. \tag{8.57b}$$

The product of voltage phasor and conjugate—current phasor gives real part proportional to active power and imaginary part proportional to reactive power:

$$\underline{u}_1(n)\underline{i}_1^*(n) = U_{1m}I_{1m}\exp[j(\varphi_{1U} - \varphi_{1I})] = 2P_1 + j2Q_1. \tag{8.58}$$

If the same product of phasors is calculated when they are represented by orthogonal components then one can get:

$$\begin{aligned}\underline{u}_1(n)\underline{i}_1^*(n) &= [u_{1C}(n) + ju_{1S}(n)][i_{1C}(n) - ji_{1S}(n)]\\ &= [u_{1C}(n)i_{1C}(n) + u_{1S}(n)i_{1S}(n)] + j[u_{1S}(n)i_{1C}(n) - u_{1C}(n)i_{1S}(n)].\end{aligned} \tag{8.59}$$

Comparing real and imaginary parts of both equations digital algorithms of power calculation are obtained:

$$P_1 = 0.5[u_{1C}(n)i_{1C}(n) + u_{1S}(n)i_{1S}(n)], \tag{8.60}$$

$$Q_1 = 0.5[u_{1S}(n)i_{1C}(n) - u_{1C}(n)i_{1S}(n)]. \tag{8.61}$$

Above equations present fundamental class of power measurement algorithms employing orthogonal signal components. More general formula can be obtained using delayed phasors as well. Using products as below one can get complex functions of active and reactive power:

$$\underline{u}_1(n)\underline{i}_1^*(n-k) = (2P_1 + j2Q_1)\exp(jk\Omega_1) \tag{8.62a}$$

$$\underline{u}_1(n-k)\underline{i}_1^*(n) = (2P_1 + j2Q_1)\exp(-jk\Omega_1) \tag{8.62b}$$

Addition and subtraction of above equations give:

$$\underline{u}_1(n)\underline{i}_1^*(n-k) + \underline{u}_1(n-k)\underline{i}_1^*(n) = 4(P_1 + jQ_1)\cos(k\Omega_1), \tag{8.63a}$$

$$\underline{u}_1(n)\underline{i}_1^*(n-k) - \underline{u}_1(n-k)\underline{i}_1^*(n) = j4(P_1 + jQ_1)\sin(k\Omega_1), \tag{8.63b}$$

Calculating left sides of the equations with use of phasors described by their orthogonal components and comparing both sides of the equations one can get after simple rearrangements:

$$P_1 = \frac{1}{2\cos(k\Omega_1)}[u_{1S}(n)i_{1S}(n-k) + u_{1C}(n-k)i_{1C}(n)], \tag{8.64}$$

$$Q_1 = \frac{1}{2\cos(k\Omega_1)}[u_{1S}(n)i_{1C}(n-k) - u_{1C}(n-k)i_{1S}(n)], \tag{8.65}$$

$$P_1 = \frac{1}{2\sin(k\Omega_1)}[u_{1S}(n)i_{1C}(n-k) - u_{1S}(n-k)i_{1C}(n)], \tag{8.66}$$

$$Q_1 = \frac{1}{2\sin(k\Omega_1)}[u_1(n-k)i_1(n) - u_1(n)i_1(n-k)]. \tag{8.67}$$

The first two out of four equations are more general versions of the fundamental relations (8.60) and (8.61). They are equal when the value of delay is equal to zero. The two remaining equations have interesting features: the second (8.67) does not require orthogonal components at all, the first, (8.66), removes or decreases low−frequency noise present in the signals including decaying DC components.

Among many other possibilities very simple algorithms of power measurements can be a result of orthogonalization by delay of a quarter of period of fundamental component:

$$\begin{aligned} u_{1C}(n - N_1/4) &= u_{1S}(n) \\ u_{1S}(n - N_1/4) &= -u_{1C}(n). \end{aligned} \tag{8.68}$$

All of these fundamental equations can be used to measure criterion values applying different methods for obtaining of orthogonal components.

Example 8.5 Assuming the same as before sampling rate (1000 Hz) specify examples of power measurement algorithms that do not require usage of orthogonal filters.

Solution Among the algorithms for power estimation only the one for reactive power, described by Eq. 8.67, does not require orthogonal components. Depending on the selected delay value one may obtain:

for $k = 1$:

$$Q_1 = k_4[u(n-1)i(n) - u(n)i(n-1)], \text{ with } k_4 = \frac{1}{2\sin(\Omega_1)} = 1.618$$

and for the delay equal to a quarter of cycle:

$$Q_1 = 0.5[u(n-5)i(n) - u(n)i(n-5)]$$

that in fact realizes signal orthogonalization by time delay. Now one can also write similar algorithm for active power measurement:

$$P_1 = 0.5[u(n)i(n) + u(n-5)i(n-5)].$$

8.2 Measurement of Protection Criterion Values

Above the fundamental versions of measurement algorithms of electrical quantities on basis of either averaging or orthogonal components have been described. Some particular solutions resulting from different methods of orthogonalization are described and analyzed below [4, 6, 8, 10].

8.2.1 Measurement of Magnitude of Voltage or Current

Magnitude measurement can be realized using the methods described in general in last paragraph. The most important are given by Eqs. 8.50, 8.53 and 8.54. It can be easily recognized that all of these algorithms can be applied when orthogonal components from rotating phasors are used, as examples in the case of FIR digital filters. In such cases it is also possible to obtain delayed signals. In contrary, for constant phasor, as in the correlation method, Eq. 8.50 can only be applied.

Considering the first case, the output signals of a pair of orthogonal FIR filters can be written in the form:

$$y_{1C}(n) = F_{1C}X_{1m}\cos(n\Omega_1 + \varphi_1 + \beta), \tag{8.69a}$$

$$y_{1S}(n) = F_{1S}X_{1m}\sin(n\Omega_1 + \varphi_1 + \beta), \tag{8.69b}$$

where $y_1(n) = \sum_{k=0}^{N-1} a(k)x_1(n-k)$ is an output signal of digital FIR filter for the input signal,

$$x_1(n) = X_{1m}\cos(n\Omega_1 + \varphi_1),$$

$$\Omega_1 = \omega_1 T_S,$$

ω_1 is an angular frequency of fundamental component (50 Hz), F are gains of filters of odd (S) and even (C) symmetry of impulse response for fundamental frequency components, β is a phase shift of the filter of even symmetry.

The orthogonal components (8.47a, b) can be obtained from (8.69a, b) just by dividing the filter outputs by filter gains:

$$x_{1C}(n) = \frac{y_{1C}(n)}{F_{1C}}, \tag{8.70a}$$

$$x_{1S}(n) = \frac{y_{1S}(n)}{F_{1S}}. \tag{8.70b}$$

Applying them in (8.50) one gets:

$$X_{1m} = \sqrt{\frac{y_{1C}^2(n)}{F_{1C}^2} + \frac{y_{1S}^2(n)}{F_{1S}^2}}. \tag{8.71}$$

This is the fundamental equation of current or voltage magnitude measurement using a pair of orthogonal filters. In practice, in most cases the orthogonal filter gains are the same and then Eq. 8.71 is simplified to:

$$X_{1m} = \frac{1}{F_1}\sqrt{y_{1C}^2(n) + y_{1S}^2(n)}. \tag{8.72}$$

In fundamental algorithm (8.50) the signal components obtained from any known method of orthogonalization can be used. For instance, in the case of orthogonalization by single delay one receives:

$$X_{1m} = \sqrt{x_1^2(n) + \left(\frac{x_1(n-k) - x_1(n)\cos(k\Omega_1)}{\sin(k\Omega_1)}\right)^2}. \tag{8.73}$$

Depending on specific value of delay we can get either very fast ($k = 1$) or very simple algorithm ($k = N_1/4$), i.e. for the latter case:

$$X_{1m} = \sqrt{x_1^2(n) + x_1^2(n - N_1/4)}. \tag{8.74}$$

It should be noticed that applying orthogonalization by delay one must be careful, since for very small delays the algorithm frequency response is bad, even with possible noise amplification.

The second group of methods of magnitude measurement is related to Eq. 8.54, which uses delayed orthogonal components. Substituting the components (8.70a, b) to (8.54) one can get:

$$X_{1m} = \frac{1}{\sqrt{F_{1C}F_{1S}\sin(k\Omega_1)}}\sqrt{y_{1S}(n)y_{1C}(n-k) - y_{1S}(n-k)y_{1C}(n)}. \tag{8.75}$$

This is an important and interesting result. There is one common gain factor, which changes according to filter gains, delay value and sampling frequency, however, the fundamental equation remains the same. This simple result is obtained even when filter gains are different and, what more, the algorithm is less sensitive to frequency deviation, especially for filters with sine and cosine windows.

When one uses correlation, methods of MSE or DFT (see Chap. 4) to get required orthogonal components, then there is much less choice of possibilities. Since orthogonal components are then of DC type instead of AC, it is impossible to use methods and algorithms with time delay. We can simply use fundamental Eq. 8.50 only. Assuming orthogonal components in the form:

$$x_{1C}(n) = X_{1C} = X_{1m}\cos(\varphi_1) \tag{8.76a}$$

$$x_{1S}(n) = X_{1S} = X_{1m}\sin(\varphi_1) \tag{8.76b}$$

the following equation of magnitude measurement is obtained:

$$X_{1m} = \sqrt{X_{1C}^2 + X_{1S}^2} \tag{8.77}$$

Example 8.6 Provide exemplary equations of the magnitude measurement algorithms with application of separate orthogonalization procedures. Assume sampling rate $f_S = 1000$ Hz.

Solution Parameters of the equations result from the sampling frequency and are equal:

$$N_1 = f_S/f_1 = 1000/50 = 20,$$

$$\Omega_1 = 2\pi/N_1 = 2\pi/20 = \pi/10,$$

$$\cos(\Omega_1) = \cos(\pi/10) = 0.951,$$

$$\sin(\Omega_1) = \sin(\pi/10) = 0.309.$$

Substituting the values to the general equations one obtains specific algorithms for magnitude measurement, and for:

- orthogonalization by single delay, $k = 1$ (8.73)

$$X_{1m} = \sqrt{x_1^2(n) + \left(\frac{x_1(n-k) - \cos(k\Omega_1)x_1(n)}{\sin(k\Omega_1)}\right)^2} = \sqrt{x_1^2(n) + \left(\frac{x_1(n-1) - 0.951x_1(n)}{0.309}\right)^2},$$

- orthogonalization by double delay, $k = 1$ (6.41)

$$X_{1m} = \sqrt{x_1^2(n-k) + \left(\frac{x_1(n-2k) - x_1(n)}{2\sin(k\Omega_1)}\right)^2} = \sqrt{x_1^2(n-1) + \left(\frac{x_1(n-2) - x_1(n)}{0.618}\right)^2},$$

- orthogonalization by single delay, $k = N_1/4$,

$$X_{1m} = \sqrt{x_1^2(n) + x_1^2(n - N_1/4)} = \sqrt{x_1^2(n) + x_1^2(n-5)}.$$

Example 8.7 Provide full-cycle magnitude measurement algorithms with application of Walsh filters of I and II order. Assume sampling rate $f_S = 1000$ Hz.

Solution For the sampling frequency $f_S = 1000$ Hz the number of samples per cycle is $N_1 = 20$. Thus the output signals $y(n)$ of respective Walsh filters are produced according to the following equations processing input values $x(n)$, (6.22), (6.25):

$$y_{1C}(n) = -\sum_{k=0}^{4} x(n-k) + \sum_{k=5}^{14} x(n-k) - \sum_{15}^{19} x(n-k),$$

$$y_{1S}(n) = +\sum_{k=0}^{9} x(n-k) - \sum_{15}^{19} x(n-k).$$

Gain values for full-cycle filters for frequency 50 Hz are identical and amount to, (6.28):

$$F_{1C} = F_{1S} = F = |W_1(j\Omega_1)| = |W_2(j\Omega_1)| = \frac{2}{\sin(\pi/N_1)} = \frac{2}{\sin(\pi/20)} = 12.78$$

therefore the magnitude measurement equation becomes:

$$X_1 = \frac{1}{F}\sqrt{y_{1C}^2(n) + y_{1S}^2(n)} = 0.0782\sqrt{y_{1C}^2(n) + y_{1S}^2(n)}.$$

If the sampling frequency equals 2000 Hz, then the number of samples per cycle is 40 and the above equations obtain the form:

$$y_{1C}(n) = -\sum_{k=0}^{9} x(n-k) + \sum_{k=10}^{29} x(n-k) - \sum_{30}^{39} x(n-k),$$

$$y_{1S}(n) = \sum_{k=0}^{19} x(n-k) - \sum_{k=20}^{39} x(n-k),$$

$$X_1 = 0.0393\sqrt{y_{1C}^2(n) + y_{1S}^2(n)}.$$

Example 8.8 For sampling frequencies 1000 and 1600 Hz derive equations of full-cycle magnitude measurement algorithms with application of full-cycle FIR filters with sine, cosine impulse responses windows.

Solution Output signals of the orthogonal filters (6.37), (6.38) are:

$$y_{1C}(n) = \sum_{k=0}^{19} x(n-k)\cos[(9.5-k)\pi/10],$$

$$y_{1S}(n) = \sum_{k=0}^{19} x(n-k)\sin[(9.5-k)\pi/10].$$

The gain coefficients for both filters for fundamental frequency are (for $f_S =$ 1000 Hz):

$$F_{1C} = F_{1S} = F = |A_C(j\Omega_1)| = |A_S(j\Omega_1)| = N_1/2 = 10,$$

thus the measurement equation becomes:

$$X_1 = \frac{1}{F}\sqrt{y_{1C}^2(n) + y_{1S}^2(n)} = 0.1\sqrt{y_{1C}^2(n) + y_1^2(n)}.$$

For the sampling frequency 1600 Hz the number of samples per cycle (also number of coefficients of the filters' impulse responses) amounts to 1600/50 = 32, and then the respective filtration and magnitude measurement equations are:

$$y_{1C}(n) = \sum_{k=0}^{31} x(n-k)\cos[(15.5-k)\pi/16],$$

$$y_{1S}(n) = \sum_{k=0}^{31} x(n-k)\sin[(15.5-k)\pi/16],$$

$$F_{1C} = F_{1S} = F = |A_C(j\Omega_1)| = |A_S(j\Omega_1)| = N_1/2 = 16,$$

$$X_1 = \frac{1}{F}\sqrt{y_{1C}^2(n) + y_{1S}^2(n)} = \frac{1}{16}\sqrt{y_{1C}^2(n) + y_{1S}^2(n)}.$$

8.2.2 Measurement of Power

Fundamental equations of power measurement with application of orthogonal components are given in Sect. 8.1.3.2. Required orthogonal components can be produced from signals on disposal, according to methods described in Chap. 6. Let us assume they are given in the form:

$$u_1(n) = U_{1m}\cos(n\Omega_1 + \varphi_{1U}) \tag{8.78a}$$

$$i_1(n) = I_{1m}\cos(n\Omega_1 + \varphi_{1I}) \tag{8.78b}$$

When orthogonal components from rotating phasor, applying, for instance, digital filters are used, the signal components have the form:

$$u_{F1C}(n) = F_{1C}U_{1m}\cos(n\Omega_1 + \varphi_{1U} + \beta), \tag{8.79a}$$

$$u_{F1S}(n) = F_{1S}U_{1m}\sin(n\Omega_1 + \varphi_{1U} + \beta), \tag{8.79b}$$

$$i_{F1C}(n) = F_{1C}I_{1m}\cos(n\Omega_1 + \varphi_{1I} + \beta), \tag{8.79c}$$

$$i_{F1S}(n) = F_{1S}I_{1m}\sin(n\Omega_1 + \varphi_{1I} + \beta). \tag{8.79d}$$

One can notice that the components were obtained using the same orthogonal filters for current and voltage, which leads to simpler form of final algorithm (in general we may use different filters for voltage and current). Substituting the components to Eq. 8.60 the active power value is reached:

$$P_1 = \frac{1}{2}\left[\frac{u_{F1C}(n)i_{F1C}(n)}{F_{1C}^2} + \frac{u_{F1S}(n)i_{F1S}(n)}{F_{1S}^2}\right] \tag{8.80}$$

This fundamental equation can be simplified when filter gains are identical: i.e. $F_{1C} = F_{1S} = F_1$ (it is true for full-cycle sine, cosine or Walsh FIR filters). And then:

$$P_1 = \frac{1}{2F_1^2}[u_{F1C}(n)i_{F1C}(n) + u_{F1S}(n)i_{F1S}(n)], \tag{8.81}$$

which presents the rescaled version of fundamental Eq. 8.60.

Substituting components (8.79a, b, c, d) to fundamental Eq. 8.61 one can get equation of reactive power:

$$Q_1 = \frac{1}{2F_{1C}F_{1S}}[u_{F1S}(n)i_{F1C}(n) - u_{F1C}(n)i_{F1S}(n)]. \tag{8.82}$$

This is a rescaled fundamental Eq. 8.61, similar to (8.81).

Among the fundamental equations of power measurements (8.64–8.67) employing delayed orthogonal components, the most important are the last two. Substituting there orthogonal components from Eqs. 8.79a, b, c, d one gets:

$$P_1 = \frac{1}{2F_{1C}F_{1S}\sin(k\Omega_1)}[u_{F1S}(n)i_{F1C}(n-k) - u_{F1S}(n)i_{F1C}(n)], \tag{8.83}$$

$$Q_1 = \frac{1}{2F_1^2\sin(k\Omega_1)}[u_{F1}(n-k)i_{F1}(n) - u_{F1}(n)i_{F1}(n-k)]. \tag{8.84}$$

It is seen in Eq. 8.84 that in this case one can use any arbitrary selected filter for voltage and current (the same for both signals) or avoid using them at all. Both algorithms are rescaled version of fundamental Eqs. 8.64 and 8.67. If the delay is equal to unity ($k = 1$), the fastest algorithm is obtained, while for the number of samples of delay being equal to a quarter of number of samples per cycle of fundamental frequency component ($k = N_1/4$), the simplest one is reached. In this case there is no need for orthogonal filters at all (similar to (8.74)):

$$P_1 = \frac{1}{2F_1^2}[u_{F1}(n)i_{F1}(n) + u_{F1}(n-N_1/4)i_{F1}(n-N_1/4)], \tag{8.85}$$

$$Q_1 = \frac{1}{2F_1^2}[u_{F1}(n-N_1/4)i_{F1}(n) - u_{F1}(n)i_{F1}(n-N_1/4)]. \tag{8.86}$$

When orthogonal components are obtained from standing phasor (correlation method, DFT and others), they can be described by the formulae:

$$u_{1C}(n) = U_{1C} = U_{1m}\cos(\varphi_{1U}), \tag{8.87a}$$

$$u_{1S}(n) = U_{1S} = U_{1m}\sin(\varphi_{1U}), \tag{8.87b}$$

$$i_{1C}(n) = I_{1C} = I_{1m}\cos(\varphi_{1I}), \tag{8.87c}$$

$$i_{1S}(n) = I_{1S} = I_{1m}\sin(\varphi_{1I}). \tag{8.87d}$$

Since these components are constant, not time-dependent, they can be used in Eqs. 8.60 and 8.61 only, i.e. in equations which apply orthogonal components at the same instant (without delay). Using components (8.87a, b, c, d) in these equations one gets:

$$P_1 = 0.5[U_{1C}I_{1C} + U_{1S}I_{1S}], \tag{8.88}$$

$$Q_1 = 0.5[U_{1S}I_{1C} - U_{1C}I_{1S}], \tag{8.89}$$

There are many methods allowing for magnitude and power measurements, as it is seen from considerations above. Some very simple algorithms use either averaging or Walsh orthogonal filters. When high immunity to noise is important then FIR sine, cosine filters should be used. When fast measurement is required Kalman filters, least square error or variable data window methods are recommended. When not only fundamental but also components of other frequencies are required, DFT is usually applied. In general one can always find a method adequate to the requirements.

Example 8.9 Permissible measurement time equals 12 ms. Algorithms for power measurement are to be proposed, with assumed sampling frequency 1000 Hz.

Solution For the assumed sampling rate the sampling period is equal to 1 ms, thus maximum window length of applied filters amounts to 12 samples. Therefore it is possible to propose the following algorithms fulfilling the requirements defined:

(a) application of FIR orthogonal filters with 12-sample window (e.g., sin, cos), algorithms (8.82) and (8.84),
(b) application of algorithms (8.83) and (8.84) with half-cycle Walsh or sin/cos FIR filters,
(c) usage of filters as in (b), application of algorithms (8.85) and (8.86), being less susceptible to frequency changes (see Chap. 9),
(d) application of only two half-cycle filters, orthogonalization by single or double delay.

The choice of one particular version depends on additional conditions that may include possible spectrum of the input signals or algorithm computational burden. The latter issue is illustrated in Fig. 8.6 where the schemes of obtaining orthogonal components of current and voltage signals are depicted. It is seen that the cases (a–c) are similar (signal delaying in principle does not introduce any computational burden), while the case (d) is the simplest since the separated orthogonalization is realized easier than any filter.

For assumed sampling frequency one gets: $N_1 = 20; N = 12; \Omega_1 = 2\pi/N_1 = 0.1\pi$.

(a) For the longest permissible window ($N = 12$) the filter gains (4.34), (4.35) are equal:

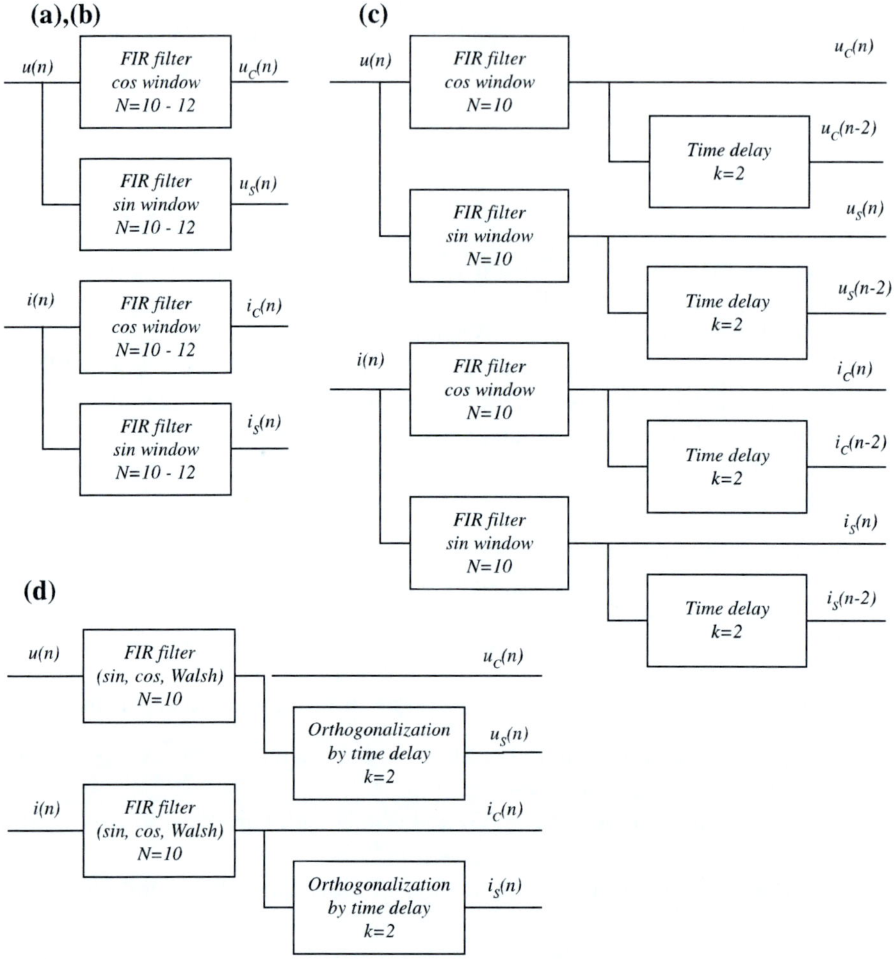

Fig. 8.6 Block schemes of signal orthogonalization for power measurement for the cases (a–d), as in Example 8.9

$$F_{1C} = 0.5\left[N + \frac{\sin(N\Omega_1)}{\sin(\Omega_1)}\right] = 0.5\left[12 + \frac{\sin(1.2\pi)}{\sin(0.1\pi)}\right] = 3.06$$

$$F_{1S} = 0.5\left[N - \frac{\sin(N\Omega_1)}{\sin(\Omega_1)}\right] = 0.5\left[12 - \frac{\sin(1.2\pi)}{\sin(0.1\pi)}\right] = 8.94$$

and after substituting them to (8.48), (8.50) one gets:

$$Q_1 = 0.0183[u_{F1S}(n)i_{F1C}(n) - u_{F1C}(n)i_{F1S}(n)],$$

where orthogonal components of voltage (similar for the current) are obtained from

$$u_{F1C}(n) = \sum_{k=0}^{11} u_1(n-k)\cos[0.1\pi(5.5-k)].$$

(b) For the half-cycle data window ($N = 10$) the filter gains are identical and equal to 5. Thus the algorithm equations take the form:

$$P_1 = 0.02[u_{F1C}(n)i_{F1C}(n) + u_{F1S}(n)i_{F1S}(n)],$$

$$Q_1 = 0.02[u_{F1S}(n)i_{F1C}(n) - u_{F1C}(n)i_{F1S}(n)],$$

where orthogonal components of voltage (similar for the current) are obtained from $u_{F1C}(n) = \sum_{k=0}^{9} u_1(n-k)\cos[0.1\pi(4.5-k)]$.

(c) In this case the filter gains are the same as in (b). Additionally one calculates: $\sin(k\Omega_1) = \sin(2 * 0.1\pi) = 0.588$ and thus

$$P_1 = 0.034[u_{F1S}(n)i_{F1C}(n-2) - u_{F1S}(n-2)i_{F1C}(n)],$$

$$Q_1 = 0.034[u_{F1}(n-2)i_{F1}(n) - u_{F1}(n)i_{F1C}(n-2)].$$

(d) Here the components of power are calculated as in (b), with different way of reaching the orthogonal components of the signals. If the single-delay method is applied (with $k = 2$ samples) the specific equations become (here—for voltage):

$$u_{F1S} = \frac{u_1(n-2) - u_1(n)\cos(2^*0.1\pi)}{\sin(2^*0.1\pi)} = \frac{u_1(n-2) - 0.809u_1(n)}{0.588}$$
$$= 1.7u_1(n-2) - 1.376u_1(n).$$

Similar equation can be used for orthogonalization of current signal.

8.2.3 Measurement of Impedance and Its Components

8.2.3.1 Application of Orthogonal Components

The algorithms allowing to measure impedance and its components as well as conductance and its components result from complex relations similar to those used before. One can write the following equations:

$$\underline{Z} = |\underline{Z}|\exp(j\varphi) = |\underline{Z}|\cos(\varphi) + j|\underline{Z}|\sin(\varphi) = R + jX. \qquad (8.90)$$

These quantities can easily be calculated from power and magnitude obtained before:

$$Z^2 = \frac{U^2}{I^2} = \frac{U_m^2}{I_m^2} \tag{8.91}$$

and

$$R = |\underline{Z}| \cos(\varphi) = \frac{U_m I_m \cos(\varphi)}{I_m^2} = \frac{2P}{I_m^2}, \tag{8.92}$$

as well as

$$X = \frac{2Q}{I_m^2}, \tag{8.93}$$

where U, I are effective values of voltage and current, U_m, I_m are their magnitudes.

These fundamental equations or similar for conductance and components can be used to find many particular algorithms using any power and any magnitude algorithms. It is reasonable, however, to use such equations for power and magnitude that constant coefficients in numerator and denominator are cancelled.

The general form of the equations uses components from the outputs of orthogonal filters, i.e. using (8.71), (8.79a, b, c, d) and (8.81). Substituting them into (8.80) equations of measurement of impedance are obtained:

$$Z^2 = \frac{\left(\frac{u_{F1C}(n)}{F_{1C}}\right)^2 + \left(\frac{u_{F1S}(n)}{F_{1S}}\right)^2}{\left(\frac{i_{F1C}(n)}{F_{1C}}\right)^2 + \left(\frac{i_{F1S}(n)}{F_{1S}}\right)^2}, \tag{8.94}$$

resistance

$$R = \frac{\frac{u_{F1C}(n) i_{F1C}(n)}{F_{1C}^2} + \frac{u_{F1S}(n) i_{F1S}(n)}{F_{1S}^2}}{\frac{i_{F1C}^2(n)}{F_{1C}^2} + \frac{i_{F1S}^2(n)^2}{F_{1S}^2}}, \tag{8.95}$$

and reactance

$$X = \frac{\frac{1}{F_{1C} F_{1S}} \left[u_{F1S}(n) i_{F1C}(n) - u_{F1C}(n) i_{F1S}(n)\right]}{\frac{i_{F1C}^2(n)}{F_{1C}^2} + \frac{i_{F1S}^2(n)}{F_{1S}^2}}. \tag{8.96}$$

When gains F_{1C}, F_{1S} are identical, they are being cancelled and simple equations using orthogonal components of voltage and current are received:

$$Z^2 = \frac{u_{F1C}^2(n) + u_{F1S}^2(n)}{i_{F1C}^2(n) + i_{F1S}^2(n)}, \tag{8.97}$$

$$R = \frac{u_{F1C}(n) i_{F1C}(n) + u_{F1S}(n) i_{F1S}(n)}{i_{F1C}^2(n) + i_{F1S}^2(n)}, \tag{8.98}$$

$$X = \frac{u_{F1S}(n)i_{F1C}(n) - u_{F1C}(n)i_{F1S}(n)}{i^2_{F1C}(n) + i^2_{F1S}(n)}, \tag{8.99}$$

These are the simple equations to measure considered quantities. Thanks to application of the filters having identical gains, the equations are gain-independent.

The second set of equations is derived using power and magnitude measurement algorithms, which are calculated by means of delayed orthogonal components. The equations for resistance and impedance are especially simple. If one takes the same filters to get orthogonal components as well as the delays are the same for all algorithms, then all coefficients are cancelled, which yields the following equations:

$$Z_1^2 = \frac{u_{F1S}(n)u_{F1C}(n-k) - u_{F1S}(n-k)u_{F1C}(n)}{i_{F1S}(n)i_{F1C}(n-k) - i_{F1S}(n-k)i_{F1C}(n)}, \tag{8.100}$$

$$R_1 = \frac{u_{F1S}(n)i_{F1C}(n-k) - u_{F1S}(n-k)i_{F1C}(n)}{i_{F1S}(n)i_{F1C}(n-k) - i_{F1S}(n-k)i_{F1C}(n)}. \tag{8.101}$$

These are important results, since very simple algorithms without any additional coefficients are obtained. Similar result cannot be obtained for reactance. When (8.74), (8.84) and (8.95) are used, one arrives at:

$$X_1 = \frac{F_{1C}F_{1S}}{F_1^2}\frac{u_{F1}(n-k)i_{F1}(n) - u_{F1}(n)i_{F1}(n-k)}{i_{F1S}(n)i_{F1C}(n-k) - i_{F1S}(n-k)i_{F1C}(n)}. \tag{8.102}$$

However, when the filter gains are the same, they are cancelled, which brings even simpler equation.

To obtain the algorithm of reactance measurement independent of filter gains Eqs. 8.100 and 8.101 can be applied and the general relation between impedance, resistance and reactance:

$$X_1^2 = Z_1^2 - R_1^2. \tag{8.103}$$

In this way one gets the set of equations of impedance and its components measurement, which are independent of filter gains. This result can be very important for frequency independent algorithms described later. The only condition is the same filters for current and voltage signals.

Using orthogonal components obtained by delay of a quarter of period of fundamental frequency component one gets:

$$Z_1^2 = \frac{u^2_{F1}(n) + u^2_{F1}(n - N_1/4)}{i^2_{F1}(n) + i^2_{F1}(n - N_1/4)}, \tag{8.104}$$

$$R_1 = \frac{u_{F1}(n)i_{F1}(n) + u_{F1}(n - N_1/4)i_{F1}(n - N_1/4)}{i^2_{F1}(n) + i^2_{F1}(n - N_1/4)}, \tag{8.105}$$

$$X_1 = \frac{u_{F1}(n - N_1/4)i_{F1}(n) - u_{F1}(n)i_{F1}(n - N_1/4)}{i_{F1}^2(n) + i_{F1}^2(n - N_1/4)}. \quad (8.106)$$

The advantage here are very simple algorithms and that one can use single filters instead of pairs of orthogonal filters. However, disadvantage is longer transient period and measurement time.

When standing phasor is used (DFT or correlation methods for instance) one obtains:

$$Z^2 = \frac{U_{1C}^2 + U_{1S}^2}{I_{1C}^2 + I_{1S}^2}, \quad (8.107)$$

$$R = \frac{U_{1C}I_{1C} + U_{1S}I_{1S}}{I_{1C}^2 + I_{1S}^2}, \quad (8.108)$$

$$X = \frac{U_{1S}I_{1C} - U_{1C}I_{1S}}{I_{1C}^2 + I_{1S}^2}. \quad (8.109)$$

Example 8.10 Derive impedance measurement algorithms with permissible measurement time equal 25, 20, 15 and 10 ms, respectively. Apply FIR sin, cos filters or signal averaging, assuming the sampling rate 1000 Hz.

Solution Except for the case of shortest measurement time 10 ms, there exist a possibility of using FIR filters with data window length corresponding to the permissible measurement time or slightly shorter, with the remaining time assigned to signal delay—in order to obtain either very simple algorithm or the one with other assumed/desirable features. This cannot apply for the case with 10 ms measurement time, where the only reasonable solution is longest possible filtration (even with 10 ms filters the effects are not perfect) and usage of measurement Eqs. 8.107–8.109. Alternatively, one can propose application of half-cycle averaging (see below). The following specific algorithms are feasible in the considered cases:

(a) measurement time ≤10 ms:

- (a1) half-cycle averaging algorithms (8.122), (8.123), $m = 1$; the reactance value can be obtained from calculated resistance and impedance,
- (a2) application of orthogonal components, algorithms (8.97), (8.98) and (8.99); the gains of half-cycle filters are identical and get cancelled in final equations.

(b) measurement time ≤15 ms:

- (b1) application of just one filter for current and voltage (10 ms, sine, cosine or other window), and 5 ms delay for signal orthogonalization (8.104–8.106),
- (b2) application of a pair of half-cycle orthogonal filters for current and voltage and then the measurement algorithms with time delay (8.100–8.102), being less susceptible to frequency deviations.

(b3) application of orthogonal of 3/4-cycle orthogonal filters and the algorithms (8.97–8.99).

(c) measurement time $\leq$20 ms (full cycle of the fundamental frequency):

(c1) application of full-cycle sin, cos filters and algorithms (8.97–8.99),
(c2) full-cycle averaging algorithms (8.122), (8.123), $m = 2$; the reactance value can be obtained from calculated resistance and impedance.

(d) measurement time $\leq$25 ms:

(d1) application of one full-cycle filter (20 ms, sine, cosine or other) for current and voltage, plus 5 ms delay for orthogonalization, (8.104–8.106),
(d2) application of a pair of half-cycle orthogonal filters for current and voltage and then algorithms (8.66) to (8.68) with 1/4-cycle delay,
(d3) application of a pair of orthogonal filters with 5/4-cycle window (different gain coefficients) and algorithms (8.94–8.96).

Selection of particular version should take into consideration all important requirements and constraints, e.g., expected frequency spectra of input signals, computational burden, frequency features of applied filters, etc. The frequency responses of various filters with windows from 1/2- to 5/4-cycle are presented in Fig. 8.7. One can see that the most favorable filtration of disturbing components in case of assumed measurement time 15 ms is obtained for the case (c), i.e., for the filter of longest possible window. However, this algorithm is the most complex one since it requires four filters, and additionally—their gains cannot be cancelled. A very simple algorithm is obtained when only one filter with 10 ms data window is applied; filter gains can now be cancelled, which leads to very simple measurement equations. Similar discussion can be done for all remaining permissible (assumed) measurement times.

The required orthogonal components needed in particular versions of algorithms can be obtained in the ways as considered in Example 8.9, according to block schemes from Fig. 8.6. It is essential to perform orthogonalization of both current and voltage signals with the same algorithm, which creates possibility of obtaining very simple measurement equations.

Since the algorithms for magnitude and power measurement have already been illustrated with numerous examples, below chosen variants only as applied for resistance measurement are presented.

Case (a2)—measurement time 10 ms ($N = 10$ for assumed sampling frequency). Orthogonal components of voltage signals can be calculated from:

$$u_{F1C}(n) = \sum_{k=0}^{9} u_1(n-k)\cos[0.1\pi(4.5-k)],$$

$$u_{F1S}(n) = \sum_{k=0}^{9} u_1(n-k)\sin[0.1\pi(4.5-k)],$$

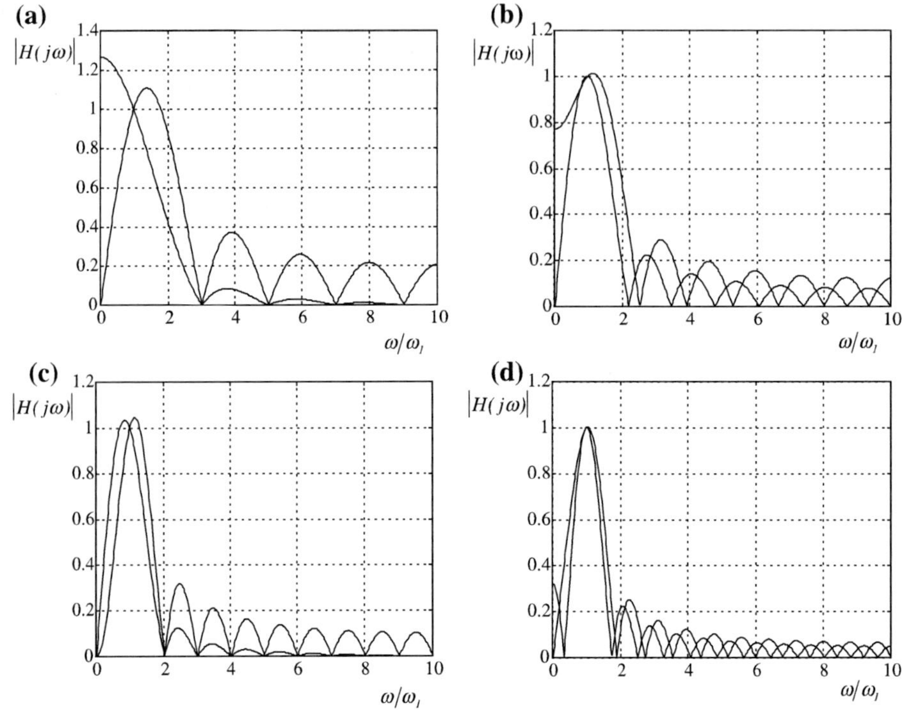

Fig. 8.7 Normalized spectra of sin, cos FIR filters for various lengths of data window: **a** 10 ms, **b** 15 ms, **c** 20 ms, **d** 25 ms

identical equations can be applied for current orthogonalization.

The resistance measurement equation becomes:

$$R = \frac{u_{F1C}(n)i_{F1C}(n) + u_{F1S}(n)i_{F1S}(n)}{i_{1F1C}^2(n) + i_{1F1S}^2(n)}$$

with subscript F standing for output signals from orthogonal filters.

Case (b3)—measurement time 15 ms, with application of 15 ms filters:

$$u_{F1C}(n) = \sum_{k=0}^{14} u_1(n-k)\cos[0.1\pi(7-k)],$$

$$u_{F1S}(n) = \sum_{k=0}^{15} u_1(n-k)\sin[0.1\pi(7-k)],$$

identical equations can be applied for current orthogonalization. Since here the filter gains are not identical:

$$F_C(j\Omega_1) = 0.5\left[\frac{15}{2} + \frac{\sin(15^*0.1\pi)}{\sin(0.1\pi)}\right] = 2.13$$

$$F_S(j\Omega_1) = 0.5\left[\frac{15}{2} - \frac{\sin(15^*0.1\pi)}{\sin(0.1\pi)}\right] = 5.37$$

the resistance measurement Eq. 8.95 becomes:

$$R_1 = \frac{0.189u_{F1C}(n)i_{F1C}(n) + 0.035u_{F1S}(n)i_{F1S}(n)}{0.189i_{F1C}(n)i_{F1C}(n) + 0.035i_{F1S}(n)i_{F1S}(n)}.$$

8.2.3.2 Impedance Calculated from Fault Loop Equations

Different class of algorithms of impedance measurement is developed on basis of the simple model of fault loop [4, 7]. Typical model of faulted line is series *R–L* circuit. Samples of current and voltages, taken from real system are just applied to calculated parameters of this model. Differential equation of the circuit is given in the well-known form (in the simplest case):

$$u(t) = Ri(t) + L\frac{\mathrm{d}i(t)}{\mathrm{d}t} \tag{8.110}$$

The unknown values of R and L are sought, which requires two sets of samples of voltages and currents. There are many methods to get these two sets from differential equation (8.110). Two of them are presented below. The first one transfers differential equation to the difference one and uses it for two discrete time instants. In the second method an integral of differential equation (8.76a, b) is used, which is in turn transferred to two difference equations for two discrete time instants.

- calculation of R, L using discrete model of (8.110) for two different time instants.

Let u_k, i_k be samples of voltage and current at instant t_k (or kT_S) where T_S is sampling period. When this period is adequately small, one may write:

$$\frac{\mathrm{d}i}{\mathrm{d}t} = \frac{i_{k+1} - i_{k-1}}{2T_S} \tag{8.111}$$

and then Eq. 8.110 can be written in the form:

$$u_k = Ri_k + L\frac{i_{k+1} - i_{k-1}}{2T_S}. \tag{8.112a}$$

The same equation but at time instant t_{k+1} has the form:

$$u_{k+1} = Ri_{k+1} + L\frac{i_{k+2} - i_k}{2T_S} \tag{8.112b}$$

Both equations can be written in matrix form:

$$\begin{bmatrix} i_k & \frac{i_{k+1}-i_{k-1}}{2T_S} \\ i_{k+1} & \frac{i_{k+2}-i_k}{2T_S} \end{bmatrix} \begin{bmatrix} R \\ L \end{bmatrix} = \begin{bmatrix} u_k \\ u_{k+1} \end{bmatrix}. \tag{8.113}$$

Having voltage and current samples, we can solve the equation to find unknown R, L:

$$R = \frac{u_k(i_{k+2} - i_k) - u_{k+1}(i_{k+1} - i_{k-1})}{i_k(i_{k+2} - i_k) - i_{k+1}(i_{k+1} - i_{k-1})}, \tag{8.114}$$

$$L = 2T_S \frac{u_{k+1}i_k - u_k i_{k+1}}{i_k(i_{k+2} - i_k) - i_{k+1}(i_{k+1} - i_{k-1})}. \tag{8.115}$$

A very interesting feature of the solution is the result independence of the possible presence of decaying DC component. This is important advantage of the method since in many algorithms it is a source of comparatively high errors that are usually difficult to remove. The disadvantage of this method is its sensitivity to high frequency noise. It can be eliminated by the method described below.

- calculation of R, L using discrete model of integral of differential equation of fault loop.

The fundamental method of suppressing oscillatory noise in the above method relies on calculating an integral of the differential equation of fault loop (8.110). Doing that for two short time periods yields two equations:

$$\int_{t_k}^{t_{k+1}} u(t)\mathrm{d}t = R\int_{t_k}^{t_{k+1}} i(t)\mathrm{d}t + L[i(t_{k+1}) - i(t_k)], \tag{8.116a}$$

$$\int_{t_{k+1}}^{t_{k+2}} u(t)\mathrm{d}t = R\int_{t_{k+1}}^{t_{k+2}} i(t)\mathrm{d}t + L[i(t_{k+2}) - i(t_{k+1})]. \tag{8.116b}$$

Calculating these integrals assuming trapezoidal integration rule one gets:

$$\int_{t_k}^{t_{k+1}} u(t)\mathrm{d}t = \frac{T_S}{2}[u(t_k) + u(t_{k+1})] \tag{8.117a}$$

and

$$\int_{t_k}^{t_{k+1}} i(t)\mathrm{d}t = \frac{T_S}{2}[i(t_k) + i(t_{k+1})]. \tag{8.117b}$$

Substituting the integrals (8.117a) and similar sum to (8.113) the matrix equation is reached:

$$\begin{bmatrix} 0.5T_S(i_{k+1}+i_k) & i_{k+1}-i_k \\ 0.5T_S(i_{k+2}+i_{k+1}) & i_{k+2}-i_{k+1} \end{bmatrix} \begin{bmatrix} R \\ L \end{bmatrix} = \begin{bmatrix} 0.5T_S(u_{k+1}+u_k) \\ 0.5T_S(u_{k+2}+u_{k+1}) \end{bmatrix}. \tag{8.118}$$

Solving it we get sought values R and L:

$$R = \frac{(u_{k+1}+u_k)(i_{k+2}-i_{k+1}) - (u_{k+2}+u_{k+1})(i_{k+1}-i_k)}{2(i_k i_{k+2} - i_{k+1}^2)}, \tag{8.119}$$

$$L = T_S \frac{(u_{k+1}+u_k)(i_{k+2}-i_{k+1}) - (u_{k+2}+u_{k+1})(i_{k+1}-i_k)}{2(i_k i_{k+2} - i_{k+1}^2)}. \tag{8.120}$$

The method is immune to high frequency noise, however, certain problem is oscillation of denominator and its small value for substantial decaying DC in current signal.

8.2.3.3 Measurement of Impedance Components Using Averaging

At the end of presentation the impedance measurement algorithms with application of averaging approach are described. They result directly from general Eqs. 8.91–8.93 and calculation of power and magnitude using averaging (8.39), (8.43), (8.45) and (8.46a, b). Assuming identical coefficients very simple equation can be reached:

$$Z_1 = \frac{\sum_{k=0}^{mN_1/2} |u_1(n-k)|}{\sum_{k=0}^{mN_1/2} |i_1(n-k)|} \tag{8.121}$$

or using (8.9a):

$$Z_1^2 = \frac{\sum_{k=0}^{mN_1/2} u_1^2(n-k)}{\sum_{k=0}^{mN_1/2} i_1^2(n-k)}. \tag{8.122}$$

Constant coefficients in numerator and denominator are cancelled thanks to the same data windows and identical algorithms for voltage and current processing.

In case of resistance and reactance the choice is smaller and the best solution is got using (8.43):

$$R_1 = \frac{\sum_{k=0}^{N-1} u_1(n-k) i_1(n-k)}{\sum_{k=0}^{N-1} i_1^2(n-k)} \tag{8.123}$$

and applying Eqs. 8.43 and 8.46a, b the reactance becomes:

$$X_1 = \frac{\sum_{k=0}^{N-1} u_1(n-k-N_1/4) i_1(n-k)}{\sum_{k=0}^{N-1} i_1^2(n-k)}. \tag{8.124}$$

Example 8.11 For the sampling frequency 1000 Hz propose: (a) possibly simplest algorithms for resistance and reactance measurement for undistorted voltage and current signals or when only harmonic distortions are expected; (b) possibly fastest algorithms insensitive to frequency deviations.

Solution

(a) the simplest algorithms for impedance components measurement result from Eqs. 8.101 and 8.102, for $k = 5$:

$$R_1 = \frac{u_1(n)i_1(n) + u_1(n-5)i_1(n-5)}{i_1(n)i_1(n) + i_1(n-5)i_1(n-5)}$$

$$X_1 = \frac{u_1(n-5)i_1(n) + u_1(n)i_1(n-5)}{i_1(n)i_1(n) + i_1(n-5)i_1(n-5)}$$

or from averaging algorithms (8.123), (8.124):

$$R_1 = \frac{\sum_{k=0}^{9} u_1(n-k)i_1(n-k)}{\sum_{k=0}^{9} i_1^2(n-k)},$$

$$X_1 = \frac{\sum_{k=0}^{9} u_1(n-k-5)i_1(n-k)}{\sum_{k=0}^{9} i_1^2(n-k)}.$$

In case when harmonic distortions may appear one can apply the same algorithms with appropriate prefiltering of the current and voltage. In case of averaging one can also apply longer averaging time (a multiple of half-cycle).

(b) the fastest possible algorithms insensitive to frequency deviations
From Eqs. 8.101–8.103 for the shortest possible delay value (1 sample) one obtains:

$$R_1 = \frac{u_{F1S}(n)i_{F1C}(n-1) - u_{F1S}(n-1)i_{F1C}(n)}{i_{F1S}(n)i_{F1C}(n-1) - i_{F1S}(n-1)i_{F1C}(n)},$$

$$Z_1^2 = \frac{u_{F1S}(n)u_{F1C}(n-1) - u_{F1S}(n-1)u_{F1C}(n)}{i_{F1S}(n)i_{F1C}(n-1) - i_{F1S}(n-1)i_{F1C}(n)},$$

$$X_1 = \sqrt{Z_1^2 - R_1^2}.$$

The resulting speed of the algorithms is dependent on the type and window length of applied orthogonal filters.

As it is seen, to measure impedance components and similarly power and magnitudes one can use either orthogonal components or averaging. Features of the algorithms will first of all depend on introductory signal processing (mainly digital filters) and in some way on the algorithms themselves. Application of identical filters leads to substantial simplification of the algorithms of impedance and its components. Insensitivity to decaying DC, when impedance is measured, can be reached using model and differential equation of fault loop.

8.2.4 Phase and Phase Shift Measurement

8.2.4.1 Signal Phase Measurement

Let the samples of a signal, voltage or current, be given by the equation:

$$x(n) = X_{1m} \sin(n\Omega_1 + \varphi), \tag{8.125}$$

where

$$\Omega_1 = \omega_1 T_S.$$

Phase is defined as argument value at the instant of measurement calculated from the reference being that moment when the function x last time crossed zero:

$$\gamma(n) = n\Omega_1 + \varphi. \tag{8.126}$$

If the magnitude of the signal (8.125) was known, it would be possible to calculate phase value using arcsin function. However, it is not convenient and the magnitude must be kept constant. This is why better solution are magnitude independent algorithms using the same components as before, i.e. orthogonal components obtained in many different ways. One should take into account, however, that till now it was assumed that there are on disposal real parts of complex phasors, and now its imaginary part is used. As a consequence, delay of the quarter of period of the signal (8.125) gives orthogonal component but with opposite sign (the other way round than before). It has no influence in case of magnitude measurement but in case of phase it has. This is why opposite signs are obtained in cases of different orthogonalization methods (single, double delay).

The simplest way of phase measurement is got using orthogonal signal components obtained by delay of quarter of period. Remembering that $N_1\Omega_1 = 2\pi$ one receives:

$$x(n - N_1/4) = X_{1m} \sin[(n - N_1/4)\Omega_1 + \varphi] = -X_{1m} \cos(n\Omega_1 + \varphi). \tag{8.127}$$

Taking into account the above and (8.125) yields:

$$\gamma(n) = -\text{arctg}\frac{x(n)}{x(n - N_1/4)} \tag{8.128}$$

To calculate phase one can also apply orthogonalization by single delay:

$$\gamma(n) = \text{arctg}\frac{x(n-k) - x(n)\cos(k\Omega_1)}{x(n)\sin(k\Omega_1)} \tag{8.129}$$

or by double delay:

$$\gamma(n-k) = \operatorname{arctg}\frac{x(n-2k)-x(n)}{2x(n-k)\sin(k\Omega_1)}. \tag{8.130}$$

To measure phase shift of the signal one can also use a pair of orthogonal filters. Assuming that input signal of orthogonal filters is equal to $x(n) = X_{1m}\sin(n\Omega_1 + \varphi)$ the output signal of one filter equals to: $y_{1C} = F_{1C}X_{1m}\sin(n\Omega_1 + \varphi + \beta)$, while the second equals $y_{1S} = F_{1S}X_{1m}\cos(n\Omega_1 + \varphi + \beta)$. Taking ratio of these signals yields:

$$\operatorname{arctg}\left(\frac{F_{1S}}{F_{1C}}\frac{y_{1C}(n)}{y_{1S}(n)}\right) = \gamma(n) + \beta = n\Omega_1 + \varphi + \beta \tag{8.131}$$

From (8.131) one can calculate phase shift, since filters parameters are known.

Example 8.12 Provide an algorithm for signal phase estimation with orthogonal components obtained from signal filtration with use of FIR sin, cos filters. Assume sampling with 1000 Hz.

Solution For given sampling rate the number of samples per cycle of 50 Hz is $N_1 = 20$ thus the filter gains are $F_{1C} = F_{1C} = N_1/2 = 10$ and their arguments amount to $\beta = -\pi + \pi/N_1 = -0.95\pi$. Taking that into account one obtains:

$$\varphi + n\Omega_1 = \operatorname{arctg}\left(\frac{F_{1S}}{F_{1C}}\frac{y_{1C}(n)}{y_{1S}(n)}\right) - \beta = \operatorname{arctg}\left(\frac{y_{1C}(n)}{y_{1S}(n)}\right) + 0.95\pi,$$

where (according to (4.30), (4.31)) the output signals of orthogonal filters are given by

$$y_{1C}(n) = \sum_{k=0}^{19} x(n-k)\cos[0.1\pi(9.5-k)],$$

$$y_{1S}(n) = \sum_{k=0}^{19} x(n-k)\sin[0.1\pi(9.5-k)],$$

for the input signal of the standard form (8.125):

$$x(n) = X_{1m}\sin(n\Omega_1 + \varphi) = X_{1m}\sin(0.1n\pi + \varphi).$$

For realizing of the signal phase measurement the input signal is to be delivered to both filters, which—besides orthogonalization function—are helpful in elimination of possible signal distortions, if any, that are not included in assumed signal model (8.127). It is good to recall that full-cycle sin, cos filters are able to completely eliminate all harmonic components, while other frequencies are only reduced, according to the filter frequency characteristics described in previous chapters.

The simplest possibility of phase shift calculation results from correlation method or DFT. Let the signal (8.125) be given in the form:

$$x(n) = X_{1m} \sin(n\Omega_1 + \varphi) = X_{1m} \cos(\varphi) \sin(n\Omega_1) + X_{1m} \sin(\varphi) \cos(n\Omega_1) \quad (8.132)$$

or

$$x(n) = X_{1C} \sin(n\Omega_1) + X_{1S} \cos(n\Omega_1) \quad (8.133)$$

Components X_{1C}, X_{1S} can be calculated in many ways, using correlation or DFT, for instance. Knowing these variables one can directly calculate phase shift of the signal:

$$\varphi = \operatorname{arctg}\left(\frac{X_{1S}}{X_{1C}}\right) \quad (8.134)$$

Example 8.13 Applying the method of full-cycle correlation calculate the phase shift of given input signal with respect to the correlating functions. Sampling frequency is equal to 1000 Hz.

Solution In considered case one can apply both non-recursive and recursive correlation procedures. In the latter version, the following values hold:

$$2/N_1 = 2/20 = 0.1; \quad \Omega_1 = 2\pi/N_1 = 0.1\pi$$

and the correlation components are calculated according to

$$\begin{aligned} X_{1C}(n) &= X_{1C}(n-1) + 2/N_1[x(n) - x(n-N_1)] \cos(n\Omega_1) \\ &= X_{1C}(n-1) + 0.1[x(n) - x(n-20)] \cos(0.1n\pi), \end{aligned}$$

$$X_{1S}(n) = X_{1S}(n-1) + 0.1[x(n) - x(n-20)] \sin(0.1n\pi).$$

The correlation components can now be directly used for calculation of the sought phase shift:

$$\varphi = \operatorname{arctg}\left(\frac{X_{1S}}{X_{1C}}\right).$$

8.2.4.2 Measurement of Phase Shift Between Two Signals

Phase shift of the signal calculated above was referenced to phase of correlation functions. One can measure the phase shift between two signals in the same way. Useful practical approach could be application of algorithm of either power or impedance components. This is unimportant whether the signals are current and voltage—one simply needs any two signals. Applying the algorithm of calculation active and reactive power one reaches:

$$\varphi = \operatorname{arctg}\left(\frac{Q}{P}\right). \quad (8.135)$$

And when algorithms of impedance components calculation are utilized, one can measure phase shift as follows:

$$\varphi = \operatorname{arctg}\left(\frac{X}{R}\right) \tag{8.136}$$

Applying these two equations one can get many particular algorithms for different orthogonalization methods.

Example 8.14 Propose a method for measurement of the mutual–phase displacement between voltage and current signals. Sampling frequency is equal to 1000 Hz.

Solution For the measurement of phase shift one can apply either Eq. 8.135 or 8.136, that are in fact equivalent when the impedance components are calculated with use of active and reactive power values obtained for example according to (8.80) and (8.82). Assuming that the orthogonal components of voltage and current are calculated with use of full-cycle sin, cos filters, the following equation for the phase shift measurement holds:

$$\begin{aligned}\varphi = \operatorname{arctg}\left(\frac{Q(n)}{P(n)}\right) &= \operatorname{arctg}\left(\frac{\frac{1}{F_{1C}F_{1S}}[u_{1S}(n)i_{1C}(n) - u_{1C}(n)i_{1S}(n)]}{\frac{1}{F_{1C}^2}u_{1C}(n)i_{1C}(n) + \frac{1}{F_1^2}u_{1S}(n)i_{1S}(n)}\right) =, \\ &= \operatorname{arctg}\left(\frac{u_{1S}(n)i_{1C}(n) - u_{1C}(n)i_{1S}(n)}{u_{1C}(n)i_{1C}(n) + u_{1S}(n)i_{1S}(n)}\right)\end{aligned}$$

where suitable cancelling of the filter gain coefficients was possible due to application of appropriate filters (identical for voltage and current, with the same gain values for 50 Hz). The orthogonal components of current/voltage signal are here defined by:

$$x_{1C}(n) = \sum_{k=1}^{19} x(n-k)\cos[0.1\pi(9.5-k)],$$

$$x_{1S}(n) = \sum_{k=1}^{19} x(n-k)\sin[0.1\pi(9.5-k)].$$

8.2.5 Measurement of Frequency

The power system frequency is almost constant at normal operating conditions. The frequency deviation from 50 Hz is then very small. If this deviation increases it could inform that some troubles appeared in the system and that is why it is important problem to measure frequency or frequency deviation with very high accuracy. Higher deviations of frequency may appear during high–power imbalance. Wide range frequency changes may appear during reversible units operation. Here frequency measurement can be less accurate, however, wider is range of frequency changes. Among many known methods three are presented below: by counting reference impulses, convolution of signal with Walsh zero–order function and application of orthogonal components [7, 8].

8.2.5.1 Digital Measurement of Frequency by Counting Reference Impulses

This is the best known method, which has been used for many years in analog technique as well. In this method the number of reference impulses between consecutive zero crossings of the signal is counted. The number is related to a half of period of the signal and must be multiplied by sampling period. Calculating reciprocal of the value one gets:

$$f_m = \frac{1}{T_m} = \frac{1}{2M_{0.5}T_S} = \frac{f_S}{2M_{0.5}}, \tag{8.137}$$

where $M_{0.5}$ is a number of reference samples during half period of the signal, f_m is a measured frequency and f_S is a sampling frequency.

Since actual zero crossing may appear at different moment than at sampling instant, it results in measurement error depending on sampling frequency. Then the measurement accuracy can be given by:

$$\delta_{0.5} = \frac{1}{M_{0.5}} = \frac{2f_m}{f_S}. \tag{8.138}$$

One can notice that the last equation results from (8.137). The equation clearly shows that the error decreases when sampling frequency increases, which is in a sense evident. It can be calculated easily that for nominal frequency of 50 Hz and for sampling frequency of 1000 Hz (typical for protection systems), 10% error of frequency measurement appears.

There are several methods to decrease the error. The most important are:

- increase of sampling frequency,
- increase of number of half periods of counting pulses (increase of measurement time) and
- better precision of zero crossing determination of the reference signal.

Basis of all his methods results from Eq. 8.138. Decrease of the error can be reached by increase in number of counted samples since the error is equal to plus or minus one sample. It is seen from (8.138) that for instance for sampling frequency of 10 kHz, i.e., ten times greater than before, 50 Hz is measured with ten times decreased error equal to 1 percent (as compared to 1 kHz of sampling frequency). Disadvantage of the method of decreasing error is necessity of application of high–sampling frequency different than typical in protection systems.

The same result can be achieved by increasing the number of counted pulses by increasing the number of half periods. Having k half periods measured frequency is:

$$f_m = \frac{kf_S}{2M_k}, \tag{8.139}$$

where M_k is a number of samples in k half-periods.

Measurement error is given by the equation:

$$\delta_k = \frac{1}{M_k} = \frac{1}{kM_{0.5}} \tag{8.140}$$

and is k-time less than in case of one–half period.

Evident disadvantage is k-time is longer measurement time.

Example 8.15 Applying the frequency measurement method by counting of impulses one should assure the error of measurement of the frequencies close to 50 Hz not higher than 1%. What should be the sampling frequency, so that the required error should not exceed during impulse counting within single half-period? How many half-periods one should consider when the sampling frequency is 1000 Hz?

Solution The required sampling frequency results directly from the error Eq. 8.138:

$$\delta_{0.5} = \frac{2f_m}{f_S} \cong \frac{2f_1}{f_S} = \frac{100}{f_S} = 0.01,$$

therefore: $f_S = 10$ kHz.

On the other hand, for the assumed sampling at 1000 Hz, where the number of samples per cycle equals 20 and for a half-cycle is 10, the frequency measurement error by counting over k half-cycles amounts to (8.140):

$$\delta = \frac{1}{10k} = 0.01,$$

thus $k = 10$. Therefore, in order to get the error not higher than 1% one should perform impulse counting over ten half-cycles, i.e. over 100 ms.

The third method of decreasing the error relies on reaching greater precision of zero crossing between sampling instants. The situation is presented in Fig. 8.8. Assuming straight line interpolation and the number of samples of the same sign equal to $M_{0.5}$ one gets more precise value of half period of the signal:

$$\frac{T_m}{2} = M_{0.5}T_S + t_{a2} - t_{b2}. \tag{8.141}$$

The equation can be rewritten in the form:

$$\frac{T_m}{2T_S} = M_{0.5} + \left(\frac{u_{k+1}}{u_{k+1} - u_k}\right)_p - \left(\frac{u_{m+1}}{u_{m+1} - u_m}\right)_{p+1}, \tag{8.142}$$

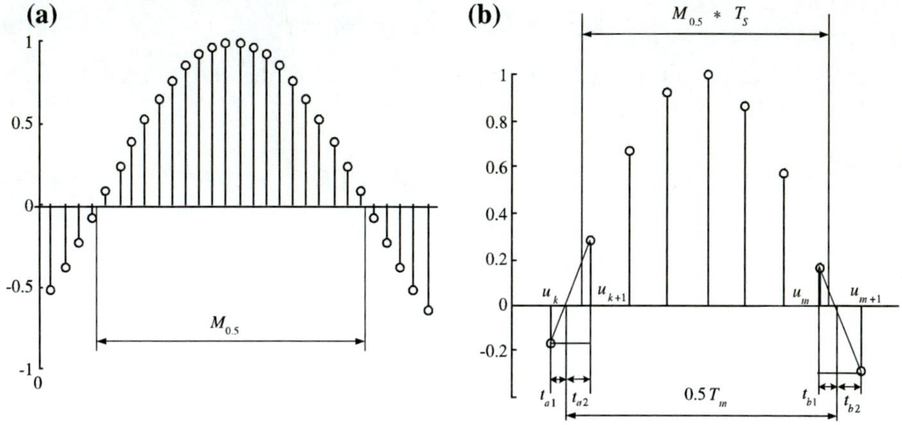

Fig. 8.8 Frequency measurement with counting of impulses: **a** standard version, **b** version with zero crossing determination

where subscripts $p+1, p$ denote zero crossings: the last and before last and subscripts of samples with +1 denote the first sample after changing the sign.

Applying the method one can reduce errors substantially and it is possible to get good results even for small sampling frequencies. Using 1000 Hz sampling frequency for instance the highest zero crossing error is equal to 0.16 percent. This results in 80 mHz error of measured frequency (nominal, i.e. 50 Hz). This is really small error for such sampling frequency and substantial improvement of the method. Additional burden are two divisions performed once at half period.

8.2.5.2 Measurement of Frequency Deviations Applying Zero-Order Walsh Function

The idea of convolution of the signal and zero-order Walsh function (averaging) results from simple observation of zero average value over one period of AC signal. If that averaging period is kept constant but the frequency changes then calculated average value is no longer equal to zero and can be the measure of frequency deviation from its nominal value. Let the signal samples on disposal be given by the equation $u(n) = U_m \cos(n\Omega_1 + \varphi)$.

Let all the samples be summed during one period of nominal frequency (20 ms for 50 Hz signal, N_1—number of samples for that period):

$$y(n) = \sum_{k=0}^{N_1-1} u(n-k).$$

Output signal of such filter is equal to zero for nominal frequency, according to Chap. 6 (i.e. $\Omega = \Omega_1$ and $\Omega_1 = 2\pi/N_1$). If the frequency is different, one observes non-zero output signal of the filter. The signal can be given by the equation:

$$y(n) = CU_m \cos(n\Omega + \varphi + \beta). \tag{8.143}$$

A ratio of output and input magnitudes is equal to the zero–order Walsh FIR filter gain for given frequency:

$$\frac{Y_m}{U_m} = C = |W_0(j\Omega)|. \tag{8.144}$$

The filter gain is given by:

$$|W_0(j\Omega)| = \left|\frac{\sin\left(N_1 \frac{\Omega}{2}\right)}{\sin\left(\frac{\Omega}{2}\right)}\right|, \tag{8.145}$$

while the frequency characteristic is shown in Fig. 6.2.

The gain is equal to zero for nominal frequency ($\Omega = \Omega_1 = 2\pi/N_1$) and the same is for magnitude of the output signal. As it is seen from Eq. 8.144 the gain value can be measured using ratio of output Y_m and input U_m magnitudes. Having it we can rearrange complex function of frequency to find it or its deviation. The last possibility leads to simpler equation. One can notice first that:

$$\frac{N_1\Omega}{2} = \frac{N_1(\Omega_1 + \Delta\Omega)}{2} = \pi + \frac{N_1\Delta\Omega}{2} \tag{8.146}$$

and substituting it to (8.145) and then to (8.144) yields:

$$C = \frac{\sin\left(\frac{N_1\Delta\Omega}{2}\right)}{\sin\left(\frac{\Omega_1+\Delta\Omega}{2}\right)}. \tag{8.147}$$

If $\Delta\Omega$ and $N_1\Delta\Omega$ are very small then one obtains:

$$\Delta\Omega = C\frac{2}{N_1}\sin\left(\frac{\Omega_1}{2}\right). \tag{8.148}$$

The measurement error is a result of simplifications made. Its value is a fraction of percent for small frequency deviations.

Simpler formula is obtained when $\Omega_1 + \Delta\Omega$ is very small (by high sampling frequency) and both sine functions in (8.147) are substituted by their arguments. Simple rearrangements give:

$$\frac{\Delta\Omega}{\Omega_1} = \frac{C}{N_1 - C}. \tag{8.149}$$

This is the simplest equation allowing to calculate small frequency deviations. To apply it is necessary to know the ratio of output and input magnitudes of zero–order Walsh filter. Having it one can calculate frequency deviation from (8.149).

For greater frequency deviations the measurement error increases significantly. One can avoid it using finite difference of output signal (8.143) at discrete time n and $n - 1$. Then:

$$\Delta y = y(n) - y(n-1) = -2CU\sin(0.5\Omega)\sin[(n-0.5)\Omega + \varphi + \beta]. \quad (8.150)$$

If the ratio of that difference and input magnitude is denoted by D ($\Delta y/U = D$), then:

$$\Delta\Omega = \frac{2}{N_1}\arcsin(D/2). \quad (8.151)$$

This is the equation of measurement of frequency deviation which does not require assumption of approximate value of frequency. When value of D is small the sine function can be substituted by its argument and then:

$$\Delta\Omega = D/N_1 \quad (8.152)$$

where $\Delta\Omega = 2\pi\frac{\Delta f}{fs}$.

Example 8.16 Calculate relative error of frequency measurement (with respect to the nominal value) realized with use of the simplified methods with Walsh zero–order function. The measured frequency is equal to 49 Hz and the input signal is sampled at 1000 Hz.

Solution According to Eq. 6.12 the ratio of output and input signals magnitudes for given signal and sampling frequencies is equal: to

$$\frac{Y_m}{U_m} = C = \frac{\sin(N_1\Omega/2)}{\sin(\Omega/2)} = \frac{\sin(N_1\pi f/f_s)}{\sin(\pi f/f_s)} = \frac{\sin(20\pi^*49/1000)}{\sin(\pi^*49/1000)} = 0.413.$$

This ratio should be measured by applying selected algorithm for magnitude measurement, the same for the Walsh filter input and output signals, which creates possibility of cancelling coefficients in numerator and denominator. On the other hand, one should remember that the separated methods of orthogonalization or orthogonalization with delay by a quarter of period cannot be used, simply because the real period of the signal is not known (it is to be determined). Taking that into account one can propose application of the following algorithms for calculation of the sought magnitudes ratio, with use of the described higher algorithms for magnitude measurement (8.55), (8.74) and (8.75):

- for any delay values (assuring non-zero values of the cancelable coefficients)

$$\frac{Y_m}{U_m} = C = \sqrt{\frac{y(n-k)y(n-m) - y(n)y(n-m-k)}{u(n-k)u(n-m) - u(n)u(n-m-k)}},$$

- for the smallest possible delay values

$$\frac{Y_m}{U_m} = C = \sqrt{\frac{y^2(n-1) - y(n)y(n-2)}{u^2(n-1) - u(n)u(n-2)}},$$

- with application of the orthogonal components and any delay values (also under condition of non-zero filter gain values, to be cancelled in numerator and denominator)

$$\frac{Y_m}{U_m} = C = \sqrt{\frac{y_S(n)y_C(n-k) - y_C(n)y_S(n-k)}{u_S(n)u_C(n-k) - u_C(n)u_S(n-k)}}.$$

Having measured the value of magnitudes ratio C the frequency deviation and the sought frequency itself can now be determined Having in mind that this value resulting from the Walsh filter characteristic is equal to 0.413 for the measured frequency of 49 Hz and substituting this value to the simpler, yet less accurate, Eq. 8.149, one obtains:

$$\frac{\Delta\Omega}{\Omega_1} = \frac{\Delta f}{f_1} = \frac{C}{N_1 - C} = \frac{0.413}{20 - 0.413} = 0.021.$$

Therefore $\Delta f = f_1 * 0.021 = 50 * 0.021 = 1.05\,\text{Hz}$ and the relative error of measurement, defined as the relative difference of real and measured values of frequencies with respect to the nominal frequency, is equal:

$$\delta f = |\Delta f_{\text{real}} - \Delta f|/f_1 = |1 - 0.5|/50 = 0.001.$$

Applying the more accurate formula (8.114) one obtains:

$$\frac{\Delta\Omega}{\Omega_1} = \frac{\Delta f}{f_1} = \frac{C\sin(\pi/N_1)}{\pi} = \frac{0.413\sin(0.05\pi)}{\pi} = 0.0205,$$

which yields $\Delta f = f_1^* 0.0205 = 50^* 0.0205 = 1.025\,\text{Hz}$ and the relative error value equal

$$\delta f = |\Delta f_{\text{real}} - \Delta f|/f_1 = |1 - 1.025|/50 = 0.0005.$$

One can see that applying more accurate Eq. 8.148 allows getting twice lower error values, at the cost of slightly higher computational complexity (necessity of calculation of the sine value). Generally, it is seen that both methods deliver quite satisfactory results, with relative errors of 0.1 and 0.05%, respectively.

8.2.5.3 Measurement of Frequency and Its Deviation Using Orthogonal Signal Components

The last from described methods adopts in a way the algorithm of magnitude measurements with application of delayed orthogonal components. Fundamental role plays the following function of orthogonal and delayed orthogonal components:

$$g_k(\Omega) = y_S(n)y_C(n-k) - y_C(n)y_S(n-k) \tag{8.153}$$

where

$$y_C(n) = |F_C|X_m \cos(n\Omega + \varphi + \beta),$$

$$y_S(n) = |F_S|X_m \sin(n\Omega + \varphi + \beta),$$

F is a filter gain, X_m is an input magnitude.

It can be shown easily, which also results from (8.74), that the function (8.153) depends on: input magnitude, gain of orthogonal filters, frequency and number of delay samples, which can be expressed in the form:

$$g_k(\Omega) = X_m^2|F_C||F_S|\sin(k\Omega) \tag{8.154}$$

If Expression (8.153) is calculated using the same as before input signals but for different value of delay k then the argument of sine function is only changed. It means that calculating ratio of function (8.154) for different delays one gets expression, which depends on frequency only (signal magnitude and filter gains are cancelled). To get simpler final expression one can choose in certain special way relationships between delays. When for instance one delay is two times greater than the other one gets:

$$\frac{g_{2k}(\Omega)}{g_k(\Omega)} = 2\cos(k\Omega) \tag{8.155}$$

Calculating that ratio with use of orthogonal components of the signal (as in (8.153)) the algorithm of frequency measurement becomes:

$$\Omega = \frac{1}{k}\arccos\left\{0.5\frac{y_S(n)y_C(n-2k) - y_C(n)y_S(n-2k)}{y_S(n)y_C(n-k) - y_C(n)y_S(n-k)}\right\}. \tag{8.156}$$

Simplified version of the method may be used for small frequency deviations from its nominal value. Such result can be obtained assuming number of delay samples giving particular delay time for instance a quarter of period of fundamental component (nominal value). For such value of k one gets:

$$\cos(k\Omega) = \cos\left[\frac{N_1}{4}(\Omega_1 + \Delta\Omega)\right] = \cos\left[\frac{\pi}{2}\left(1 + \frac{\Delta\Omega}{\Omega_1}\right)\right] = -\sin\left(\frac{\pi}{2}\frac{\Delta\Omega}{\Omega_1}\right) \cong -\frac{\pi}{2}\frac{\Delta\Omega}{\Omega_1} \tag{8.157}$$

and then:

$$\frac{g_{N_1/2}(\Omega)}{g_{N_1/4}(\Omega)} \cong -\pi\frac{\Delta\Omega}{\Omega_1}.$$

Finally, simplified equation of measurement of frequency deviation is as follows:

$$\frac{\Delta\Omega}{\Omega_1} = \frac{1}{\pi}\left\{\frac{y_S(n)y_C(n - N_1/2) - y_C(n)y_S(n - N_1/2)}{y_S(n)y_C(n - N_1/4) - y_C(n)y_S(n - N_1/4)}\right\}. \tag{8.158}$$

Described method can and may be used to measure both frequency and frequency deviations. The latter can use simplified equations giving reasonable accuracy for frequency deviations that are met in practice.

Example 8.17 Provide examples of simplified and accurate algorithms of frequency measurement with use of orthogonal components. Assume sampling at 1000 Hz. For measured frequencies 48 and 45 Hz determine relative measurement error of the simplified method (related to the nominal value of 50 Hz). Draw the course of this error in the frequency range from 45 to 55 Hz.

Solution Measurement of frequency can be done according to Eq. 8.156. Assuming possibly shortest delay ($k = 1$) one obtains:

$$\Omega = \frac{2\pi f}{f_S} = \arccos\left\{0.5\frac{y_S(n)y_C(n-2) - y_C(n)y_S(n-2)}{y_S(n)y_C(n-1) - y_C(n)y_S(n-1)}\right\},$$

whereas the orthogonal signal components for the adopted sampling frequency are calculated from:

$$y_C(n) = \sum_{k=0}^{19} x(n-k)\cos[0.1\pi(9.5-k)],$$

$$y_S(n) = \sum_{k=0}^{19} x(n-k)\sin[0.1\pi(9.5-k)].$$

Instead of function arccos one can apply the simplified version, under the condition that the delay is equivalent to a quarter of signal cycle ($k = N_1/4 = 5$), which gives:

$$\frac{\Delta\Omega}{\Omega_1} = \frac{\Delta f}{f_1} = \frac{1}{\pi} * \frac{y_S(n)y_C(n-10) - y_C(n)y_S(n-10)}{y_S(n)y_C(n-5) - y_C(n)y_S(n-5)}$$

and finally the frequency deviation can be obtained from:

$$\Delta f = -\frac{2f_1}{\pi}\cos\left(\frac{\pi f_{md}}{2f_1}\right).$$

The error of measurement resulting from the adopted simplification is equal:

$$\Delta = f_1 + \Delta f - f_{md} = 50 + \Delta f - f_{md}$$

where f_1 is a nominal power frequency (50 Hz), Δf is a frequency deviation calculated with the simplified method, f_{md} is an accurate value of the measured

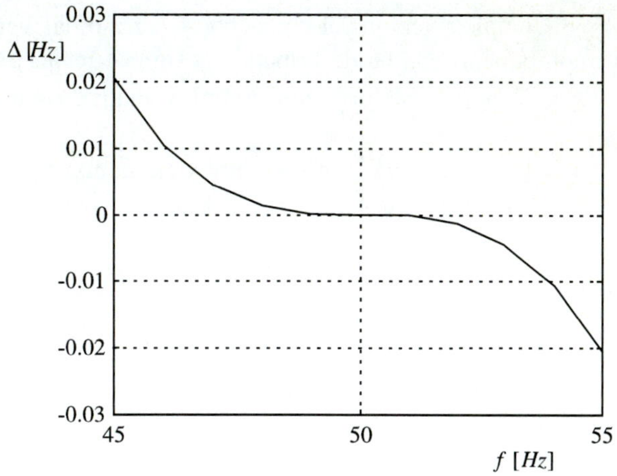

Fig. 8.9 Relative error of frequency measurement for Example 8.17

frequency. When the real value of frequency amounts to 48 Hz the measured frequency deviation is:

$$\Delta f = -\frac{100}{\pi}\cos\left(\frac{48\pi}{100}\right) = -1.99866$$

and for $f_{\text{md}} = 45$ Hz:

$$\Delta f = -\frac{100}{\pi}\cos\left(\frac{45\pi}{100}\right) = -4.979.$$

Therefore one gets:

$$\Delta_{48} = f_1 + \Delta f - f_{md} = 50 - 1.99866 - 48 = 0.00234\,\text{Hz}$$

$$\Delta_{45} = 50 - 4.979 - 45 = 0.021\,\text{Hz}$$

The course of this error as a function of measured frequency is shown in Fig. 8.9.

8.2.6 Filtering of Symmetrical Components

Transformation of three phase sinusoidal variables to symmetrical components is realized according to known matrix equation of complex variables:

$$\underline{\boldsymbol{x}}_{012} = \underline{\boldsymbol{S}}\underline{\boldsymbol{x}}_{L1L2L3}, \tag{8.159}$$

where $\underline{\boldsymbol{x}}_{L1L2L3} = \left[\underline{x}_{L1}, \underline{x}_{L2}, \underline{x}_{L3}\right]^{\text{T}}$ is a vector of three phase sinusoidal variables, $\underline{\boldsymbol{x}}_{012} = \left[\underline{x}_0, \underline{x}_1, \underline{x}_2\right]^{\text{T}}$ is a vector of symmetrical components, i.e., zero, positive and negative sequence components.

Complex vectors of three phase sinusoidal variables as well as symmetrical components can be described as a sum of respective orthogonal components. i.e., $\underline{x} = x^C + jx^S$. In turn, the complex matrix of transformation is the following: $\underline{S} = \frac{1}{3}\begin{bmatrix} 1 & 1 & 1 \\ 1 & \underline{a} & \underline{a}^2 \\ 1 & \underline{a}^2 & \underline{a} \end{bmatrix}$,with complex elements:

$$\underline{a} = \exp(j2\pi/3) = -0.5 + j\sqrt{3}/2,$$

$$\underline{a}^2 = \exp(j4\pi/3) = -0.5 + j\sqrt{3}/2.$$

Transformation of three phase voltages or currents into their symmetrical components and the other way round can be done in digital technique in many ways. Below principles of two of them will be presented, applying orthogonal signal components and signal delays.

8.2.6.1 Filters of Symmetrical Components Using Orthogonal Signal Components

The matrix of transformation $\underline{S}$ as a complex matrix can be given using its real and imaginary parts:

$$\underline{S} = S_R + jS_I = \frac{1}{3}\begin{bmatrix} 1 & 1 & 1 \\ 1 & a_R + ja_I & a_R - ja_I \\ 1 & a_R - ja_I & a_R + ja_I \end{bmatrix}, \tag{8.160}$$

where

$$a_R = -0.5; \quad a_1 = \sqrt{3}/2.$$

If phasors of three phase sinusoidal variables and symmetrical components are presented using orthogonal components, the equation of transformation (8.159) can be given in the form:

$$\underline{x}^C_{012} + j\underline{x}^S_{012} = (S_R + jS_I)\left(\underline{x}^C_{L1L2L3} + j\underline{x}^S_{L1L2L3}\right). \tag{8.161}$$

Performing multiplication of the components at the right−hand side of the equation and comparing real and imaginary parts of both equations sides yields:

$$\underline{x}^C_{012} = S_R\underline{x}^C_{L1L2L3} - S_I\underline{x}^S_{L1L2L3} \tag{8.162}$$

and

$$\underline{x}^S_{012} = S_R\underline{x}^S_{L1L2L3} + S_I\underline{x}^C_{L1L2L3}. \tag{8.163}$$

Each of these equations can be used to get symmetrical components. Equation 8.162, for instance, can be written in the following form:

$$\begin{bmatrix} x_0^C \\ x_1^C \\ x_2^C \end{bmatrix} = \frac{1}{3}\begin{bmatrix} 1 & 1 & 1 \\ 1 & a_R & a_R \\ 1 & a_R & a_R \end{bmatrix}\begin{bmatrix} x_{L1}^C \\ x_{L2}^C \\ x_{L3}^C \end{bmatrix} - \frac{1}{3}\begin{bmatrix} 0 & 0 & 0 \\ 0 & a_I & -a_I \\ 0 & -a_I & a_I \end{bmatrix}\begin{bmatrix} x_{L1}^S \\ x_{L2}^S \\ x_{L3}^S \end{bmatrix}. \quad (8.164)$$

The equation can be implemented directly using orthogonal components of three phase sinusoidal variables obtained applying any methods of orthogonalization.

8.2.6.2 Symmetrical Component Filters Using Signal Delays

Simpler filters of symmetrical components can be realized using signal delays. Simplicity of such a solution results from the simplest operation in a digital system—delay by certain number of samples. To get adequate shape of transformation matrix one can use transformation operators $\underline{a}$ and $\underline{a}^2$ in the form:

$$\underline{a} = \exp(j2\pi/3) = \exp(j\gamma) = \exp(-j4\pi/3) = \exp(-j2\gamma), \quad (8.165)$$

$$\underline{a}^2 = \exp(j4\pi/3) = \exp(j2\gamma) = \exp(-j2\pi/3) = \exp(-j\gamma). \quad (8.166)$$

Using these expressions the filters of symmetrical components can be written in the form similar to (8.159):

$$\begin{bmatrix} \underline{x}_0 \\ \underline{x}_1 \\ \underline{x}_2 \end{bmatrix} = \frac{1}{3}\begin{bmatrix} 1 & 1 & 1 \\ 1 & \exp(-j2\gamma) & \exp(-j\gamma) \\ 1 & \exp(-j\gamma) & \exp(-j2\gamma) \end{bmatrix}\begin{bmatrix} X_{L1}\exp\{j(n\Omega_1 + \varphi_{L1})\} \\ X_{L2}\exp\{j(n\Omega_1 + \varphi_{L2})\} \\ X_{L3}\exp\{j(n\Omega_1 + \varphi_{L3})\} \end{bmatrix}. \quad (8.167)$$

If one multiplies matrices of right–hand side of the above equation and then compares either real or imaginary components of both sides one can get resulting equations of transformation, which need addition of adequately delayed three phase system components. Comparing real components one gets:

$$x_0^C(n) = \frac{1}{3}\{X_{L1}\cos(n\Omega_1 + \varphi_{L1}) + X_{L2}\cos(n\Omega_1 + \varphi_{L2}) + X_{L3}\cos(n\Omega_1 + \varphi_{L3})\}, \quad (8.168a)$$

$$x_1^C(n) = \frac{1}{3}\{X_{L1}\cos(n\Omega_1 + \varphi_{L1}) + X_{L2}\cos(n\Omega_1 + \varphi_{L2} - 2\gamma) + X_{L3}\cos(n\Omega_1 + \varphi_{L3} - \gamma)\}, \quad (8.168b)$$

$$x_2^C(n) = \frac{1}{3}\{X_{L1}\cos(n\Omega_1 + \varphi_{L1}) + X_{L2}\cos(n\Omega_1 + \varphi_{L2} - \gamma) + X_{L3}\cos(n\Omega_1 + \varphi_{L3} - 2\gamma)\}. \quad (8.168c)$$

The delay angles are equal either one–third or two–thirds of period of fundamental frequency component. It means that signals should be delayed by equivalent number of samples. It also means that the number of samples in one period of fundamental component should be dividable by three. Then Eqs. 8.168a, b, c can be written as follows:

$$x_0(n) = \frac{1}{3}\{x_{L1}(n) + x_{L2}(n) + x_{L3}(n)\}, \tag{8.169a}$$

$$x_1(n) = \frac{1}{3}\{x_{L1}(n) + x_{L2}(n - 2N_1/3) + x_{L3}(n - N_1/3)\}, \tag{8.169b}$$

$$x_2(n) = \frac{1}{3}\{x_{L1}(n) + x_{L2}(n - N_1/3) + x_{L3}(n - 2N_1/3)\}. \tag{8.169c}$$

These are the simplest filters of symmetrical components. The equations are realized in digital technique directly, as it is given above.

8.3 Summary, Conclusions and Recommendations

In this chapter the methods and specific algorithms of processing the input samples of currents and voltages, filtered currents and voltages or their orthogonal components into the sought criteria signals are described. All (almost all) the discussed measurement algorithms have nonlinear character, which makes assessment of their quality and resulting errors quite difficult. Therefore their features are summarized here only partially, the other part of their assessment can be done basing on the simulation investigations, which is presented later.

The measurement algorithms commit transformation of adequately selected signal components into the measured variables. The basic and most popular algorithms of protection criteria measurement apply orthogonal components of currents and voltages. Certain specific features of measurement one can also obtain applying algorithms with delayed orthogonal components.

As one could see, there are lots of basic methods and algorithms, and their multitude results from a number of ways of obtaining orthogonal components, which is described in Chap. 10. Those methods are responsible for the resulting algorithm features, that is their dynamics, ability of suppressing signal distortions and noise, as well as computational complexity.

One can say that the DC (or decaying DC) component present in measured signals (especially currents) may be a cause of significant, difficult to remove errors of measurement. Similar conclusions may also apply for other distorting components, depending on the algorithm at hand. It means that possible distortions of the input signals should be appropriately minimized or eliminated before the actual measurement algorithm is applied. Crucial are in this respect properly selected filters or other methods of digital signal processing.

The above-described algorithms can be categorized as follows.

(A) *Classification*

- Basic classification of the protection criteria measurement algorithms related to the principle of measurement allows distinguishing methods applying: signal averaging, orthogonal components and delayed orthogonal components. Other special approaches met include counting of impulses and solving of the fault circuit equation.
- Very similar methods and principles as well as adopted requirements apply to algorithms of current and voltage magnitude and power measurement.
- The algorithms for estimation of impedance, phase shift and some of the frequency measurement equations differ from the ones for magnitude and power in the sense that they are quotient type, however, they are also similar in the sense that the variables in the numerator and denominator are determined with the algorithms from previous point, i.e., they are usually powers and magnitudes of certain signals.
- Slightly different and more diverse are some of the algorithms for frequency measurement and symmetrical components filtration.

(B) *Dynamics*

- Dynamic features of measurement are dependent on applied filters and other methods of introductory digital signal processing.
- A bit part in resulting dynamics play signals delaying being applied in some algorithms (most often few times lower that the filter window length); the total duration of measurement transients is a sum of filter window length and delay value.
- Dynamics of the averaging methods depends clearly on the period of signal averaging.

(C) *Filtering efficiency of the algorithms*

- Filtering, elimination or suppression of signal distortions is determined mainly by the frequency characteristics of applied filters or other signal processing methods. The algorithm itself has generally little influence in this context. Sparse examples are exceptions to that rule; e.g., the algorithm of magnitude estimation employing averaging of squared sample values (being a kind of filter, incidentally). If one uses orthogonal components obtained from signal delaying (without filtration), the resulting algorithms perform in fact only spectral transformation of the processed AC components.

(D) *Computational complexity*

- Decisive for ease or difficulty of numerical implementation of the measurement algorithms are again applied filters. Complexity of the algorithms is rather small, since they usually base on simple mathematical operations like additions, multiplications, divisions or squaring of numbers.

(E) *Recommendations*

On the strength of the above remarks one may try to recommend some of the very many−described algorithms for practical applications, those with the most favorable resulting virtues, fulfilling the standard or more specific requirements.

- Magnitude and power measurement algorithms
  Typical example of this kind applies full-cycle FIR filtration (sin/cos windows) for both elimination of distortion and realization of signal orthogonalization. The filters' gain values are identical and thus the final algorithm has the simplest possible form. The recursive filtration or correlation procedures assure effective filtration of unwanted signal components, with the resulting estimation time equal 20 ms (a cycle of fundamental frequency component). If the signal distortions are small, one can recommend Walsh filters. In case when shorter measurement time is of importance, the methods based on least mean square error, Kalman filter, shorter window orthogonal filters or methods employing separated orthogonalization can be applied. When the permissible measurement time is 25 ms or longer one can apply single filter plus orthogonalization with time delay by a quarter of cycle. Sometimes it may be important to use algorithms employing delayed orthogonal components, characterized by decreased susceptibility to small frequency distortions around nominal value. Their further usage in the quotient type algorithms of other criteria measurement may also be advantageous. All the considered algorithms perform well independently of the sampling frequency level. Also in this respects there are no visible reasons for the utilization of averaging algorithms.
- Quotient type impedance measurement algorithms
  The discussion from previous point applies also to the algorithms of impedance components estimation that employ standard algorithms for powers and signal magnitudes as intermediate variables. It is recommended to use the same filters for the measured variables in the numerator and denominator. The best effects are obtained with application of algorithms based on delayed orthogonal components, which offer practical independence of the frequency deviations and comparatively simple final equations, with the filter gain coefficients cancelled from the numerator and denominator.
- Frequency measurement by impulse counting
  This algorithm differs from all others by the general principle of measurement. The accuracy of measurement can be increased by changing the sampling rate; however, for most of power system protection and control applications the resulting precision is not sufficient. Necessary for getting better results is an extension of the time of counting (over multiple cycles) or application of improved method with zero−crossing correction. Advantage of this approach is its simplicity, but one should remember that the result of measurement is delivered every half-cycle only (not a continuous process).

Nevertheless, when higher accuracy is required, other methods, including those based on orthogonal components, are recommended.

- Impedance estimation based on fault circuit model
  This approach is characterized by full independence of the decaying DC component (this feature is inherited in the algorithm, being taken into account in the circuit model). One of the disadvantages in certain susceptibility to higher frequency oscillating components. This shortcoming can be limited by applying appropriate input filters, what on the other hand increases the total measurement time.
- Filtration of symmetrical components
  This operation is most frequently realized by applying phase signals delaying or by use of their orthogonal components. The first method is very simple, but requires sampling frequencies dividable by 3 (e.g., 1200 Hz) and phase signals devoid of distortions. The second one is more complex, but thanks to signal filtration one gets elimination of distortions and orthogonal components that can also be used for measurement of other criteria signals. Thus, usage of this method seems justified to be recommended for practical applications.

References

1. Advancements in microprocessor based protection and communication (1997) IEEE Tutorial Course, New York, NY, USA, Sachdev MS (Coordinator), Publication No. 97TP120-0
2. Brooks AW (1994) Distance relaying using least-squares estimates of voltage, current and impedance. IEEE Comput Appl Power 7:8–13
3. Johns AT, Salman KS (1983) Digital protection for power systems. Peter Peregrinus Ltd., on behalf of IEE PtC, Stevenage
4. Rosolowski E (1992) Analysis of digital power system protection algorithms. Sci Works of the Institute of Electrical Power Engineering, WUT, Wroclaw, Poland. Monographs, No 27 (in Polish)
5. Sachdev MS, Giray MM (1985) A least squares technique for determining power system frequency. IEEE Trans PAS 104:437–444
6. Szafran J (1990) Signal recognition in digital protection. Sci Works of the Institute of Electrical Power Engineering, WUT, Wroclaw, Poland. Monographs, No 23 (in Polish)
7. Szafran J, Rebizant W (1998) Power system frequency estimation. IEE Proc GTD 145:578–582
8. Szafran J, Wiszniewski A (2001) Measurement and decision algorithms of digital protection and control. WNT, Warszawa (in Polish)
9. Ukil A (2007) Intelligent systems and signal processing in power engineering. Springer, Berlin
10. Ungrad H, Winkler W, Wiszniewski A (1995) Protection techniques in electrical energy systems. M Dekker, New York

Chapter 9
Characteristics of Measurement of Criterion Values and Adaptive Algorithms

Standard control systems are designed and evaluated taking into account dynamics and required accuracy for various input signals. Digital protection systems should also be evaluated in a similar way, however, the fundamental parameter is time needed to take decision concerning the state of protected element of the power system. Sometimes it is as short as a fraction of period of the fundamental frequency component. This decision time, different for various power system elements, is usually fixed to avoid possible extended damage of protected plants. For some criterion values and some decision-making methods the final decision is to be done even during transient states of measurement when one can still expect substantial errors criteria values determination. This is why while designing a protection system one must take dynamic features of measurement algorithms into consideration and perform optimization of its elements and the entire system as a decision-making device.

9.1 Dynamics of the Measurement Process

Processes of measurement and decision-making in protection devices are realized under heavy dynamic conditions [9]. Signals, usually voltages and currents, change rapidly during faults, causing also changes of the measured criterion values. In order to reach adequately fast decision concerning the state of protected element it is necessary to have properly short transient period between normal and faulty steady states of measurement.

Applied measurement algorithms can be divided for either operating continuously or starting its operation when fault appears. In many protection systems both types are applied, however, general principle is that in the fastest protection systems the second ones must be used. Examples could be systems using Kalman filters or mean square error (MSE) methods. But even in such a case it is necessary

W. Rebizant et al., *Digital Signal Processing in Power System Protection and Control,* Signals and Communication Technology,
DOI: 10.1007/978-0-85729-802-7_9,

to use continuous measurement to identify the fault and to estimate its inception instant. The resulting conclusion is the following: either continuous measurement only or mixed continuous and starting its operation at the fault instant are used in contemporary protection systems.

From the viewpoint of final protection decision-making of importance is not only the transient time but its trajectory as well. Monotonic, increasing or decreasing trajectory is advantageous since then the number of wrong decisions could significantly be minimized. These trajectories are not always monotonic and that is why delays must sometimes be used to avoid over-/under-tripping. Deep analysis clearly shows that transients of measurement depend on signal transient and noise which must be removed by adequate filters, the criterion values trajectory, on the other hand, depends on features of applied filters and measurement algorithms.

Among the filters and similar digital processing methods applied one can distinguish: finite impulse response (FIR) filters, infinite impulse response (IIR) filters, Kalman filters and other signal processing (not filters), including discrete Fourier transform, correlation and MSE methods. Standard IIR digital filters are used in protection systems rather rarely. Disadvantages of the filters in that application are: nonlinear phase, troubles to get orthogonal components and very long lasting filter transient. An exception could be Kalman filter. It has short and monotonic transient when model of the process is well defined and close to reality. The most universal and frequently used are finite impulse filters (FIR) and similar correlation methods. Their advantages are: sharply-defined period of transient, linear phase and easy available orthogonal filters as well as the resulting orthogonal signal components. Transients of the FIR filters are analyzed below.

A FIR filter processes input signal samples according to the equation:

$$y(n) = \sum_{k=0}^{N-1} a(k)x(n-k), \tag{9.1}$$

where $a(k)$ is a filter coefficients, and x is an input signal.

Such simple equation can be used and is valid in the steady state only when the signal parameters are constant within the entire filter window (N samples). However, when sudden change of the signal appeared within the filter window, say at instant m (see Fig. 9.1), then the output signal of the filter should be expressed in the form including samples of "old" and "new" signal:

$$y(n) = \sum_{k=0}^{n-m} a(k)x_2(n-k) + \sum_{k=n-m+1}^{N-1} a(k)x_1(n-k), \tag{9.2}$$

which is valid for $0 \le n - m \le N - 1$.

When time is passing n increases the number of samples of new signal (x_2) observed in the filter data window also increases whereas the number of samples of old signal (x_1) decreases. After N samples counted from the fault inception (change of signal parameters), there are samples of "new" signal only inside the filter

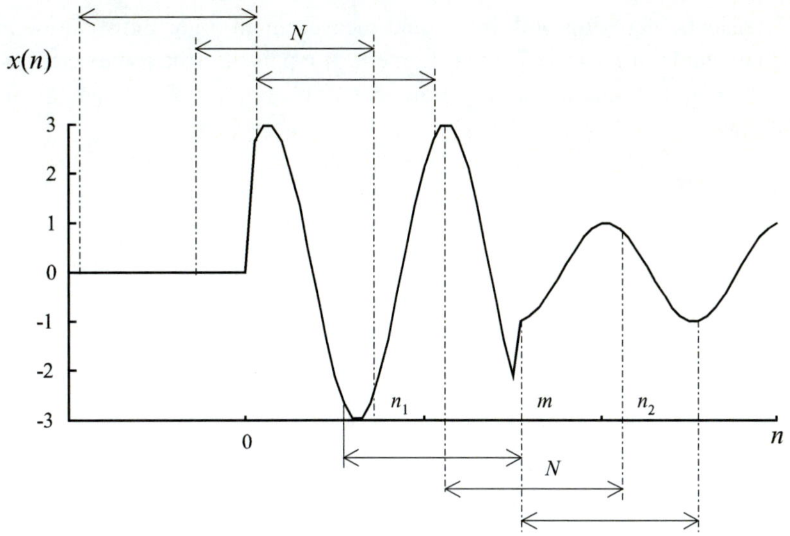

Fig. 9.1 Signal waveshape after measurement start ($n = 0$) and change of parameters

window and one reaches the actual steady state which will last until new disturbance appears. It means that filter transient lasts for the period equivalent to filter window length. The transient period can last shorter when shorter filter window are applied.

The situation is a little bit different when a filter starts its operation at the fault instant. In such a case zero initial conditions may be assumed, i.e., signal samples are equal to zero before the filter starts its operation (at the instant $m = 0$, Fig. 9.1). The second sum in Eq. 9.2 is then equal to zero and output signal of the filter is given by the equation:

$$y(n_1) = \sum_{K=0}^{n_1} a(k)x(n_1 - k), \tag{9.3}$$

for $0 \le n_1 \le N - 1$.

It is seen (9.3) that at the very beginning the first term in (9.2) has the initial sum limit n_1 which is increasingly growing along the filtration process. The growth lasts as long as n_1 reaches the value equal to filter window length (N) and then this value becomes constant. At that time instant the new steady state of measurement begins. Examinations and simulations concerning filters and measurement algorithm transients are easier to realize in the latter case since there is less parameters affecting transients that should be considered. Though the measurement algorithms are usually nonlinear, it is possible to come to general conclusions important for some arbitrary dynamical states. To reach this one may notice that each AC signal is determined by three parameters: magnitude, frequency and phase shift. For constant frequency of signals (as it is in power system) transients trajectory will depend mainly on their phases since magnitude has influence on scaling factors only.

Transients including both filters and measurement algorithms can be analyzed for an example of magnitude measurement. If a pair of orthogonal filters is applied then their output signals (during transients, i.e., for $n \leq N - 1$) are given by the equations:

$$y_C(n) = \sum_{k=0}^{n} a_c(k)x(n-k), \tag{9.4a}$$

$$y_S(n) = \sum_{k=0}^{n} a_s(k)x(n-k). \tag{9.4b}$$

Let input signal of the filters be given by:

$$x(n) = X_{1m} \cos(n\Omega_1 + \varphi_1), \tag{9.5}$$

where

$$\Omega_1 = \omega_1 T_S = 2\pi f_1/f_S = 2\pi/N_1,$$

f_1 is a frequency of fundamental component, f_S, T_S are sampling frequency and period, and N_1 is a number of samples in the period of fundamental frequency component.

Delayed signal samples in equations of the filters (9.4a, b) can be written in the form:

$$x(n-k) = X_{1m} \cos[(n-k)\Omega_1 + \varphi_1] = b_c(k)x_C(n) + b_s(k)x_S(n), \tag{9.6}$$

where

$$x_C(n) = X_{1m} \cos[(n+0.5)\Omega_1 + \varphi_1],$$

$$x_S(n) = X_{1m} \sin[(n+0.5)\Omega_1 + \varphi_1],$$

$$b_c(k) = \cos[(k+0.5)\Omega_1],$$

$$b_s(k) = \sin[(k+0.5)\Omega_1].$$

Substituting (9.6) into (9.4a, b) one obtains:

$$y_C(n) = d_{CC}(n)x_C(n) + d_{CS}(n)x_S(n), \tag{9.7a}$$

$$y_S(n) = d_{SC}(n)x_C(n) + d_{SS}(n)x_S(n), \tag{9.7b}$$

where

$$d_{CC}(n) = \sum_{k=0}^{n} a_c(k)b_c(k),$$

$$d_{CS}(n) = \sum_{k=0}^{n} a_c(k)b_s(k),$$

$$d_{SC}(n) = \sum_{k=0}^{n} a_s(k)b_c(k),$$

$$d_{SS}(n) = \sum_{k=0}^{n} a_s(k)b_s(k).$$

For given orthogonal filters, i.e., for their impulse responses being known, it is possible to calculate the above coefficients as a function of actual window length n. If for instance full-cycle sine, cosine orthogonal filters are used, their impulse responses are given by:

$$a_c(k) = \cos[(k + 0.5)\Omega_1] = b_c(k), \tag{9.8a}$$

$$a_s(k) = \sin[(k + 0.5)\Omega_1] = b_s(k), \tag{9.8b}$$

the coefficients of equations (9.7a, b) are following:

$$d_{CC}(n) = \sum_{k=0}^{n} \cos^2[(k + 0.5)\Omega_1] = \frac{n+1}{2} + \frac{\sin(n\Omega_1)}{2\sin(\Omega_1)}\cos(n\Omega_1), \tag{9.9a}$$

$$d_{SS}(n) = \sum_{k=0}^{n} \sin^2[(k + 0.5)\Omega_1] = \frac{n+1}{2} - \frac{\sin(n\Omega_1)}{2\sin(\Omega_1)}\cos(n\Omega_1), \tag{9.9b}$$

$$d_{CS}(n) = d_{SC}(n) = \sum_{k=0}^{n} \cos[(k + 0.5)\Omega_1]\sin[(k + 0.5)\Omega_1] = \frac{\sin(n\Omega_1)}{2\sin(\Omega_1)}\sin(n\Omega_1). \tag{9.9c}$$

Substituting the above coefficients into output signals of orthogonal filters (9.7a, b) and applying them in standard algorithm of magnitude measurement (sum of squared orthogonal components) after simple rearrangements one obtains:

$$X_{1m}^2(n) = X_{1m}^2 \frac{4}{N_1^2}\left\{\left(\frac{n+1}{2}\right)^2 + \left(\frac{\sin(n\Omega_1)}{\sin(\Omega_1)}\right)^2 \right. \\ \left. +(n+1)\frac{\sin(n\Omega_1)}{2\sin(\Omega_1)}\cos[(n+1)\Omega_1 + 2\varphi_1]\right\}. \tag{9.10}$$

According to expectations the measured magnitude depends on initial phase shift of the signal. Transients and steady state of magnitude measurement is shown

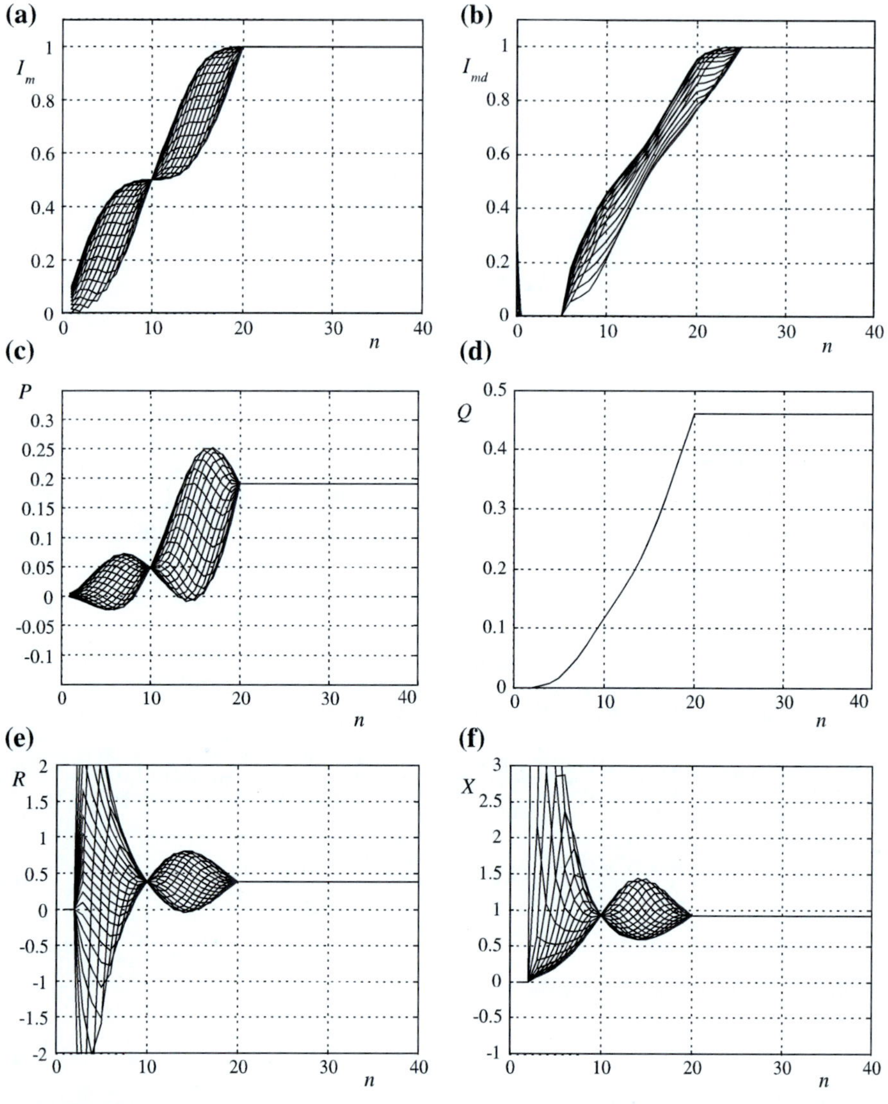

Fig. 9.2 Transient response of the measurement algorithms for zero initial conditions: **a**, **b** measurement of current magnitude, **c**, **d** active and reactive power, **e**, **f** impedance components; algorithms: (8.37), (8.41), (8.46a, b), (8.48), (8.64), (8.65)

in Fig. 9.2a. It is seen that steady state is reached after the time equivalent to filter window, in this case after N_1 equal to 20 samples (here and in many other cases the sampling frequency was assumed 1000 Hz, which means that sampling period is equal to 1 ms). At instant equal to half filter window and for time greater than filter window length one observes that they have values independent of the initial phase shift. At other instants during transients measured values are different, depending on initial phase shift. Since it is unknown, the actual magnitude cannot be estimated then.

Equations similar to (9.10) can be obtained for other algorithms of magnitude measurement (with application of different filters, their window lengths, etc.). Procedure of calculation of final result is either similar or identical. The steps are following: calculation of output signals of orthogonal filters (9.7a, b) with non-stationary coefficients (similar to (9.9a, b)), substituting them into given algorithm, rearranging and simplifying. The calculations are sometimes complex, chosen results are shown in Fig. 9.2.

Transients of magnitude measurements according to algorithm with delayed orthogonal components are shown in Fig. 9.2b. It is seen that the set of trajectories has a little bit different shape, however, one important feature is the same as before—there is no overshoot of measured value. This virtue is extremely important with the viewpoint of decision-making process in power system protection.

Active power measurement transients are different than before (Fig. 9.2c). Now the transients that are non-monotonic and overshoot depending on power angle appears. It is seen even when the phase angle is equal to $\pi/2$, i.e., when active power steady state value is equal to zero, which may be a cause of problems during decision-making. On the other hand advantageous is the shape of transient of reactive power (Fig. 9.2d). Here, the course of transient is not only monotonic but it also does not depend on initial phase shift. This is a very important result allowing for easier decision-making as well as for dynamical correction of protection criterion values measurement.

Transients of impedance components are shown in Fig. 9.2e, f. It is seen now that changes of measured values are tremendous, which is a result of measurement algorithms being a ratio type, where the algorithm transients depend on transients of the numerator and denominator. Here, information during transients is not useful and one should wait until steady state is reached to avoid wrong decisions. It could be said that during transients of impedance components measurement the decision-making should be off.

The examples discussed before presented trajectories of measurement during transients that were obtained with use of above algorithms under assumptions of zero initial conditions, what is the case by starting the measurement at all or by restarting it (reinitiating) after disturbance inception. The signals observed during continuous measurement without filters' restarting look quite differently. Selected exemplary results of simulations are presented in Fig. 9.3. As one can see, the differences are significant, however, the general picture is common: the magnitude measurement is always monotonic, while the other criteria values may have non-monotonic trajectory during transients. One can note even sign changes and overshoots, especially big for resistance and reactance measurement, especially when the signal phase changes due to disturbance are high. It is therefore to be stressed that detailed analysis of transient behavior of the measurement algorithms is extremely important for selection of appropriate procedures both for criteria estimation and final protection decision-making.

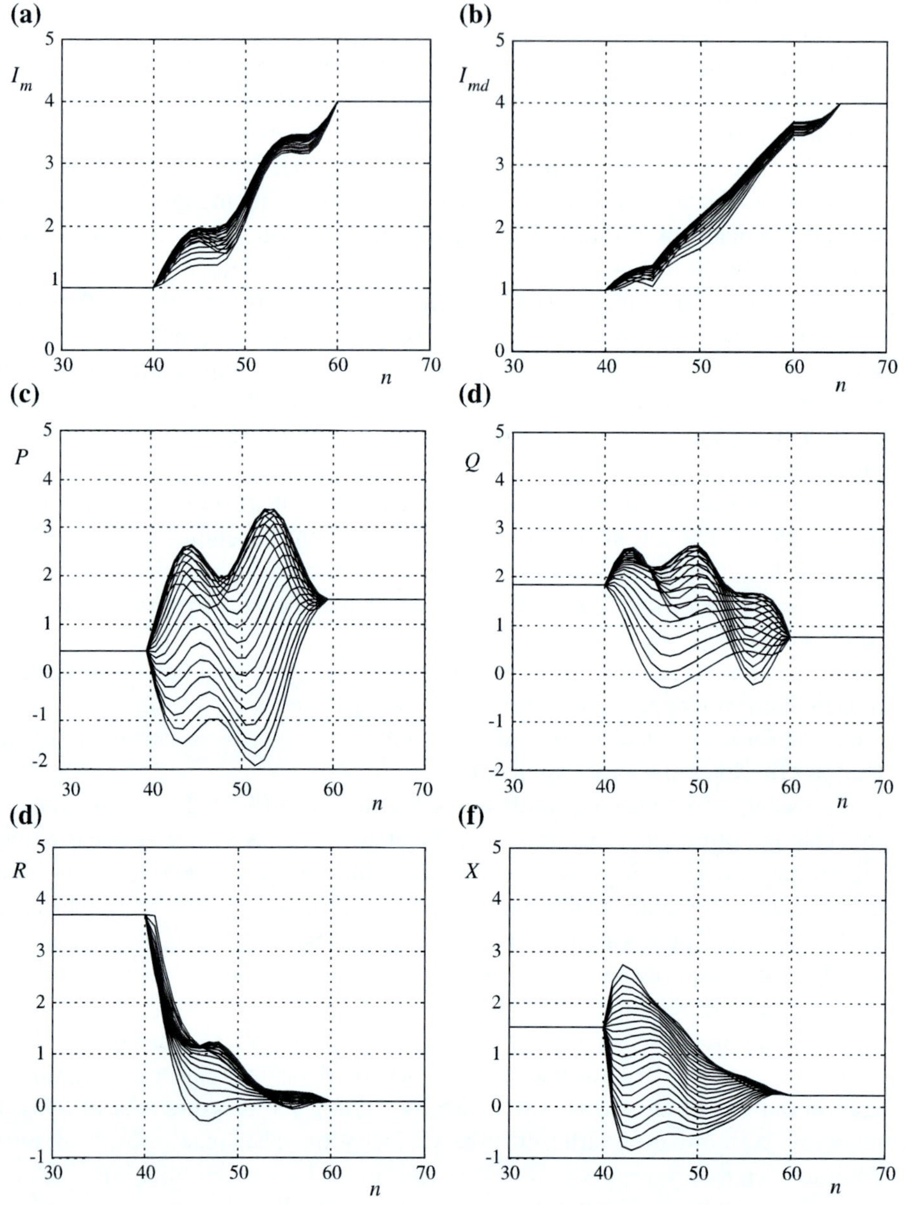

Fig. 9.3 Measurement trajectories for transient initiated at $n = 40$; algorithms as in Fig. 9.2

9.2 Dynamical Correction of Measurement of Criterion Values

Analysis of transients using theory and selected simulation runs made it possible to notice at least two possibilities of dynamical correction [1, 2, 7, 8]. The first one can rely on calculation of orthogonal signal components on basis of the filter outputs during transients, the second one can use unique transient response of reactive power and its application in development of corrected algorithms for other criterion values. Both methods require zero initial conditions and precise estimation of fault instant. With the viewpoint of protection systems that instant should be estimated as soon as possible.

The basis for the first method is Eqs. 9.7a, b which describes output signals of a pair of orthogonal filters (during transients) as a function of variable coefficients and orthogonal components of the signal. The pair of non-stationary equations allows to calculate orthogonal signal components:

$$x_C(n) = \frac{d_C(n)}{d(n)} \qquad x_S(n) = \frac{d_S(n)}{d(n)}, \tag{9.11}$$

where

$$d(n) = \begin{vmatrix} d_{CC}(n) & d_{CS}(n) \\ d_{SC}(n) & d_{SS}(n) \end{vmatrix} = d_{CC}(n)d_{SS}(n) - d_{CS}(n)d_{SC}(n),$$

$$d_C(n) = \begin{vmatrix} y_C(n) & d_{CS}(n) \\ y_S(n) & d_{SS}(n) \end{vmatrix} = d_{SS}(n)y_C(n) - d_{CS}(n)y_S(n),$$

$$d_S(n) = \begin{vmatrix} d_{CC}(n) & y_C(n) \\ d_{SC}(n) & y_S(n) \end{vmatrix} = d_{CC}(n)y_S(n) - d_{SC}(n)y_C(n).$$

These orthogonal components known during transients allow to calculate magnitude of the signal using standard algorithm:

$$X_{1m}^2 = x_C^2(n) + x_S^2(n). \tag{9.12}$$

The trajectory of magnitude measurement applying this method of dynamical correction is shown in Fig. 9.4. Substantial speed-up of reaching the steady state result can be observed. It should be added here that both calculations and simulations were made for pure sinusoidal signals without noise. When noise is present in the signal then the result changes oscillatory around steady state and variance of error decreases until filter window length is reached.

It is evident that in the same way dynamical correction of the other criterion values can be arranged.

The basis of the second method of dynamical correction is the algorithm of measurement of reactive power. It was mentioned before that any algorithm of reactive power has unique, phase independent transient response. It can be easily proved for standard algorithm:

$$Q_1 = \frac{1}{2F_{1C}F_{1S}}\left[u_{F1S}(n)i_{F1C}(n) - u_{F1C}(n)i_{F1S}(n)\right], \tag{9.13}$$

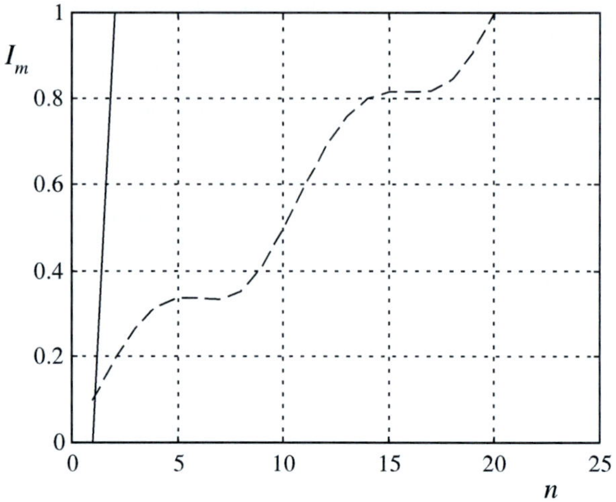

Fig. 9.4 Signal magnitude measurement with correction acc. to (9.11); standard algorithm (*dashed line*), alg. with correction (*solid line*)

Output signals of orthogonal filters of voltage and current during transients can be written similarly to (9.7a, b) in the form:

$$\begin{aligned} u_{F1C}(n) &= d_{CC}(n)u_C(n) + d_{CS}(n)u_S(n) \\ u_{F1S}(n) &= d_{SC}(n)u_C(n) + d_{SS}(n)u_S(n) \\ i_{F1C}(n) &= d_{CC}(n)i_C(n) + d_{CS}(n)i_S(n) \\ i_{F1S}(n) &= d_{SC}(n)i_C(n) + d_{SS}(n)i_S(n), \end{aligned} \tag{9.14}$$

where

$$u_C(n) = U_m \cos[(n + 0.5)\Omega_1 + \varphi_U],$$

$$u_S(n) = U_m \sin[(n + 0.5)\Omega_1 + \varphi_U],$$

$$i_C(n) = I_m \cos[(n + 0.5)\Omega_1 + \varphi_I],$$

$$i_S(n) = I_m \sin[(n + 0.5)\Omega_1 + \varphi_I],$$

φ_U is an initial phase shift of voltage, and φ_I is an initial phase shift of current.

Substituting the values into (9.13) after simple rearrangements yields:

$$Q_1(n) = \frac{d_{SS}(n)d_{CC}(n) - d_{CS}(n)d_{SC}(n)}{2F_{1C}F_{1S}} Q, \tag{9.15}$$

where

$$Q = 0.5U_m I_m \sin(\varphi_U - \varphi_I).$$

Assuming application of sine, cosine full-cycle orthogonal FIR filters one obtains values of reactive power during transient period:

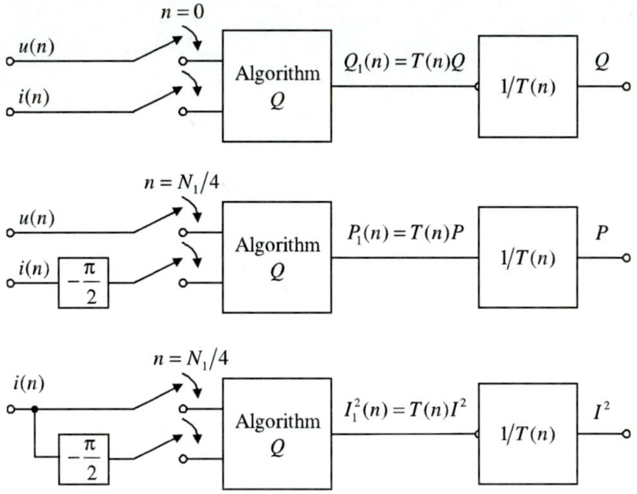

Fig. 9.5 Block scheme of *P*, *Q* and *I* measurement with correction

$$Q_1(n) = \frac{1}{N_1^2}\left[(n+1)^2 - \frac{\sin^2(n\Omega_1)}{\sin^2(\Omega_1)}\right]Q, \tag{9.16}$$

which can be written in a shorter form:

$$Q_1(n) = T(n)Q. \tag{9.17}$$

Knowing time instant n and function $T(n)$ the steady state value Q

$$Q = \frac{1}{T(n)}Q_1(n) \tag{9.18}$$

can be calculated much earlier, at the very beginning, shortly after fault inception.

Since algorithms of measurement of reactive power only have unique, phase independent transient response there appears a problem of dynamical correction of the remaining criterion values. The simplest solution of the problem is application of $\pi/2$ delay of current signal (equivalent number of delay samples). Then an algorithm of reactive power measurement becomes an algorithm of active power since $\sin[\varphi_U - (\varphi_I - \pi/2)] = \cos(\varphi_U - \varphi_I)$. The block scheme of active power measurement with correction is presented in Fig. 9.5. Realization of zero initial conditions requires starting the algorithm after the time needed to realize the current signal delay. In the simplest case it means the number of samples equivalent to a quarter of period of the fundamental frequency component. It is also possible to use the methods of orthogonalization by single or double delay. They also provide the same delay, however, here additional speed-up measurement is obtained since delayed current is given as early as after one or two samples.

It is also evident that the same algorithm (of active power) can be used to measure magnitudes of voltages and currents.

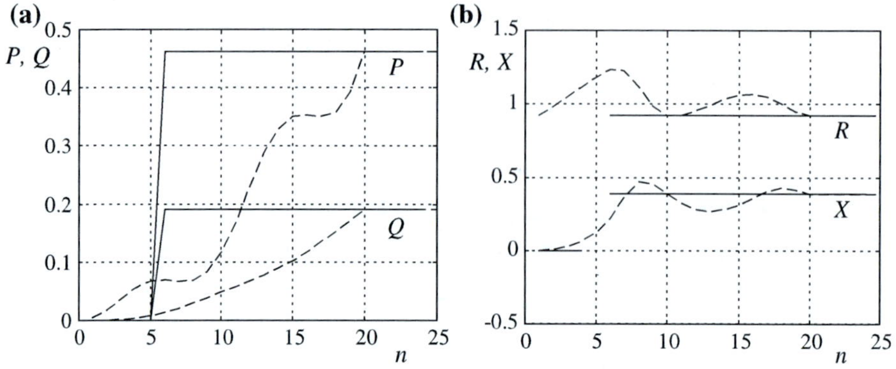

Fig. 9.6 Transient behavior of the algorithms without (*dashed line*) and with correction (*solid line*) for: **a** power measurement, **b** impedance measurement

Having dynamically compensated algorithm measuring signal magnitudes and power it is also possible to get algorithms for measurement of impedance or admittance and their components. Very attractive solution appears when one synchronizes the delays for the algorithms where it is required. Then all algorithms will start their operation at the same instant having the same function of dynamical correction $T(n)$. The functions will be identical in numerator and denominator of impedance and its components algorithms, which will allow canceling them as it is shown below:

$$R(n) = \frac{2T(n)P}{T(n)I^2} = R, \tag{9.19a}$$

$$X(n) = \frac{2T(n)Q}{T(n)I^2} = X. \tag{9.19b}$$

It means that thanks to proper synchronization dynamically corrected algorithms of impedance and its components without correction of calculated results at consecutive samples are obtained. Impedance measurement algorithms (9.19a, b) are thus the most attractive ones.

Simulation results of power and impedance measurement using presented method are shown in Fig. 9.6 (solid line). It is seen that we can get steady state of measurement much faster than with standard methods (dashed line). It should be added that the delay of five samples can be further reduced even to one sample delay, but at the cost of worse response to noise.

9.3 Frequency Characteristics of Measurement Algorithms

Frequency characteristics of protection measurement algorithms are of interest at least for a couple of reasons. Firstly, for most of cases constant frequency of the current and voltage signals is assumed, usually equal 50 or 60 Hz. In reality, however, the frequency of the power system may change, with the deviations as high as a few percent noted during power imbalance cases and close generator faults. Crucial for proper operation of digital relays is susceptibility of employed algorithms to frequency deviations. Secondly, sometimes it is needed to apply the protection algorithms in wide range of frequencies, e.g., for reversible machines in power plants. Thirdly, the input current and voltage signals may contain components of various levels and frequencies. All these factors should be taken into account during design and evaluation of the protection algorithms [5].

The frequency spectra of measurement algorithms are directly associated with applied filters and the shape of measurement equations themselves. The filters, usually used as pairs of orthogonal ones, are linear units of well-known frequency responses. Unfortunately, the measurement algorithms mostly employ nonlinear operations like multiplications, squaring, divisions or calculating of absolute values; therefore one should be very careful by drawing suitable conclusions or making generalizations.

In order to enable analysis of frequency features of measurement algorithms three types of characteristics will be determined: sensibility characteristics for small frequency deviations, algorithms' spectra in full range of considered frequencies (up to half of sampling frequency) and error characteristics for additive distortions. The considerations and analyses will have both theoretical and simulative character. One should, however, remember that all of the analyzed algorithms are nonlinear and thus concluding with respect to their advantages and drawbacks are to be conducted with high cautiousness.

The algorithms of the averaging family (Sect. 8.1.2) apply either averaging of absolute or squared values of signals samples for magnitude estimation and averaging of current and voltage samples products for measurement of active and reactive powers. For the sake of preserving the clarity of conducted analyses the considerations for algorithms with absolute values averaging will be omitted. It is justified also by the fact that their accuracy is also dependent on sampling frequency, which would make analysis doubly difficult.

Let the current and voltage signals for further considerations be given in the form:

$$u(n) = \cos(n\Omega + \varphi_U) + g_k \cos(n\Omega_k + \varphi_{Uk}), \tag{9.20a}$$

$$i(n) = \cos(n\Omega + \varphi_I), \tag{9.20b}$$

where

$$g_k = U_{kM}/U_{1m}.$$

The analysis will be performed for two exemplary protection criterion signals—magnitude and active power. Application of averaging principle requires calculation of the components of addition that in considered cases for the signals (9.20a, b) are:

$$
\begin{aligned}
u^2(n) = & \ 0.5[1 + \cos(2n\Omega + 2\varphi_U)] + 0.5g_k^2[1 + \cos(2n\Omega_k + 2\varphi_{Uk})] \\
& + g_k\{\cos[n(\Omega_k + \Omega) + \varphi_{uk} + \varphi_U] + \cos[n(\Omega_k - \Omega) + \varphi_{Uk} - \varphi_U]\},
\end{aligned}
\tag{9.21}
$$

$$
\begin{aligned}
u(n)i(n) = & \ 0.5[\cos(\varphi_U - \varphi_I) + \cos(2n\Omega + \varphi_U + \varphi_I)] \\
& + 0.5g_k\{\cos[n(\Omega_k + \Omega) + \varphi_{Uk} + \varphi_I] + \cos[n(\Omega_k - \Omega) + \varphi_{Uk} - \varphi_I]\}.
\end{aligned}
\tag{9.22}
$$

The above components are summed up for a number of samples being a multiple of signal half-cycle. One can see that the estimation error may result from either non-zero sub-sums of AC components or the existing constant parts of (9.21), (9.22). It is worth to add that during analyses of the algorithm spectra, either for small or bigger frequency deviations, the additive distortions should be assumed zero ($g_k = 0$). Determination of the characteristic for distortion influence is done under assumption of signal frequency being equal to nominal value ($\Omega = \Omega_1$) and the magnitude of distortion components equal to unity ($g_k = 1$). Changing the frequency of distortion component Ω_m from zero to half of the sampling frequency one gets measurement oscillations around accurate value, with the magnitude dependent on distortion frequency.

In the following considerations calculation of the sums of cosine components, being typical components of Eqs. 9.21 and 9.22, could be quite useful. Simple transformations, facilitated with application of exponential complex functions, yield:

$$
\sum_{n=0}^{N-1} \cos(n\Omega + \varphi) = \cos[0.5(N-1)\Omega + \varphi] \frac{\sin(0.5N\Omega)}{\sin(0.5\Omega)}
\tag{9.23}
$$

The above expression can be utilized for determination of algorithm's spectra, as well as specific conditions for zeroing of error components. One can conclude that for the frequency characteristics the source of error of magnitude and power measurement is possible non-zero sum of the second harmonic. This sum is always zero when the signal frequency is a full multiple of the fundamental frequency, i.e., for $\Omega = k\Omega_1$. From (9.23) results that the algorithms susceptibility for small frequency deviations can be quite significant.

The above observations are confirmed by the simulative investigations, illustrated in Figs. 9.7 and 9.8.

Considering the errors due to oscillating distorting components one should assume the values $\Omega = \Omega_1$ and $g_k = 1$. The sums of second harmonic are here equal to zero and do not cause errors. On the other hand, deviations from accurate measurement values result from non-zero sums of components of frequencies $\Omega_k, \Omega_k + \Omega_1, \Omega_k - \Omega_1$ and additionally non-zero sum of constant component in voltage signal ($0.5g_k^2$). For certain frequencies at least the sums resulting from AC components are zero, it is so for instance for the third harmonic ($\Omega_k = 3\Omega_1$) and then the measurement error is minimal, though also quite significant.

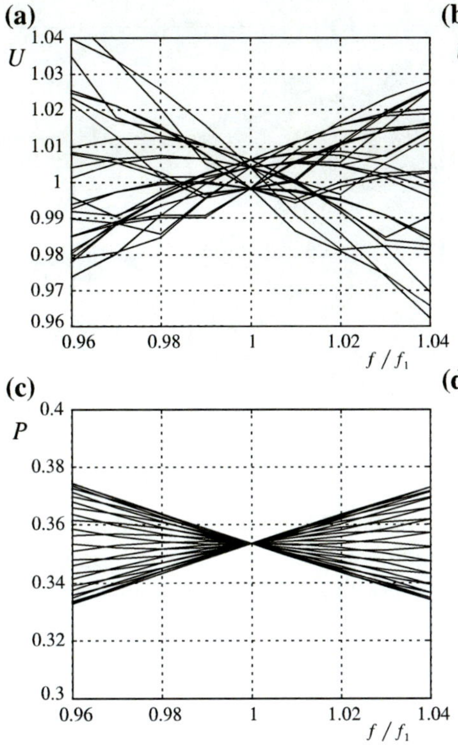

Fig. 9.7 Spectra of selected averaging measurement algorithms for **a**, **b** voltage magnitude and **c**, **d** power components; algorithms: (8.3a, b), (8.9a, b, c), (8.11), (8.12)

Fig. 9.8 Spectra of averaging algorithms in wide-frequency range; algorithms: (8.9a, b, c), (8.11), (8.12)

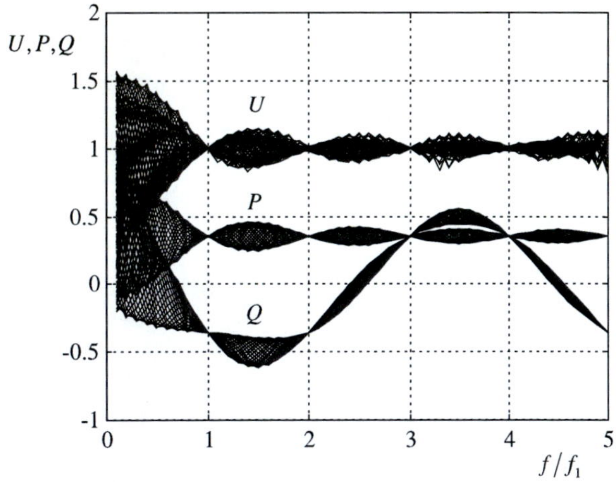

The simulative analysis results shown in Fig. 9.9 confirm the above, presenting also the characteristics for reactive power (not discussed previously). The above observations lead to two general conclusions: the algorithms considered are highly

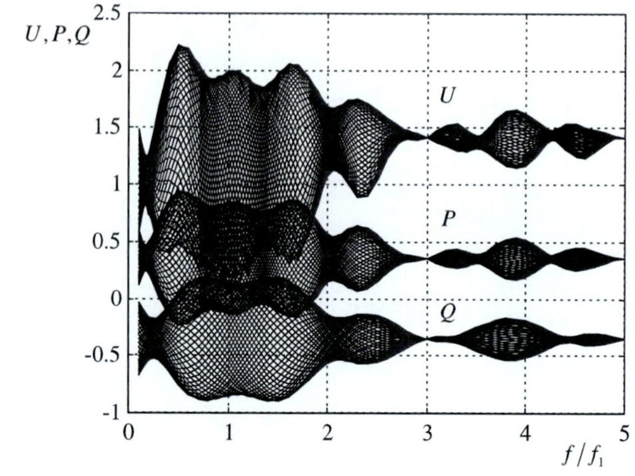

Fig. 9.9 Algorithms spectra for additive distortions; algorithms as in Fig. 9.8

susceptible to frequency deviations, both of the changing fundamental frequency component and oscillating distortion components. This feature significantly decreases general good opinion about the algorithms, whose undisputed advantage is their numerical simplicity.

For frequency analysis of the measurement algorithms applying orthogonal filters one can utilize current and voltage signals of the form:

$$\frac{u_{FC}}{F_C} = h_C \cos(n\Omega + \varphi_U) + g_{Ck} \cos(n\Omega_k + \varphi_{Uk}), \tag{9.24a}$$

$$\frac{u_{Fc}}{F_S} = h_S \sin(n\Omega + \varphi_U) + g_{Sk} \sin(n\Omega_k + \varphi_{Uk}), \tag{9.24b}$$

$$\frac{i_{FC}}{F_C} = h_C \cos(n\Omega + \varphi_I), \tag{9.25a}$$

$$\frac{i_{FS}}{F_S} = h_S \sin(n\Omega + \varphi_I), \tag{9.25b}$$

where

$$h_C = \left|\frac{H_C(j\Omega)}{H_C(j\Omega_1)}\right|,$$

$$h_S = \left|\frac{H_S(j\Omega)}{H_S(j\Omega_1)}\right|,$$

$$g_{Ck} = \left|\frac{H_C(j\Omega_k)}{H_C(j\Omega_1)}\right| \frac{U_{km}}{U_{1m}},$$

$$g_{Sk} = \left|\frac{H_S(j\Omega_k)}{H_S(j\Omega_1)}\right| \frac{U_{km}}{U_{1m}},$$

$$F_C = |H_C(j\Omega_1)|,$$

$$F_S = |H_S(j\Omega_1)|.$$

Similarly as before for analysis of algorithms' spectra it is assumed that $g_{Ck} = g_{Sk} = 0$, while h_C and h_S are functions of Ω. For analysis of additive distortion influence one should assume $\Omega = \Omega_1$, which induces unity values of the coefficients $h_C = h_S = 1$.

One has to mention that for investigations discussed here additive distortions in one of the signals only (voltage) were introduced. Moreover, since g_{Ck} and g_{Sk} are very small, the errors resulting from products of various distortion components can also be neglected. Thus, assuming free change of distortion components phase shifts one can conclude that the errors due to a number of components are roughly the same as the sum of errors for particular components separately. With this in mind the conclusions from consideration for single distortion components can be extended accordingly.

It is essential that the orthogonal FIR filters, having linear phase, deliver at their outputs orthogonal components of input signal of any frequency, i.e., this holds both for fundamental frequency component and all distortion components. This also explains the form of Eqs. 9.24a, b.

Basing on the above models of current and voltage signals one can determine sought spectra of the measurement algorithms, for assumed zero level of additive distortions ($g_{Ck} = g_{Sk} = 0$). Considering standard algorithms for signal magnitude, active and reactive power (8.37), (8.47a, b), (8.48), after simple transformations one gets:

$$U_m^2 = 0.5(h_C^2 + h_S^2) + 0.5(h_C^2 - h_S^2)\cos(2n\Omega + 2\varphi_U), \tag{9.26}$$

$$P = 0.25(h_C^2 + h_S^2)\cos(\varphi_U - \varphi_I) + 0.25(h_C^2 - h_S^2)\cos(2n\Omega + \varphi_U + \varphi_I), \tag{9.27}$$

$$Q = 0.5h_C h_S \sin(\varphi_U - \varphi_I). \tag{9.28}$$

It is seen that measurement result of both magnitude and active power contains, apart from constant value proportional to measured variable, also the second harmonic term, being a source of error. The latter one appears always when the normalized filter gains h_C and h_S differ one from another (for given frequency). Depending on the initial phase of the signals the maximum error values of magnitude measurement may vary between the values of h_C and h_S. When the filter gains are identical, which is a case for frequencies being integer multiple of the base frequency of applied orthogonal filters, this error disappears.

The result of reactive power contains only a constant term proportional to the measured value, multiplied by the product of filter gains. The measured value differs from accurate one when this product differs from unity.

The calculated measurement algorithms' spectra (signal magnitude, active and reactive power) in wide range of frequency changes are presented in Fig. 9.10. In the vicinity of the fundamental frequency the highest deviations of the measured

Fig. 9.10 Spectra of the measurement algorithms without time delay: **a** U (8.37), **b** P (8.47a, b), **c** Q (8.48)

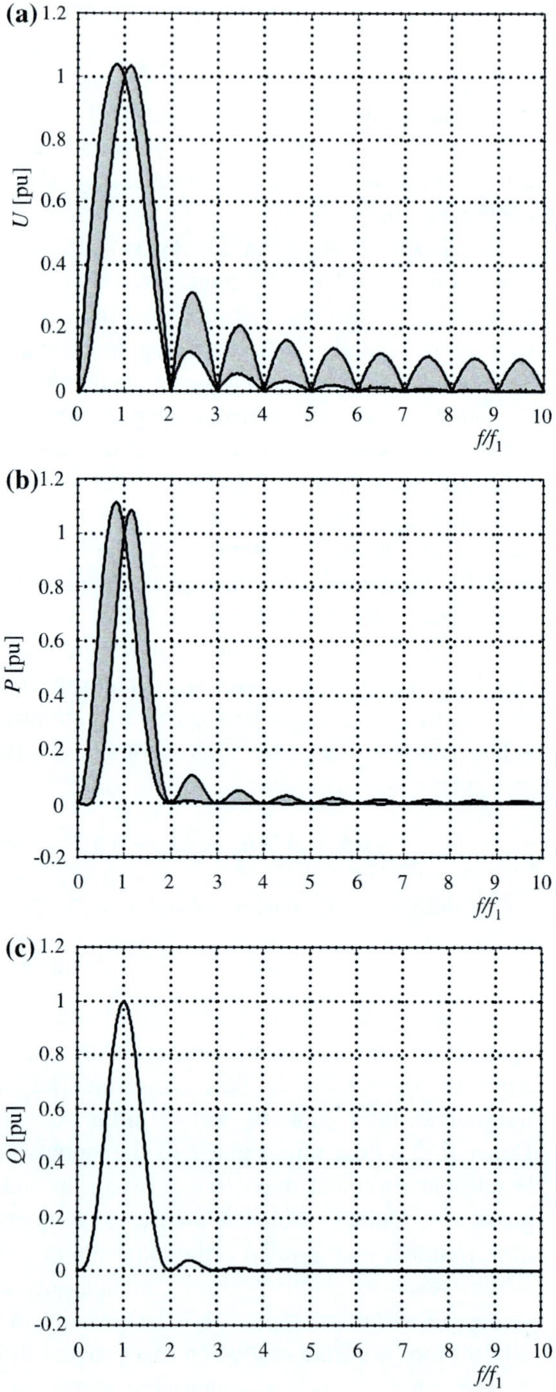

value are observed for active power, moderate—for voltage magnitude, and the lowest and unique (phase independent) for reactive power. Such results are a derivative of the normalized filters' characteristics (see Fig. 9.7b). Since one of the filter gain coefficients is slightly higher and the other one slightly lower than unity, their product is close to unity and this is why we have almost constant level of reactive power (Fig. 9.10c). On the other hand, squared values of gains differ much more from the gains themselves (in considered range of frequency), therefore deviations for active power are higher than that for magnitude. Considering the spectra in wider range one can conclude that the results for higher frequencies are different than before. The values of active and reactive power are lower than for magnitude, since for low values of filter gains their squared values or products are much lower than the gains themselves.

Simulation studies have also been performed for the algorithms with time delay (8.41), (8.49) and (8.50), with application of the sine/cosine filters for signal orthogonalization and sine filter only in case of the algorithm for reactive power estimation. The results are presented in Fig. 9.11. Characteristic here is phase independent shape of spectra (unique curve) for all considered algorithms, which may allow for eventual adaptation of estimators to varying signals frequency. One can see that the signal magnitude U estimator spectrum has the side bands at the level between the values resulting from the sine/cosine filters frequency responses. The spectra of both P and Q estimators exhibit only the main band, whereas the signal frequencies above 100 Hz are almost fully suppressed, which is a significant advantage over the algorithms without time delay. One should remember, however, that this superiority is achieved at the cost of elongated response in time domain, due to introduction of additional time delay.

Analytical evaluation of the accuracy of standard measurement algorithms of magnitude, active and reactive power for additive distortion components can be performed when one substitutes the voltage and current terms according to (9.24a, b) and (9.25a, b) to the equations of particular measurement equations, assuming constant value of fundamental frequency ($\Omega = \Omega_1$). Measured quantities are now described as follows:

$$U_m^2 = 1 + g_{Ck}^2\cos^2(\beta) + 2g_{Ck}\cos(\alpha)\cos(\beta) + g_{Sk}^2\sin^2(\beta) + 2g_{Sk}\sin(\alpha)\sin(\beta), \tag{9.29}$$

$$P = 0.5[\cos(\varphi_U - \varphi_I) + g_{Ck}\cos(\gamma)\cos(\beta) + g_{Sk}\sin(\gamma)\sin(\beta)], \tag{9.30}$$

$$Q = 0.5[\sin(\varphi_U - \varphi_I) + g_{Sk}\cos(\gamma)\sin(\beta) - g_{Ck}\sin(\gamma)\cos(\beta)], \tag{9.31}$$

where

$$\alpha = n\Omega_1 + \varphi_U,$$

$$\beta = n\Omega_k + \varphi_{Uk},$$

$$\gamma = n\Omega_1 + \varphi_I.$$

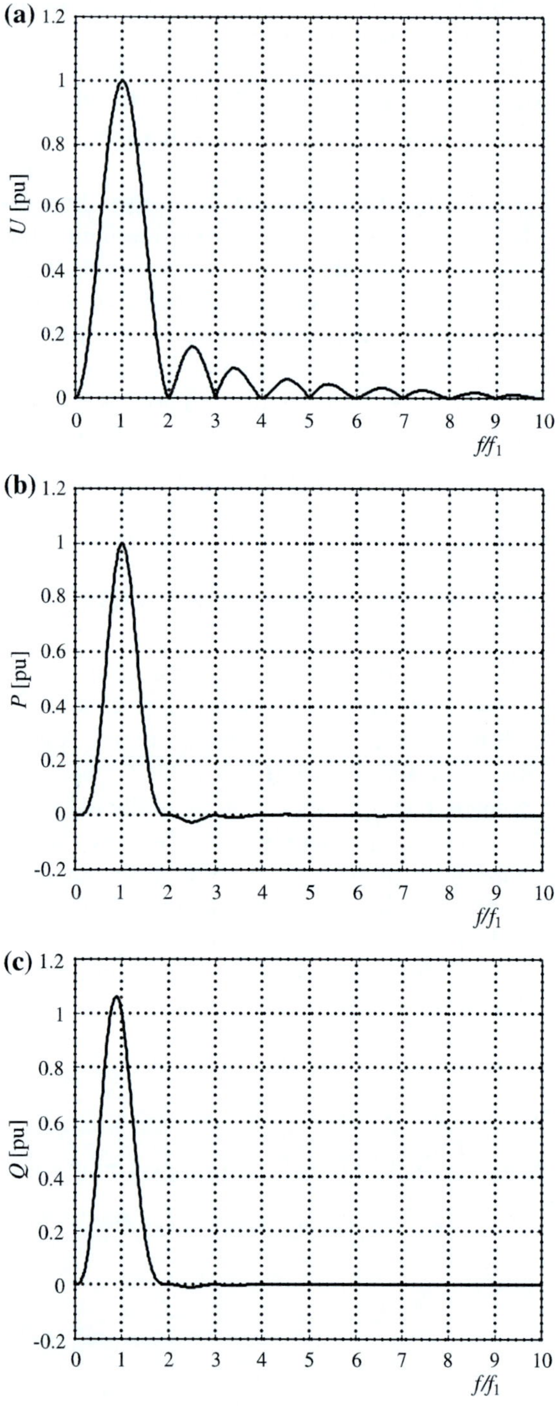

Fig. 9.11 Spectra of the measurement algorithms with time delay ($k = N_1/4$): **a** U (8.41), **b** P (8.49), **c** Q (8.50)

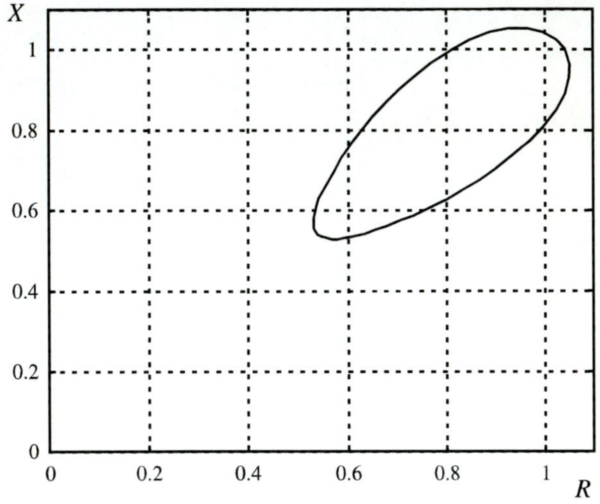

Fig. 9.12 Complex characteristic of impedance measurement for additive component of frequency $f/f_1 = 2.5$

From (9.29) to (9.31) follows that the highest measurement errors are approximately equal to higher value of filter gains for given frequency. The investigation results shown in Fig. 9.11 also confirm this observation. Generally, one can say that the shape and level of deviations can be identified with the characteristics of applied orthogonal filters. One can also notice that the relationship between the algorithms' spectra and characteristics for additive distortions is rather weak. It should be mentioned that the measurement errors due to distortion components, here quite high because of assumed level of additive components, will be much lower in reality, where expected harmonics and other components are usually much smaller than the fundamental frequency component.

There exist also other possibilities of presenting expected measurement errors. For instance, in order to show both module and phase error of a complex variable one can present the results on a plane R–X or P–Q. An example of such a characteristic is shown in Fig. 9.12 for selected impedance measurement algorithm and frequency of the current/voltage signals. The accurate value amounts $R = X = 0.707$, whereas the measured values surround this point along an ellipse, showing both module and argument deviations. Quite big size of the ellipse results from relatively low frequency of the additive component and its unit magnitude. In practice the size of the curve should always be much more limited.

The frequency characteristics of measurement algorithms presented in this chapter illustrate the frequency features of considered algorithms. The way of algorithms evaluation can be further transmitted to assessment of other algorithms, e.g. for impedance measurement. Interested reader can perform similar analyses himself, one can also reach for available references, e.g., [5, 11].

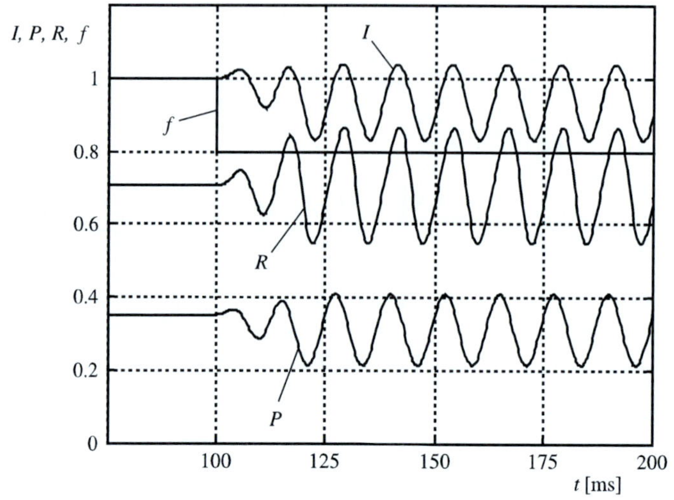

Fig. 9.13 Waveshapes of selected criteria values measured during sudden change of frequency at $t = 100$ ms

9.4 Adaptive Frequency Insensitive Estimators

Measurements of criterion values of power system protections use the signals that in most cases have constant frequency. However, there are some situations, for instance during great power imbalance, when frequency changes by as much as a few Hz. Then some problems arise concerning accuracy of applied measurement algorithms since the filters are tuned and matched to constant nominal frequency (50 or 60 Hz). Protection relay are also used in reversible generation units where the range of frequency changes is wide and under such conditions the protection criteria values should be measured. This is why one must also consider designing protection systems, which either can operate during wide range of frequency changes or which are insensitive to small frequency deviations, depending on application.

It can be proved that deviations of signal frequency cause errors of measured criteria values. In some cases these errors depend on signals phase shifts (magnitude and active power). When a frequency deviation occurs, second harmonic of the signals as well as constant value proportional to measured quantity may appear. This is illustrated in Fig. 9.13. After step change of frequency from 50 to 40 Hz (this is simulation only, such change must not appear in practice) measured current magnitude, power and resistance (standard algorithms applied) exhibit unsymmetrical oscillations around accurate values. The frequency of oscillations is two times higher (2*40 Hz) than signal frequency. The errors are too big to accept them in practice.

However, beside standard algorithms one has on disposal many others that are less or even much less sensitive to frequency deviations. Responses of two types of algorithms of active power and resistance measurement are presented in Fig. 9.14

Fig. 9.14 Comparison of measurement results of **a** active power, **b** resistance for various algorithms: 1—standard, 2—with time delay, 3—with adaptation

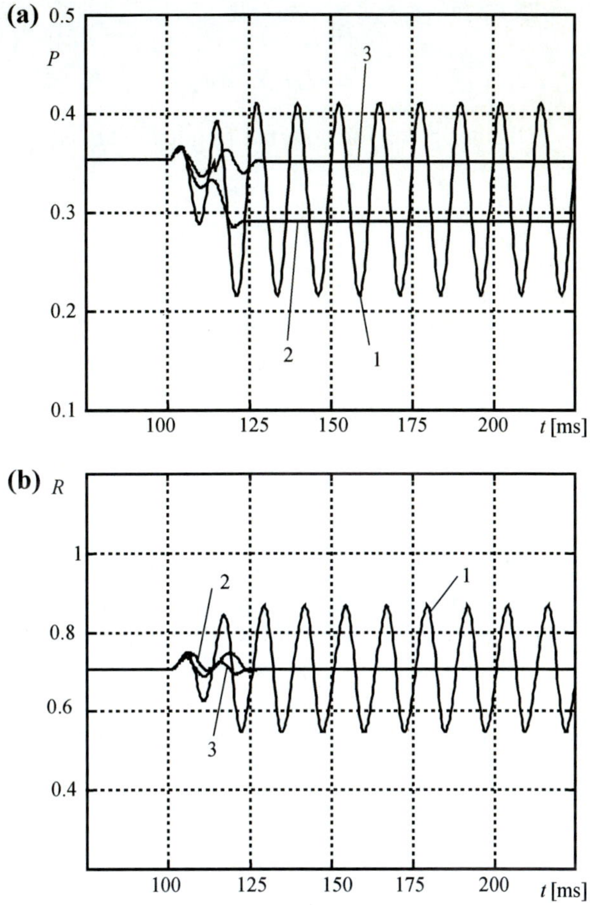

for four percent of frequency deviation. It is seen that algorithms applying delayed orthogonal components are much less sensitive to frequency deviations and, what is very important for adaptive solutions errors, their responses are unique, phase independent. Theoretical analysis of the situation may be interesting.

It is known that FIR orthogonal filters have output signals (for the same input) which phase angles differ by $\pi/2$ at any arbitrary frequency. Then at given frequency Ω the filter outputs can be expressed in the form:

$$y_{FC}(n) = |H_C(j\Omega)| y_C(n), \tag{9.32a}$$

$$y_{FS}(n) = |H_S(j\Omega)| y_S(n). \tag{9.32b}$$

Let us calculate the function appearing in magnitude and active power algorithms using outputs of the filters (9.32a, b) (subscript F denotes output of the filter, C, S stand for sine, cosine filter window, respectively):

$$\begin{aligned} & u_{FS}(n)i_{FC}(n-k) - u_{FS}(n-k)i_{FC}(n) \\ & \quad = U_m I_m \cos(\varphi_U - \varphi_I)|H_C(j\Omega)||H_S(j\Omega)| \sin(k\Omega). \end{aligned} \tag{9.33}$$

If filter gains for nominal frequency Ω_1 are:

$$\begin{aligned} |H_C(j\Omega_1)| &= F_C \\ |H_S(j\Omega_1)| &= F_S \end{aligned} \tag{9.34}$$

then one can calculate active power using algorithms with delayed orthogonal components. To do that Eq. 9.33 should be divided by constant coefficient $(2F_C F_S \sin(k\Omega_1))$ obtaining:

$$P(\Omega) = \frac{u_{FS}(n)i_{FC}(n-k) - u_{FS}(n-k)i_{FC}(n)}{2F_C F_S \sin(k\Omega_1)} = h_C h_S \frac{\sin(k\Omega)}{\sin(k\Omega_1)} P = CP \tag{9.35}$$

where h_C, h_S are normalized filter gains (ratios of gain at given frequency and nominal frequency).

The conclusion is that during frequency variation measured active power is equal to the accurate value multiplied by a constant factor. When the frequency is equal to nominal value the coefficient C is equal to unity, otherwise it is different giving certain measurement error.

In the same way using Eq. 9.33, substituting current instead of voltage samples, one obtains an algorithm of current magnitude measurement depending on frequency variations:

$$I_m^2(\Omega) = \frac{i_{FS}(n)i_{FC}(n-k) - i_{FS}(n-k)i_{FC}(n)}{2F_C F_S \sin(k\Omega_1)} = h_C h_S \frac{\sin(k\Omega)}{\sin(k\Omega_1)} I^2 = CI^2 \tag{9.36}$$

with coefficient C depending on frequency being the same as in (9.35).

For standard algorithm of reactive power we get in turn:

$$Q(\Omega) = h_C h_S Q \tag{9.37}$$

However, in this case one obtains different factor depending on frequency (there is no sine function).

Time responses of standard algorithms and those using delayed orthogonal components are shown in Fig. 9.14. According to the above considerations in case of in standard algorithms the second harmonic component appears additionally when frequency changes, however, for algorithms with delayed orthogonal components a constant deviation of measured values is observed. In case of resistance measurement it is even better since after transient period disappears the algorithm is again accurate, irrespective of signal frequency. This is very important and advantageous result, which can be proved easily. The resistance is calculated as a ratio of active power (9.35) and current magnitude squared (9.36):

$$R(\Omega) = \frac{2P(\Omega)}{I_m^2(\Omega)} = \frac{2CP}{CI_m^2} = R, \tag{9.38}$$

and

$$R = \frac{u_{FS}(n)i_{FC}(n-k) - u_{FS}(n-k)i_{FC}(n)}{i_{FS}(n)i_{FC}(n-k) - i_{FS}(n-k)i_{FC}(n)}. \tag{9.39}$$

Since the same factor C appears in numerator and denominator it is evident that the resistance measurement is completely independent of frequency (except for C equal to zero).

Similarly, writing Eq. 9.37 for measurement of voltage magnitude and dividing voltage by current, after canceling the same constant factors in numerator and denominator one gets an accurate frequency independent impedance algorithm:

$$Z^2(\Omega) = \frac{U_m^2(\Omega)}{I_m^2(\Omega)} = \frac{CU_m^2}{CI_m^2} = Z^2 \tag{9.40}$$

Unfortunately, there is no similar result for reactance algorithms. However, possible solution is calculation of reactance using measured impedance and resistance:

$$X(\Omega) = \sqrt{Z^2(\Omega) - R^2(\Omega)} \tag{9.41}$$

Since both quantities of the right-hand side of the equation are frequency independent then the same is valid for the left side, i.e. reactance measurement is also frequency independent.

Equations 9.36–9.41 describe the algorithms that are less sensitive or insensitive to frequency changes. The algorithms of impedance components (9.39–9.41) are completely frequency independent. All of the algorithms can be applied with good result and small errors to measurements of criterion values during small frequency deviations in the range ±2.5 Hz around nominal frequency.

When frequency of the signals changes in wider range (protections of reversible generation units) then even those algorithms less sensitive to frequency deviations may give too big errors. Measured active power and magnitude differ more from their accurate value since coefficient C differs much more from unity. Even in measured impedance some errors may appear due to another reason not taken into account till now. This is noise which can cause errors not only during transients but also during steady state. Simply during substantial frequency changes one must take into consideration reshaping of the filters frequency responses. The situation is shown in Fig. 9.15 presenting a part of frequency response of full period sine, cosine orthogonal filters. If the filters were designed for nominal frequency 50 Hz but actual frequency is equal, say, 25 Hz (1h′ in Fig. 9.15) then normalized cosine filter gain is equal to 0.4 for this frequency and the 2nd harmonic is gained 2.5 times, approximately (2h′). In general, quite good band pass filter becomes substantially worse when frequency changes in wider range.

Fig. 9.15 Sine/cosine filters' frequency responses with marked frequencies and their harmonics when fundamental frequency changes

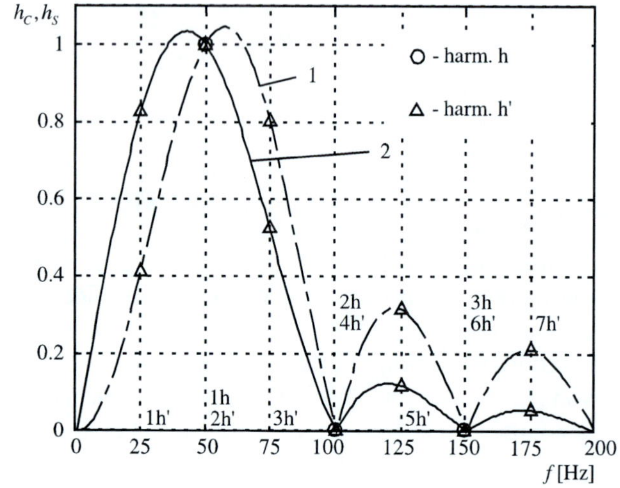

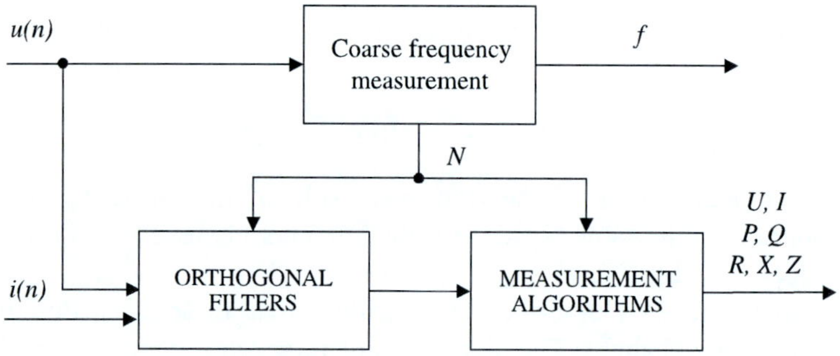

Fig. 9.16 Block scheme of the frequency-adaptive measurement procedure

One can avoid the above effect using adaptive solutions, where the filters are being matched to actual frequency of the signals, as well as applying correction of coefficient C according to actual, measured frequency [3, 4, 6, 10]. The solutions can be outlined with the block scheme depicted in Fig. 9.16. Important part of the scheme is a coarse frequency measurement, which need not be very accurate if applied algorithms of criterion values measurement are insensitive to small frequency deviations.

Modification of the filters is easy. One period FIR sine or cosine filter with the window N_0 has its basic angular frequency equal to $2\pi/N_0$. To get adaptive solution one should use filter window as close to the period of fundamental components of signals as possible. To obtain that one must measure actual frequency of signals and control whether the frequency deviation is greater than assumed.

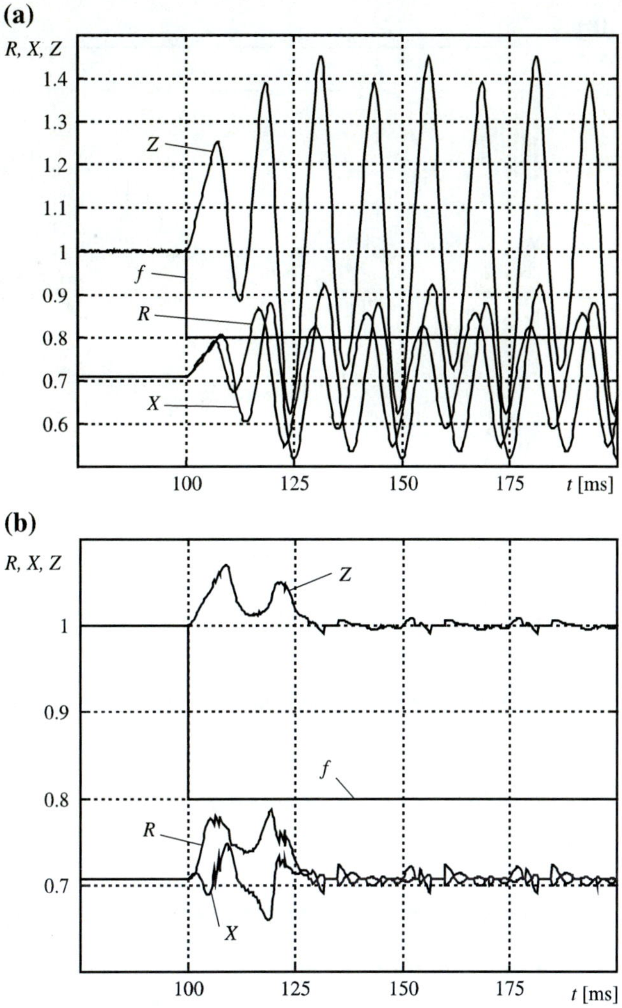

Fig. 9.17 Impedance measurement with **a** standard and **b** adaptive algorithms; signals contaminated by noise, step change of frequency from 50 to 40 Hz

From the filter features (sine, cosine full-cycle filters) it is known that filters having window length N_0 have a gain for frequency $2\pi/N_0$ equal to $N_0/2$. It means that knowing signals frequency it is possible to realize adaptive procedure and calculate new filter gains immediately.

Calculation of frequency deviation can be realized in many ways. Observing the methods described in Sect. 8.2.5 devoted to frequency measurement one can derive the following frequency dependent equation:

$$\cos(k\Omega) = 0.5\frac{u(n-2k)u(n-k) - u(n)u(n-3k)}{u(n-k)u(n-k) - u(n)u(n-2k)}. \tag{9.42}$$

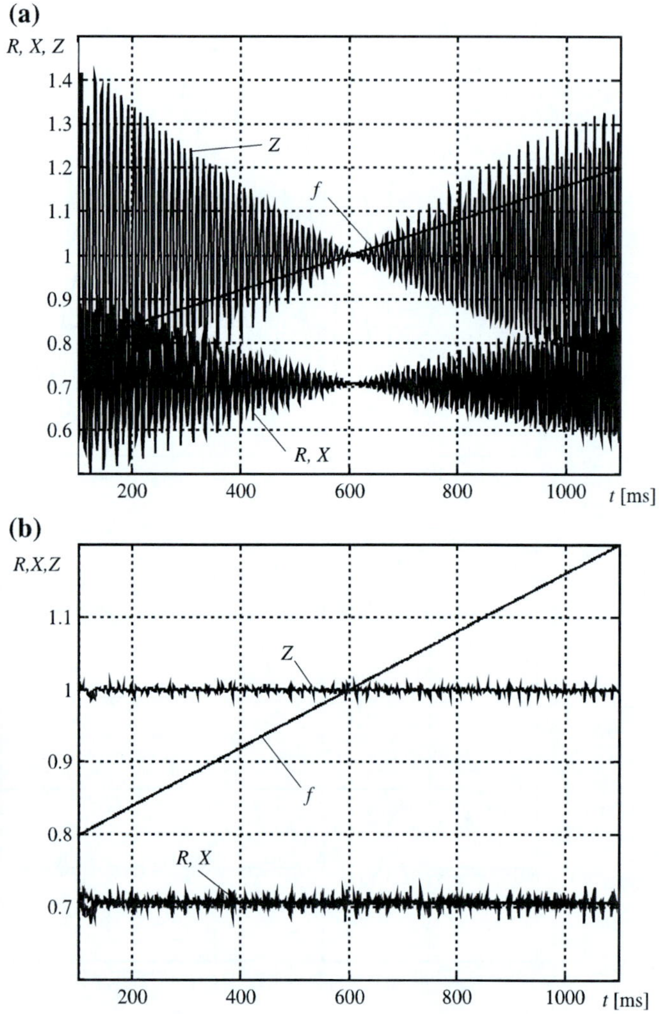

Fig. 9.18 Impedance measurement with **a** standard and **b** adaptive algorithms; signals contaminated by noise, linear change of frequency from 40 to 60 Hz

If now the delay k is equal to a quarter of fundamental frequency cycle then one gets simple equation allowing to find frequency deviation for given k:

$$\cos(k\Omega) = \cos\left(\frac{N_1}{4}\Omega\right) \cong -\frac{\pi}{2}\frac{\Delta\Omega_1}{\Omega_1} \cong -k\Delta\Omega_1 \tag{9.43}$$

Increasing or decreasing filter window length can be realized according to:

$$\text{if}\left\{\begin{array}{c} -k\Delta\Omega_1 < -\varepsilon \\ |-k\Delta\Omega_1| \leq \varepsilon \\ -k\Delta\Omega_1 > \varepsilon \end{array}\right\}\text{then}\left\{\begin{array}{c} \text{increase } k+1 \\ \text{no change } k \\ \text{decrease } k-1 \end{array}\right\} \tag{9.44}$$

It is evident that when frequency measurement shows its increase it means that the signal period decreases and filter window should be shorter, and the other way round when frequency decreases.

Results of the adaptive solution application for measurement of active power and resistance are shown in Fig. 9.14. For resistance nothing has really changed (accurate results were observed also without adaptation on), but now also active power is measured accurately.

Further advantages of the adaptive methods are visible when some noise is added to input signals (2nd and 4th harmonic, 0.15 per unit each, Figs. 9.17 and 9.18 for step and linear change of signal frequency). Analyzing the results for impedance measurement without and with adaptation, one can easily notice additional effect of decreasing of the errors due to noise for the adaptive solution. The measurement results are not perfect (the filters might not be exactly matched to actual cycle of the signals), however, they oscillate quite closely around accurate values of R, X and Z.

References

1. Bąk D, Szafran J, Rebizant W (2005) New method of dynamical correction of frequency insensitive impedance measuring algorithms. In: Proceedings of the 15th power systems computation conference, Liege, Belgium, paper 79
2. Moftah AM (1997) New dynamically corrected fast estimators of protection criterion values. PhD dissertation, Wroclaw University of Technology, Wroclaw, Poland
3. Rebizant W, Rusek B (2003) Efficient frequency-adaptive amplitude tracking algorithms for protection purposes. In: Proceedings of the IEEE PowerTech conference, Bologna, Italy, paper BPT03-12
4. Rebizant W, Szafran J (1998) Adaptive wide frequency range estimators of protection criterion values. In Proceedings of the 33rd UPEC conference, Edinburgh, UK, vol 1, pp 1–4
5. Rebizant W, Szafran J (2002) Frequency responses of relaying algorithms based on orthogonal components. In Proceedings of the 13th international conference on power system protection, Bled, Slovenia, pp 161–166
6. Rebizant W, Szafran J, Michalik M (1999) Frequency insensitive symmetrical components estimators for power system relaying. In: Proceedings of the 13th power systems computation conference, Trondheim, Norway, pp 705–712
7. Rosolowski E, Szafran J (1993) Fast estimation of protection criterion values using dynamical correction. In: Proceedings of the 11th power systems computation conference, Avignon, France, vol 2, pp 805–811
8. Rosolowski E, Szafran J (1995) Dynamically corrected fast estimators of current and voltage magnitude. IEE Proc Gener Transm Distrib 142:310–316
9. Szafran J (1990) Signal recognition in digital protection, Sci Works of the Institute of Electrical Power Engineering, WUT, Wroclaw, Poland. Monographs, no 23 (in Polish)
10. Szafran J, Rebizant W, Michalik M (1999) Adaptive measurement of power system currents, voltages and impedances in off-nominal frequency conditions. In: Proceedings of the 16th IEEE instrumentation and measurement technology conference, Venice, Italy, vol 2, pp 801–806
11. Wang L (1999) Frequency responses of phasor based microprocessor relaying algorithms. IEEE Trans Power Deliv 14:98–109

Chapter 10
Decision Making in Protective Relays

Having in mind that protective relays are devices that are supposed to evaluate the state of the protected plant and to react properly in case of abnormal operating conditions, the relay final output is usually the command to trip the protected element or to raise an alarm, whenever it is necessary. Before a tripping signal to the associated circuit breaker or an alarm is issued the decision is to be taken that is based on the locally measured criteria values and additional information from other protective relays and/or control centers. While in preceding chapters the measurement techniques and algorithms were presented, here the possible approaches to the decision making are outlined and the detailed decision procedures are described.

The very basic classification of the decision making philosophies that lead to particular decision procedures are the following:

- deterministic decision making (according to the specified recipe, with fixed or adaptive decision boundaries),
- statistic decision making (that mostly reduces to testing of statistical hypotheses related to the state of protected plant).

In many cases, especially for the protected plants of higher complexity or significant importance for the power system operation (e.g. synchronous machines or power transformers), the final relay output is worked out taking into consideration multiple criteria that bring information about respective types of phenomena or events, which may occur in or around the protected object. Aggregation of particular sub-decisions in the final output is not trivial and may be done in many ways.

Below the selected decision making procedures are described, with an example of the advanced decision scheme for generator protection, closing the chapter.

W. Rebizant et al., *Digital Signal Processing in Power System Protection and Control,* Signals and Communication Technology,
DOI: 10.1007/978-0-85729-802-7_10, © Springer-Verlag London Limited 2011

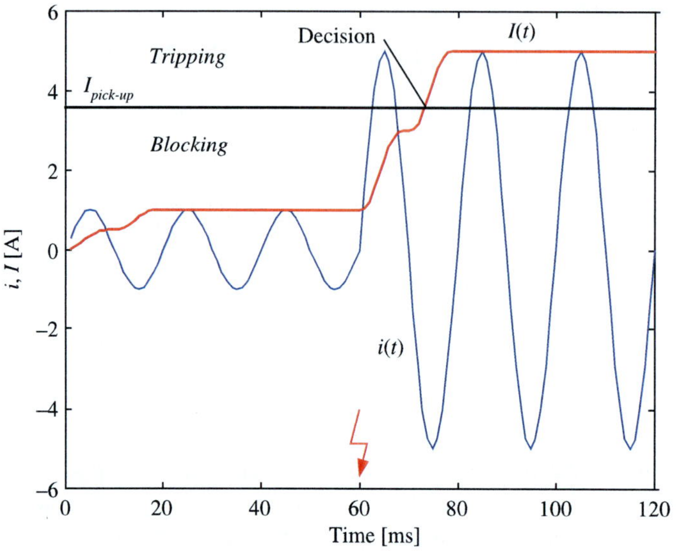

Fig. 10.1 Illustration of the decision making in overcurrent protection

10.1 Deterministic Decision Making

The simplest and, at the same time, the most frequently used approach to the decision making is the deterministic one. Deterministic approach means that the algorithm to issue a decision is precisely defined, whereas the decision thresholds may be fixed or being changed according to the prescribed scenario.

Generally speaking, with the deterministic approach the decision concerning state of the protected plant is taken by comparing the estimated/measured criteria values with appropriately set thresholds or characteristics. For the single-dimension decision problem (single criterion taken into account) the discrimination is made when the criterion signal crosses or exceeds the pre-defined border value (threshold) separating two classes of events to be distinguished, e.g. normal operation versus fault conditions. Depending on the problem the discrimination may be of one of the two types:

- overreaching, when the criterion value is higher than the threshold, e.g. for overcurrent protection, see illustration in Fig. 10.1, or
- underreaching, when the criterion value is lower than the threshold, e.g. for undervoltage protection.

With reference to Fig. 10.1, the decision to trip the protected element is taken when the measured current (instantaneous value or signal magnitude) exceeds the pick-up value:

$$I(n) > I_{\text{pick-up}}. \tag{10.1}$$

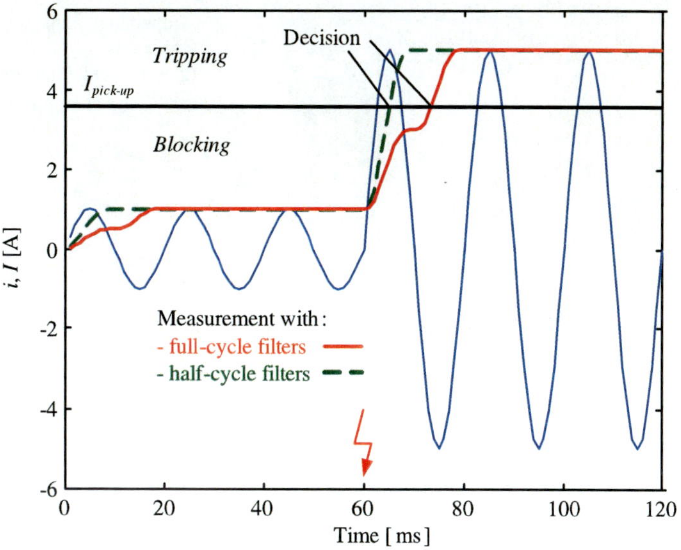

Fig. 10.2 Decision speed versus measurement data window

The decision threshold is to be set with care, i.e. taking into account the maximum expected load current and minimum fault current in the protected element:

$$I_{L\_\max} < I_{\text{pick-up}} < I_{f\_\min}. \tag{10.2}$$

The speed of the overcurrent protection is closely related to the algorithm of the current magnitude measurement. The shorter the data window of applied filters (generally—data processing window of the algorithm), the faster the relay decision can be made, Fig. 10.2.

The measurement algorithm immunity to signal contamination, especially to decaying DC component, is also important. Possible transient overestimation of signal magnitude (overshoot) can lead to false decisions, if appropriate time delay is not applied, which was already discussed in Chap. 8.

Another situation exists when the applied criterion is not a single-dimension variable but is a complex vector $\underline{Z} = R + \mathrm{j}X$, as in case of fault distance estimation in distance relays. In such a case the decision is not taken by simple checking of (10.2) but by comparing mutual location of the measured complex variable with appropriately set characteristic.

With reference to Fig. 10.3, the protection decision is made (fault confirmed) when the measured impedance vector trajectory exceeds the Mho characteristic and its locus is seen within the prescribed curve. The Mho curve should embrace the region of fault loop impedance for in-zone fault cases, with possible fault resistance, as shown in Fig. 10.3. The trip decision is taken when

$$Z \text{ is within the Mho curve.} \tag{10.3}$$

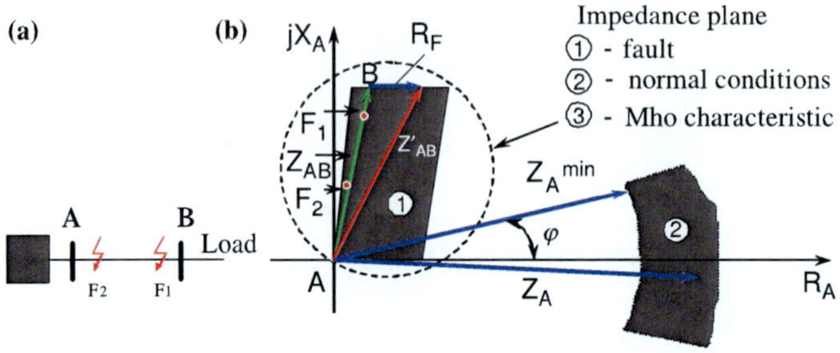

Fig. 10.3 Illustration of the decision making in distance protection: **a** simple system with faults, **b** impedance plane

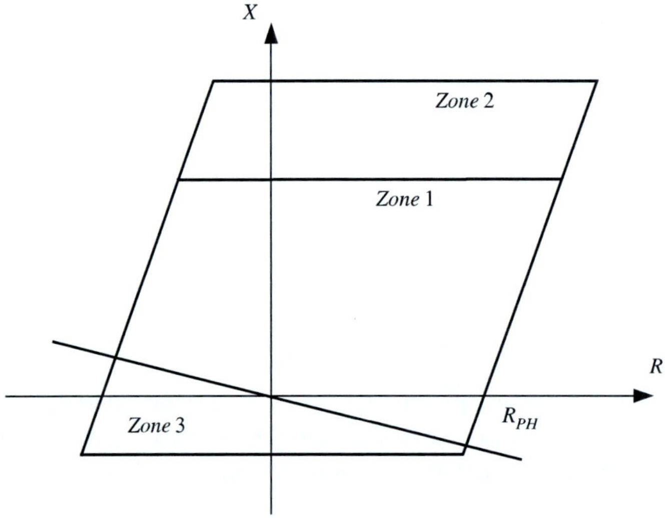

Fig. 10.4 Polygonal shape of distance characteristic

The relay manufacturers offer protection devices with distance function of various pick-up characteristics. Except Mho type the polygonal shapes are also in use, one of them is shown in Fig. 10.4. Independent of the impedance characteristic the condition (10.3) is checked in numerical way, by analysis of the equations describing the curve.

It is also a frequent practice to apply protection schemes where one criterion quantity is confronted with another one on a common decision plane. A representative example of such a scheme is the differential protection of transmission line, transformer, etc. Independent of the type of object to be protected the trip decision is met when the trajectory of measurement loci on the Diff-Bias plane (differential versus stabilization currents) exceeds the pick-up characteristic (see Fig. 10.5). After the consecutive points of the curve enter the tripping region located above the section-wise characteristic the relay picks up, and eventually the procedure of checking additional criteria (second harmonic level—for excluding

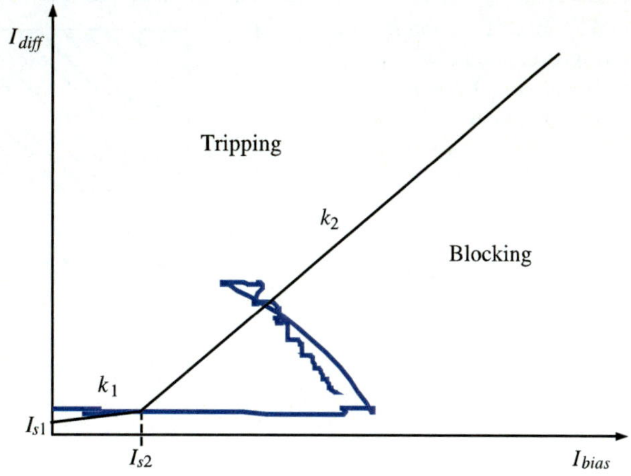

Fig. 10.5 Differential relay characteristic with sample current trajectory during fault

inrush conditions in power transformer, etc.) can be fired, before the final protection decision is taken.

In all the above-presented cases the decisions were taken without any intentional time delay. Usually some decision redundancy is introduced, whereas not just a single time but for a number of times (for consecutive time steps resulting from sampling process) the decision is repeated. Such a measure makes the relay less dependent on signal noise and sudden signal jumps or peaks, which results in more reliable final decisions.

Intentional delays are introduced in some cases, when the time dependency is needed, mainly for time grading, which should assure protection coordination, e.g. in:

- overcurrent protection of distribution networks,
- distance protection, with appropriate tripping time set for particular protection zones.

10.2 Statistical Hypotheses Testing

The deterministic approach to decision making is fully justified and should bring erroneous protection performance provided that:

- the protection criteria are well defined to assure proper distinguishing of the phenomena under investigation,
- one has significant confidence in the accuracy of protection criteria measurement (in other words—the expected measurement errors, both transient and steady-state, are not too high),
- the decision boundaries, thresholds and/or characteristics, are unambiguously specified and enable reliable division of the decision space.

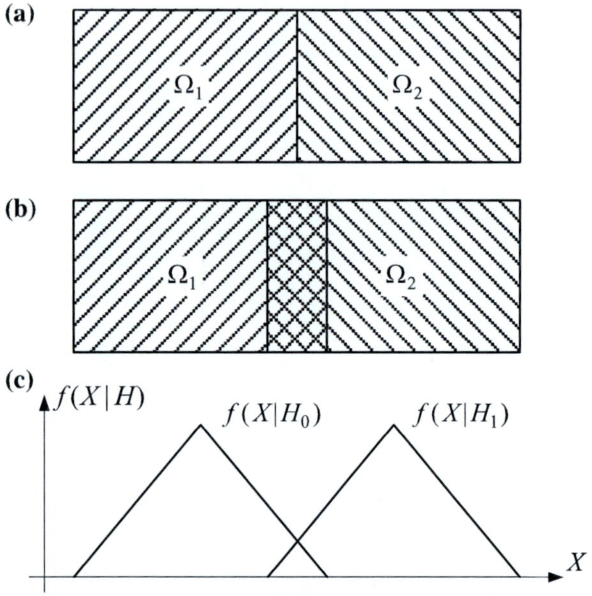

Fig. 10.6 Decision space division: **a** ideal situation, **b** real case, **c** statistic interpretation

In real life all the above conditions are rarely fulfilled in 100% situations. An illustration of the decision space division for distinguishable and non-distinguishable problems is shown in Fig. 10.6. There are cases when the decision subareas corresponding to classes of events to be perceived overlap each other. It is so for example when overloads and high impedance faults are to be distinguished. The levels of current amplitudes in both cases are similar and proper setting of the required threshold meets considerable difficulties.

Very helpful can be then the methods belonging to the Artificial Intelligence family (e.g. fuzzy logic-based reasoning systems, see Chap. 11) or statistical decision making procedures, one of which is presented below. Additional motivation for their application in power system protection is related to the fact that the fault (any abnormal situation) occurrence time and parameters are of probabilistic nature; thus some of the decision tasks may and can be solved on this basis.

The statistical decision theory interprets overlapping of decision areas corresponding to two different classes of events as overlapping of conditional probability density functions (PDFs) of the decision vector (Fig. 10.6c). Statistical approach to the decision-making problem in digital protection assumes that criterion values can be considered as random variables and that required conditional statistics are known. Probabilistic nature of the decision vector is a result of random localization of fault or such other conditions as fault resistance, fault angle, pre-fault load, etc. [7].

With statistical approach to the decision-making problem various algorithms of hypothesis testing can be applied. The fundamental decision theory with probabilistic roots is the Bayesian approach [3]. For practical technical problems the application of methods based on statistical hypothesis testing is proposed, where the hypotheses advanced could represent normal and faulty/abnormal operating conditions. When distinguishing between two hypotheses is required, one can use the Sequential Probability Ratio Test (SPRT) [12]. The approach can also be adapted for multiple hypotheses testing; then so-called Multi-hypotheses Sequential Probability Ratio Test (MSPRT) can be used [1]. The SPRT methods use conditional PDFs of the decision vector (for given classes of events) to generate the decision.

The algorithm of SPRT to be used in power system relaying was first proposed by Sakaguchi in 1980. His work [12] may be treated as first attempt to solve a decision-making problem with probabilistic method. The author applied the method to decide whether a phase is faulted or not, by using differential instantaneous signal as criterion value (d'Alembert relay). It must be said that the criterion signal was chosen arbitrary and the required PDFs were estimated very roughly (with 32 fault cases only). In testing phase their further approximation with normal distribution was used. The paper showed only possibilities of the SPRT application in digital relaying without any optimization suggestions.

SPRT belongs to decision methods in which number of samples of the decision vector necessary to issue the decision is not pre-defined. The algorithm of the SPRT can be written in the form:

$$\Theta_k = \prod_{l=1}^{k} \frac{f_l(\mathbf{X}_l|H_1)}{f_l(\mathbf{X}_l|H_0)}, \tag{10.4}$$

$$\text{If} \left\{ \begin{array}{c} \Theta_k \geq A \\ A > \Theta_k > B \\ \Theta_k \leq B \end{array} \right\} \text{then} \left\{ \begin{array}{c} \text{stop \& accept } H_1 \\ \text{continue test} \\ \text{stop \& accept } H_0 \end{array} \right\}, \tag{10.5}$$

$$A = \frac{1-\varepsilon_1}{\varepsilon_0}, \quad B = \frac{\varepsilon_1}{1-\varepsilon_0}, \tag{10.6}$$

where:

$f_l(\mathbf{X}_l\|H_i)$	PDFs of the decision vector $\boldsymbol{X}$ at the instant l after test starting for hypothesis H_i,
Θ_k	test index at the instant k,
A, B	probabilistic thresholds,
$\varepsilon_0, \varepsilon_1$	assumed values of the first and second order error probabilities.

The above procedure minimizes values of the first and the second type error probabilities $\varepsilon_0, \varepsilon_1$ [3]. The first type error comprises situations of recognizing hypothesis H_1 instead of H_0 (underfunction), while the second type error corresponds to the decision H_0 instead of H_1 (overfunction). Flow chart of the SPRT algorithm is shown in Fig. 10.7.

Application of the SPRT method is possible when the following conditions are satisfied:

Fig. 10.7 Flow chart of the Sequential Probability Ratio Test

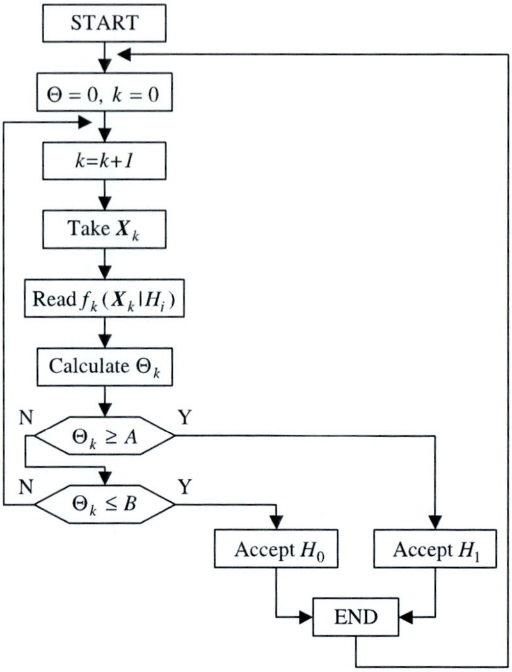

- random decision quantities are at two different time steps stochastically independent,
- conditional PDFs of the decision vector are known.

The required distributions $f_l(\mathbf{X}_l|H_i)$ may be calculated on basis of simulation cases or estimated taking into account distributions of all factors possibly influencing system behavior for given hypothesis (with the last method sought functions can be assessed rather roughly, especially when dynamic distributions are to be found). Accuracy of PDF's calculation is a nonlinear function of number of simulation runs. Basing on some statistics referring to the mean value and variance of distribution (assuming that a PDF is of normal type [3]) it can be found that about 1,000 simulation cases are needed to calculate required PDF with 5% accuracy (with probability level 0.95).

Another question is how to choose a proper decision signal for SPRT-based decision making. It seems evident that if a criterion value allows to make successful reasoning with deterministic approach, its PDFs for hypothesis H_0 and H_1 (probabilistic method) should differ one from another to such an extent that it would be possible to perform statistical inference with good results. An ideal criterion value (vector) for SPRT would be such which conditional PDFs for both hypotheses would be completely separated one from another in the steady-state as well as in the transient interval. In such case a decision would be issued always immediately (in first sample) and with 100% of probabilistic confidence. For proper (from SPRT application point of view) choice of a decision value among a set of possible ones for given decision problem, a value of distance Δ between both PDFs may be used, defined as [7]

$$\Delta_1 = \sum_{i=1}^{p} [f(x_i|H_0) - f(x_i|H_1)]^2 \tag{10.7}$$

or

$$\Delta_2 = \sup_{\substack{d_i \in D_X \\ i=1,\dots,p}} [F(x_i|H_0) - F(x_i|H_1)], \tag{10.8}$$

where:

D_X range of possible values of the decision random variable X,
f, F probability density and distribution functions for both hypotheses,
p number of intervals of discretized range D_X.

The best results of probabilistic decision making will be achieved for decision variable with highest value of distance Δ.

The application of the SPRT algorithm for fault detection and fault type identification [10] may serve as an example how the probabilistic technique can be used for decision making. The block scheme of proposed fault detection and classification algorithm is shown in Fig. 10.8. A set of SPRT units supplemented with additional logic is proposed to realize both detection and classification of fault events. Three simultaneously operating SPRT blocks (SPRT1–SPRT3) are applied in each phase to indicate which of them are involved in a fault (SPRT equipped with PDFs referring to normal operation state and fault conditions). The fourth SPRT block (SPRT4) discriminates whether a ground fault occurred (PDFs for ground and isolated faults).

The Logic and Control Unit (LCU) realizes:

- determination of the fault instant and type (on basis of the four SPRT blocks outputs),
- coordination of SPRT blocks operation (commands to block and restart the sequential testing procedures),
- on-line correction of SPRT parameters (setting the time instant in blocks remaining in testing transient state).

The proposed algorithm operates as follows. In normal operation state, when none of the testing blocks indicates a fault, all the SPRT units work independently one from another. Their operation is automatically restarted after a decision of unfaulty condition was taken. If one of the testing blocks indicates a fault in corresponding phase (e.g. in phase L1), the LCU is activated. If any of SPRT blocks remains in transient interval of testing, the LCU prevents restarting of those SPRTs which have already completed their work. This allows avoiding problems with convergence of the method in cases when particular SPRT blocks issue their decisions with some delay (not synchronously). For those which are in operation a

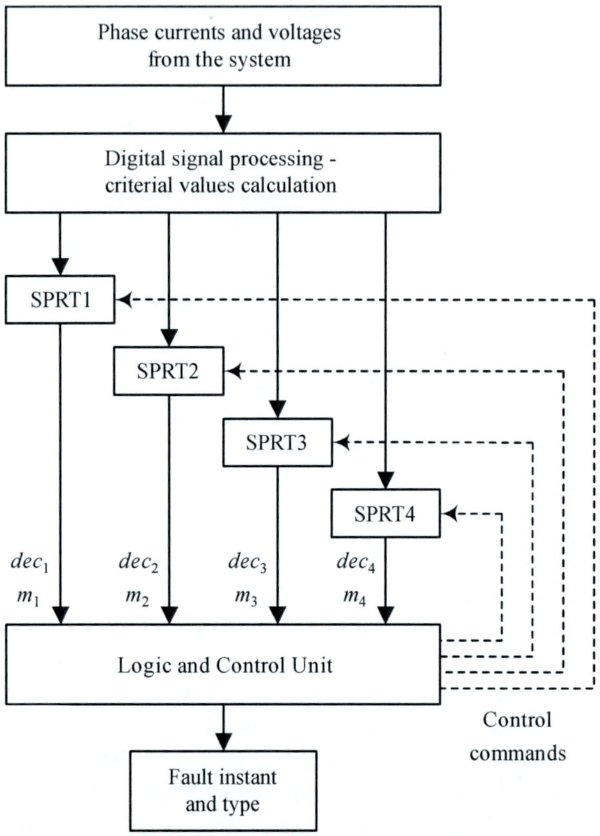

Fig. 10.8 Block scheme of proposed fault detection and classification algorithm

new corrected time instant l is set (increased by number of samples m that passed from activating of the SPRT block which first indicated a fault). The formula (10.4) changes to

$$\Theta_k = \prod_{l=1}^{k} \frac{f_{l+m}(\mathbf{X}_l | H_1)}{f_{l+m}(\mathbf{X}_l | H_0)}. \tag{10.9}$$

This adaptation tool speeds up the fault classification procedure. When all of particular SPRT decisions are made, the LCU discriminates the fault type and restarts the entire procedure.

The conditional PDFs of the decision variable $X = I_0$ (zero-sequence current amplitude) for successive samples k after fault inception are shown in Fig. 10.9. It is seen that the PDFs change with time and their divergence for both hypotheses (ground versus isolated fault) increases sample by sample which means that more and more information is delivered to the decision algorithm. For time instant $k = 6$ first signs of CT's saturation can be seen (due to inaccuracy of current transformation some zero-sequence current is generated).

The simulative tests confirmed that the SPRT algorithm demonstrated excellent ability of fault detection. All cases were detected properly, mostly in the first or

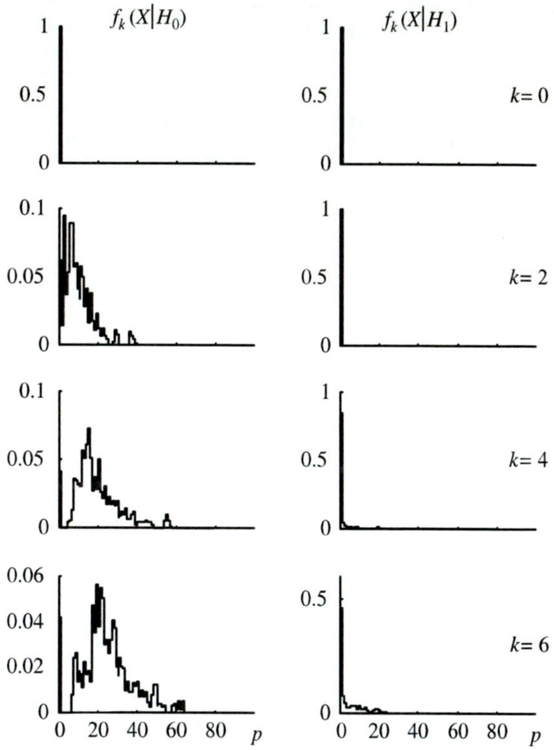

Fig. 10.9 PDFs for hypothesis H_0—ground fault, H_1—isolated fault

second sample after fault inception (in rare cases only more time was needed). Fault classification was made with exactitude of 97.2% and average time of 3 ms, which should be considered as a good result.

In [5] promising results of the SPRT application for determination of fault direction and location detection are presented. This proves that with use of probabilistic approach many protection problems may be solved and essential improvement of selectivity and self-time of protection devices may be reached.

One has to admit that the application of SPRT-based decision methods requires of course intensive simulative investigations and calculations to be done (conditional PDFs for given hypothesis must be known). Similar introductory work (off-line) is also necessary if other novel approaches (with use of Kalman filtering or neural networks) are to be applied. On-line computational complexity of SPRT algorithms is relatively small; therefore they may be implemented in real time protection systems.

10.3 Decision Making with Multiple Criteria

Depending on the plant to be protected the analysis of its state that is performed by protective relays can be done on the strength of a single criterion signal or with support of information from additional criteria. The first option is widely used for

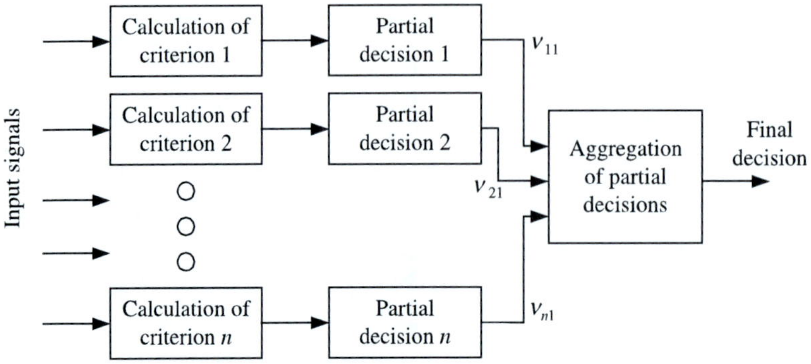

Fig. 10.10 Decision making with multiple criteria

plants where the number of expected phenomena and operating conditions is not very big; thus with single criterion discrimination of the plant state is quite possible. It is so for instance for transmission and distribution lines that are usually protected with application of differential, distance or overcurrent functions (either as main or backup protection), i.e. by analyzing the course and values of single criterion signal (current amplitude, impedance locus). For the plants with complicated structure or installed at a point where the operating conditions might be tough and/or difficult to be analyzed, the second option with multiple criteria is usually applied. Since additional information is delivered and processed by the relay, the multi-criteria decision making brings about improvement of protection reliability in terms of:

- increased decision confidence,
- elimination of wrong decisions,
- higher speed of decision making.

A very basic scheme of multi-criteria decision making is shown in Fig. 10.10. The final decision is worked out in several steps including:

- calculation of particular criteria values,
- generation of partial decisions for all criteria (e.g. by comparing their values with some thresholds or characteristics, as described above),
- aggregation of partial decisions.

For particular criteria any of the decision making approaches i.e. deterministic or probabilistic one can be applied.

Aggregation of partial decisions can be done with use of:

- simple Boolean logic operators (AND, OR),
- the strategy some-out-of-all (the decision is taken when not less than the minimum number of partial decisions support one of the possible protection options),
- weighting factor method (confidence coefficients for particular criteria are used to express their relative strength or quality and to calculate the weighted support for final decision),

- decision rules of the IF … THEN form (certain knowledge is required to set up the rules),
- decision trees (equivalent to structurised logic with a combination of AND/OR operands),
- more complex decision schemes (Fuzzy Inference Systems or Expert Systems, Chaps. 11 and 14),
- Artificial Neural Networks (Chaps. 12),
- other techniques.

Deterministic aggregation with weighting factors can be expressed as follows:

$$\delta_1 = w_1 v_{11} + w_2 v_{21} + \cdots + w_n v_{n1}, \tag{10.10}$$

$$\delta_2 = 1 - \delta_1, \tag{10.11}$$

$$\delta_1 - \delta_2 > \Delta \rightarrow Dec_1 \text{ confirmed}, \tag{10.12}$$

where:

v_{i1} partial support values for the decision Dec_1 from the ith criterion,
w_i weighting factors for particular criteria,
δ_1 total support for the decision Dec_1,
δ_2 support for the opposite decision Dec_2,
Δ discrimination threshold.

The partial decision support values v_{i1} may be crisp (taking values 0 or 1, where 0 stands for no support and 1—for full support) or fuzzy (here any value between 0 and 1 is permissible, the larger the value, the higher the support).

The representative example of protection function, where multiple criteria decision making is realized, is the differential protection of power transformer. The differential protection reaches its basic conclusion by locating the consecutive points of Diff-Bias trajectory with respect to appropriately set characteristic (example—see Fig. 10.5). Nevertheless, additional information is needed to distinguish the cases of internal faults when immediate tripping is required from those cases when significant differential current may appear due to some other non-fault reasons, which may lead to undesired tripping decision and relay malfunction. The weak points of the standard transformer differential protection are as follows [16]:

- differential currents may be caused by a variation of the transformation ratio due to the on load tap changing;
- differential currents may be caused by the magnetizing currents, inrush ones (caused by the sudden increase of the transformer terminal voltage) and stationary ones (caused by the overexcitation of the transformer core);
- current transformer errors due to core saturation in transient states during external short circuits may cause spurious differential currents;

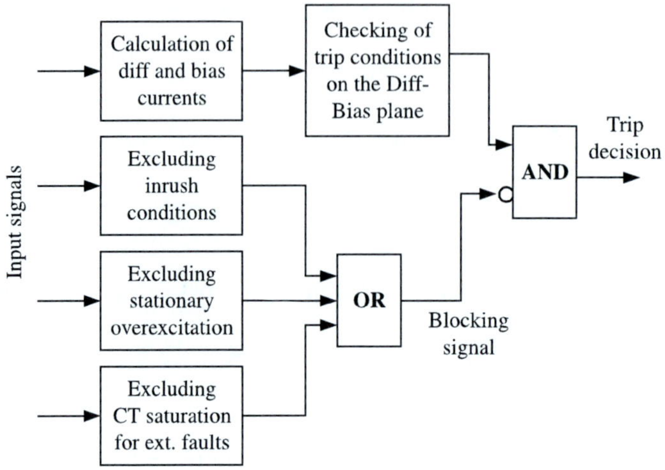

Fig. 10.11 Asserting the tripping decision in transformer differential protection

- current transformer errors due to saturation in transients and in steady sates during internal short circuits at the transformer terminals may greatly distort waveshapes of the differential currents.

Whatever the reason of an inadvertent tripping, it brings the following consequences. First, the power system stability margin may be drastically reduced, what may lead to the developing disturbance. Second, the operating personnel becomes confused, suspecting an internal fault in the transformer, what may cause hesitation whether to reclose the transformer or not.

To prevent malfunction of the differential relays the standard ways of percentage and harmonic restraint are adopted [14]. In most of the relays, if the second harmonic ratio measured in the differential current exceeds 15–20% of the fundamental harmonic level, the relay is blocked. It is sufficient in preventing maloperation during external short circuits, and in the majority of inrush cases. The situations of stationary overexcitation can be excluded through analysis of the fifth harmonic content in the differential current. If the ratio of fifth harmonic to the first harmonic is higher than 30%, the relay should remain blocked. Checking, if there was no CT saturation during external fault, can also be done by applying the so-called "add-on stabilization" [6], where the course of trajectory on the Diff-Bias plane is observed to discriminate the event.

Whatever specific criteria and algorithms are adopted, the blocking signal of the relay results from aggregation of sub-decisions for protection stabilization against above-mentioned inrush, external faults with CT saturation and overexcitation conditions. Simple logic with OR, AND blocks, as presented in Fig. 10.11, is sufficient to assert the final protection decision. Transformer tripping is not permitted, if the blocking signal coming from any of the three stabilization units is issued.

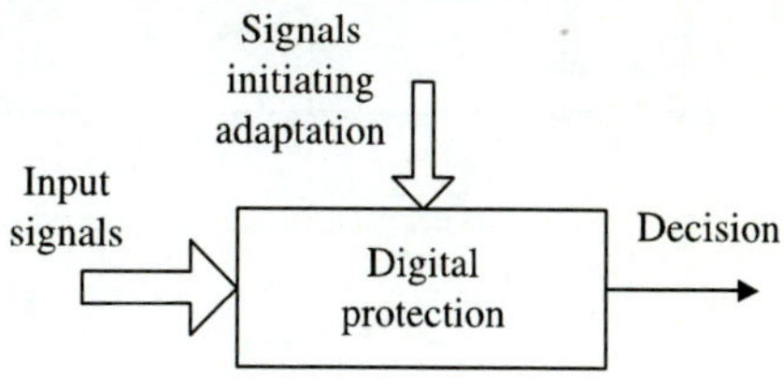

Fig. 10.12 Protective relay with adaptation initiated from outside

10.4 Adaptive Decision Schemes

Adaptation in a power system protection unit can be understood as a process of automatic adjusting to changing power system conditions, so that some functional algorithms and internal logic are transformed, in order to achieve possibly highest level of protection in all predictable abnormal operation cases [4, 8, 13, 15]. Essential condition for realization of an effective adaptation is a supply of the relay (as a whole or to its particular units) with the signals that would initiate the procedure at appropriate time instant (Fig. 10.12). Well designed adaptation should always result in improved sensitivity and selectivity, thanks to departure form conservative assumptions prescribing the relay to operate by constant (and not always optimal) settings for all possible power system disturbances.

Taking the origin of initiating signals into consideration, the adaptation procedures can be classified as external or internal ones. In the first case the initiating signal is delivered to the protection from outside, i.e. from another relay or from supervisory control. In case of internal version both the cause and the effects of adaptation are located inside of the same protection relay.

As regards the area and range of adapted functions one can distinguish:

- local adaptation, that concerns changes within the single protective unit,
- system (global) adaptation, resulting in mutual coordination of settings, activated protection functions, but also appropriate matching of control functions, automation and system stability supervision.

Global adaptation is usually executed in an integrated system of wide-area protection and control [2, 11]. In this chapter main attention is paid to some aspects and examples of local adaptation that is related to functionality of a stand-alone protective relay.

From the functional viewpoint the adaptation may be related to some changes in one or more of the following protection units (Fig. 10.13):

- analog filtration (switching to another set of filters),
- analog-to-digital conversion (change of sampling frequency),
- digital signal processing (selection of the processing algorithm, modification of the filters' frequency characteristics),
- measurement of protection criteria (change of the algorithm parameters and/or type),

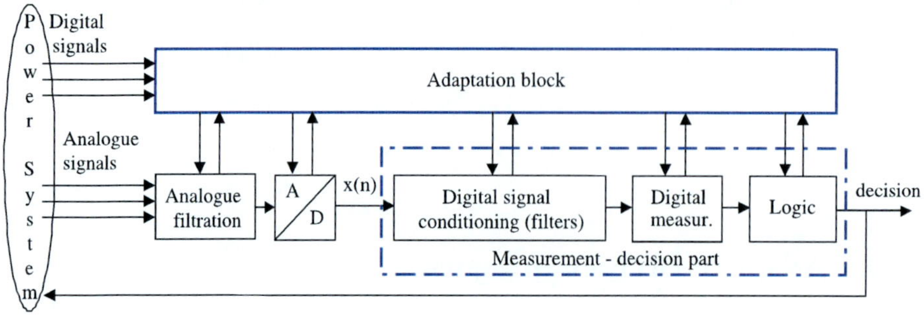

Fig. 10.13 Block scheme of an adaptive protection

- decision making (change of thresholds and decision characteristics, selection of new or additional criteria and logic signals).

Changes initiated by the adaptation block may include not only certain modification of selected algorithms but also their exchange for the ones better suited for given power system operating conditions.

The typical and widely accepted examples of adaptive schemes include:

- time-dependent overcurrent relay (operation dependent on both time and current amplitude, according to a standardized function),
- percentage differential scheme (decision threshold for differential current being a function of the bias current),
- adaptive under-frequency load shedding (load shedding steps depending on actual load measured at particular feeders),
- adaptive distance overreaching/underreaching schemes (first zone extended or shortened according to the information gained from the opposite line terminals), etc.

Below an example of adaptation in digital part of the protection relay is presented, that is related to the decision making performed in the differential protection of synchronous machines.

Differential relay adaptation

Even though the differential relays are considered quite reliable and robust, there are situations when they may malfunction. Contradictory requirements of high sensitivity and selectivity cannot reconcile in cases of faults with small through currents which may evoke over-tripping if the CTs saturate due to content of DC component with high value of decaying time constant. Such conditions may arise e.g. during external faults close to generator-transformer units when this time constant is as high as 200–300 ms. Contrary to high amplitude faults when the CTs saturate within first full-cycle of the current signal after fault inception, the saturation effects for DC-induced cases appear only few cycles later and have somewhat different nature. In [9] the idea of adaptive differential relay is presented with the aim of avoiding false tripping for such external faults.

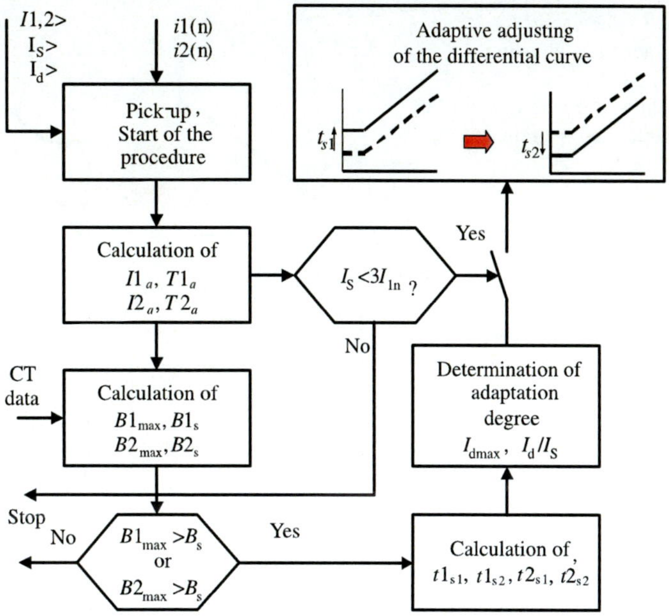

Fig. 10.14 Block scheme of the adaptive differential protection proposed

The developed procedure of adaptive adjusting of the relay differential characteristic consists of the following steps (Fig. 10.14):

- calculation of the initial value (amplitude) I_a and time constant T_a of the decaying DC component as well as the amplitude of the current fundamental frequency component I_1 for the time instant just after fault inception,
- determination (prediction) of the flux density value B_s leading to CT saturation, maximal expected flux density level $B_{\max}$ in the CT core and comparison of the values,
- calculation of the expected CT saturation period, i.e. time instants t_{s1} and t_{s2} of saturation beginning and fizzling out,
- determination of the necessary level of adaptation for given fault case (shifting up of the differential curve or slope changing of the stabilizing section),
- measurement of the through current amplitude values (differential curve adjusting takes place for fault currents not higher than three times rated machine current),
- execution of the on-line adaptation of the differential curve for the time period $[t_{s1}, t_{s2}]$.

It is worth to notice that the calculation of above-mentioned values is performed for the signals from CTs at both sides of the protected plant. The adaptation is carried into effect slightly before the predicted time instant of CT saturation (earlier of both calculated values is accepted for this purpose).

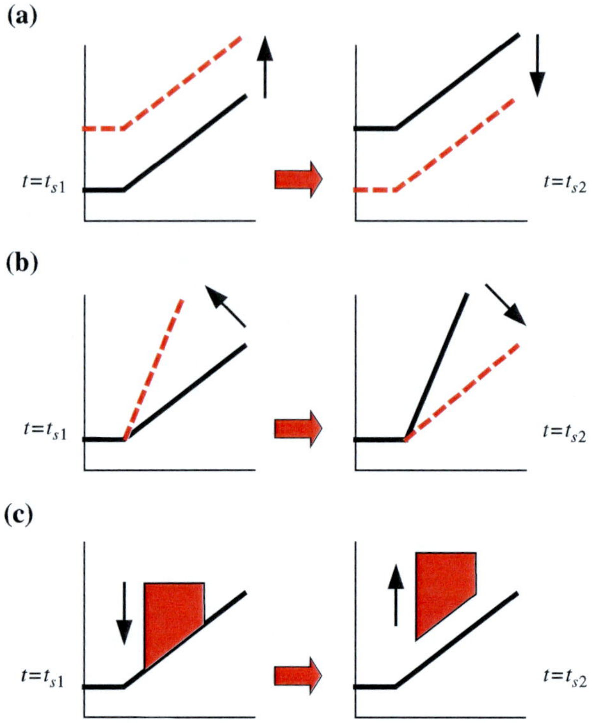

Fig. 10.15 Conceivable versions of adaptive adjusting of the differential curve: **a** vertical shifting **b** slope change **c** additional restrain area

The design of the adaptive differential protection included also research on appropriate versions and level of adaptation. Adaptive adjusting of the differential characteristic may be realized according to one of the following scenarios (see Fig. 10.15):

(a) shifting up the stabilizing curve before the saturation began, then shifting it down to the standard position after CT saturation disappeared,
(b) changing the slope m_1 of the differential curve (stabilizing section) to a higher value, then return to standard settings,
(c) forming an additional transitory restraining area above the stabilizing curve e.g. polygonal shape (covering the area of foreseeable location of current trajectory), then removing the area.

Versions (a) and (b) were examined in more detail, for both of them appropriate formulae for the required level of adaptation have also been determined [9].

In Fig. 10.16 an example of the adaptive protection operation for a real external fault close to generator terminals is presented. The fault was accompanied with DC-induced CT saturation. In Fig. 10.16a the current signals from both sides of the protected generator (phase L1) are shown. The effects of adaptive adjusting of the differential curve can be seen in Fig. 10.16b, c. Appropriate changes (version *a*—shifting up, version *b*—setting new value of the stabilizing slope) were introduced at time instant $t_{s1} = 62.3$ ms, i.e. before the non-adaptive relay picks up (curve Out (no ad.) in Fig. 10.16b). The adaptation by shifting up (version *a*) was not effective

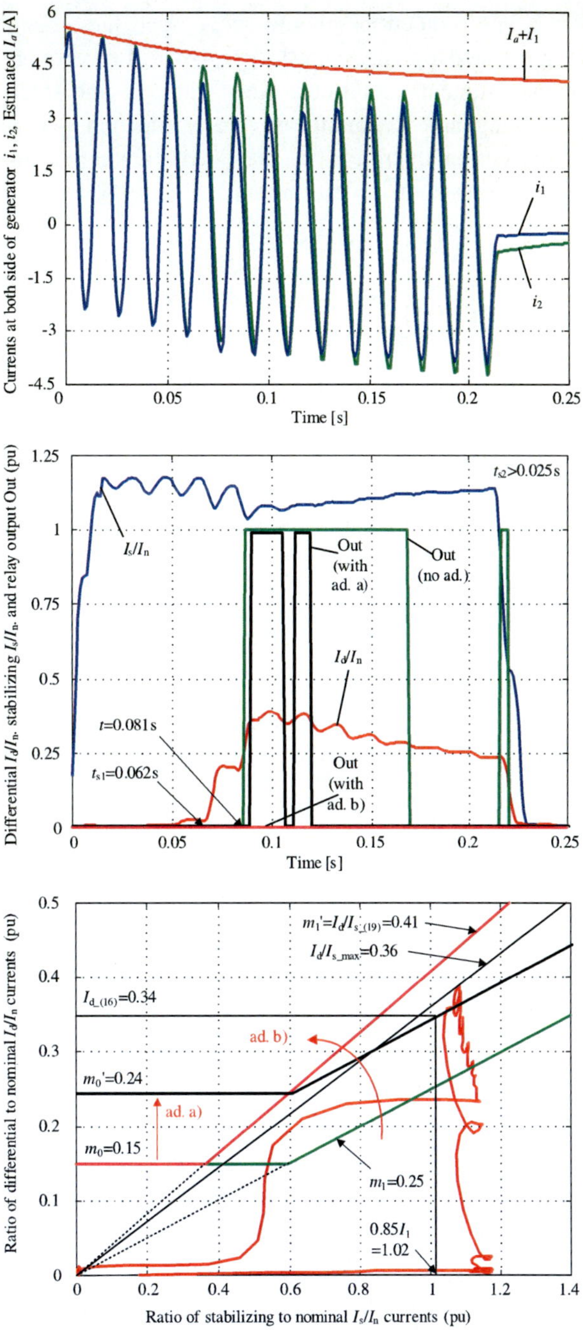

Fig. 10.16 Operation of the adaptive differential protection for a real-world external fault case: **a** current signals and estimated DC component **b** differential and stabilizing currents, protection output signals **c** adaptive adjustment of the differential curve in the $I_d = f(I_s)$ plane

enough in the considered case, the differential curve shifting was too small and the relay picked up. The newly set stabilizing slope (version *b*), on the contrary, has fulfilled the expectations, i.e. the stabilization conditions have been significantly improved and unwanted tripping for the external fault was avoided.

The protection schemes developed have been thoroughly tested with EMTP-generated signals as well as real-world recordings, having proved better operation than standard solutions. New adaptive and intelligent approaches to signal processing and reasoning are pretty much hoped to contribute to improvement of protection operation in many cases where traditional algorithms show certain weaknesses or limitations.

References

1. Baum CW, Veeravalli VV (1994) A sequential procedure for multi-hypothesis testing. IEEE Trans Inf Theory 40:1994–2007
2. Cholley P et al (2000) System protection schemes in power networks. CIGRE Study Committee Task Force SCTF 38.02.19. Final draft v5.0
3. Dowdy S, Wearden S (1983) Statistics for research. Wiley, New York
4. Jampala AK, Venkata SS, Damborg MJ (1989) Adaptive transmission protection: concepts and computational issues. IEEE Trans Power Deliv 4:177–185
5. Kasztenny B, Rebizant W, Szafran J (1994) Probabilistic approach to estimation of criterion values and decision making process of power system protection. Adv Model Anal 31:31–40
6. Numerical differential protection relay for transformers, generators, motors and mini busbars (2006) 7UT613/63x V.4.06 Instruction Manual, Order No. C53000-G1176-C160-2, SIEMENS AG
7. Rebizant W (1995) Probabilistic characteristics of signals and criterion values of power system protections and their application. PhD dissertation, Wroclaw University of Technology, Wroclaw, Poland (in Polish)
8. Rebizant W (2008) Adaptive and intelligent approaches in contemporary protection systems. In: Proceedings of the international summer CRIS workshop on distributed and renewable power generation, O-v-G Universitaet Magdeburg, Germany, pp 142–147
9. Rebizant W, Feser K, Hayder T, Schiel L (2004) Differential relay with adaptation during saturation period of current transformers. In: Proceedings of the 14th international conference on power system protection, Bled, Slovenia, pp 124–129
10. Rebizant W, Szafran J (1999) Power system fault detection and classification using probabilistic approach. Eur Trans Electr Power 9:183–191
11. Rehtanz C (2001) Online stability assessment and wide area protection based on phasor measurements. In: Proceedings of the 5th symposium on bulk power system dynamics and control, Onoomichi, Japan
12. Sakaguchi T (1980) A statistical decision theoretic approach to digital relaying. IEEE Trans Power Apparatus Syst 5:1918–1926
13. Thorp JS, Horowitz SH, Phadke AG (1988) The application of an adaptive technology to power system protection and control. In: Proceedings of the CIGRE session, Rep. 34–03
14. Ungrad H, Winkler W, Wiszniewski A (1995) Protection techniques in electrical energy systems. Marcel Dekker, New York
15. Winkler W, Wiszniewski A (1995) Adaptive protection—potential and limitation. In: Proceedings of the CIGRE SC34 colloqium, Stockholm, Rep. 34–206
16. Wiszniewski A, Rebizant W, Schiel L (2006) Modified strategy for protection of power transformers. In: Proceedings of the international symposium on modern electric power systems, Wroclaw, Poland, pp 337–341

Chapter 11
Elements of Fuzzy Logic in Protective Relays

As mentioned, e.g. in the preceding chapter on decision making, the protective relay equipped with standard, deterministic signal processing algorithms (measurement with fixed data window, discrimination with fixed thresholds or characteristics) may not perform properly for all conceivable power system operating conditions. The problems may arise mainly due to:

- transient inaccuracy of measurement with commonly used deterministic algorithms (here especially the measurement overshoot is dangerous for relay selectivity),
- difficulties in setting of the relay (especially when the regions corresponding to the classes to be distinguished overlap each other).

In this chapter the main features of fuzzy logic (FL) based systems are presented that are hoped to improve protection operation for unclear cases, situations when some data are missing or corrupted or when the measurement unit may deliver criteria signal estimates obtained with low level of confidence. Below the fundamentals of the fuzzy sets and logic are outlined, which is followed by the application hints and selected examples of this technique for power system protection purposes.

11.1 Fuzzy Sets and Fuzzy Numbers

Most of the concepts describing the problems of real life are not very precise. It is difficult to assess many things quantitatively with high degree of precision, especially when the situation is dynamic (well known Heisenberg's uncertainty principle related to measurement) or influenced by a number of factors, some of which are weakly defined. To describe certain concepts, e.g. people's height or distance to destination, the linguistic statements like "high", "short", "far away", etc. are usually used. The reasons for this are, among others, that

W. Rebizant et al., *Digital Signal Processing in Power System Protection and Control,* Signals and Communication Technology,
DOI: 10.1007/978-0-85729-802-7_11,

- high precision is not always needed and such fuzzy statements are good enough to represent the main meaning of the statement,
- accurate measurement of certain variables is difficult, costly and may take a lot of time,
- precise measurements bring not much when the model of the plant to be controlled/protected is very complex,
- the decision is sometimes to be made quickly, even if the measurement of criteria signals is not very precise.

Therefore, especially when the final result/decision is of classification type (like an assessment of the state of protected plant by the protective relays), precise statements are not better than the ones that are not that accurate (fuzzy). Moreover, a very precise model may in some cases cause difficulties when some information is missing or becomes corrupted or contaminated by noise. The well known Zadeh's principle of incompatibility [21]: "As the complexity of a system increases, our ability to make precise and yet significant statements about its behavior diminishes until a threshold is reached beyond which precision and significance (or relevance) become almost mutually exclusive characteristics", leads to the conclusion that the usage of fuzzy sets/FL may become a mechanism for abstraction of unnecessary or too complex details.

Before the application of fuzzy sets and logic for protection and control are outlined, let us consider/recall the basic definitions and rules of the related theory. Coming out from linguistic variables that correspond to the imprecise linguistic statements, let us define the fuzzy set as follows. A fuzzy set A in the universe of discourse X is expressed as a set of ordered pairs [5, 22]

$$A = \{(x, \mu_A(x)) | x \in X\}, \quad X :\rightarrow [0, 1] \tag{11.1}$$

where $\mu_A(x)$ is a membership function (MF), $\mu_A(x) \in [0, 1]$.

A fuzzy set is totally characterized by its MF. It brings an information about the degree (between 0 and 1, inclusive) to which an element belongs to the set A.

In many applications fuzzy sets provide an interface between a numerical scale and a symbolic scale (composed of linguistic terms). The elements of X may be of various natures, while the X itself may be discrete (with limited or countable number of elements) or continuous. For example, the MF for the fuzzy set defined linguistically as "set of people being around 50 years old" may be expressed by

$$\mu_A(x) = \frac{1}{1 + \left(\frac{x-50}{10}\right)^4}. \tag{11.2}$$

which is illustrated graphically in Fig. 11.1.

The notion of fuzziness may be characterized as the object similarity to imprecisely defined properties. Looking at Fig. 11.1 one can conclude that a person belongs to the set "around 50 years old" with the degree of membership higher than 0.5 only when his/her age is in the range (40, 60). According to (11.2) a person being 65 years old belongs to A only with the degree of ca. 0.16, which is

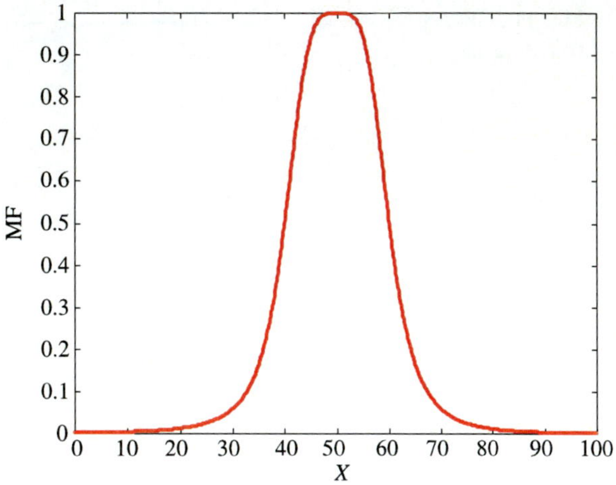

Fig. 11.1 Membership function of the set of people "around 50 years old"

rather low. In other words, we would say that he/she "rather does not belong to the fuzzy set A".

One may enumerate a lot of examples of engineering terms that are quite often used as fuzzy instead of providing precise/measured values. They are, e.g.,

- bandwidth: narrowband, broadband,
- correlation: low, medium, high, perfect,
- errors: large, medium, small, a lot of, not so great, very large, very small, almost zero,
- frequency: low, high, ultra-high,
- resolution: low, high,
- sampling: low-rate, medium-rate, high-rate,
- stability: stable (lightly damped, highly damped, over-damped, critically damped), unstable.

With regard to fuzzy sets the following set-theoretical operations can be defined [5]:

- subset (containment of A in B)

$$A \subseteq B \Leftrightarrow \mu_A(x) \leq \mu_B(x), \quad x \in X. \tag{11.3}$$

- complement (not A, defining how much do elements not belong to the set)

$$\overline{A} = X - A \Leftrightarrow \mu_{\overline{A}}(x) = 1 - \mu_A(x). \tag{11.4}$$

- union (sum of sets, determining how much of the element is in either set)
  - logical sum

$$C = A \cup B \Leftrightarrow \mu_C(x) = \max[\mu_A(x), \mu_B(x)] = \mu_A(x) \vee \mu_B(x). \tag{11.5}$$

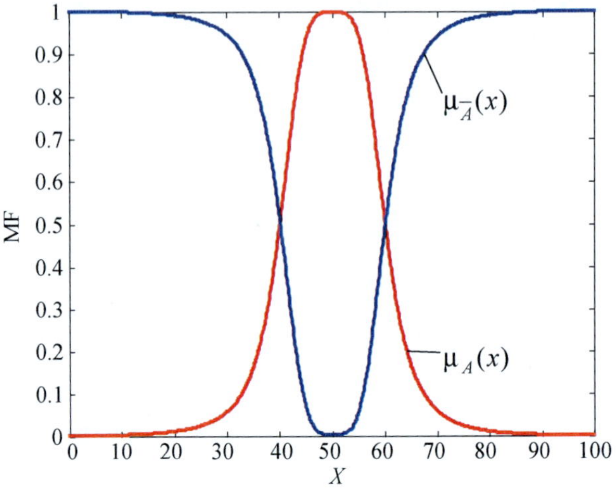

Fig. 11.2 Fuzzy set and its complement

– or algebraic sum

$$C = A \cup B \Leftrightarrow \mu_C(x) = \mu_A(x) + \mu_B(x) - \mu_A(x)\mu_B(x). \tag{11.6}$$

- intersection (product of sets, describing how much of the element is in both sets)
  – logical product

$$C = A \cap B \Leftrightarrow \mu_C(x) = \min[\mu_A(x), \mu_B(x)] = \mu_A(x) \wedge \mu_B(x). \tag{11.7}$$

  – or algebraic product

$$C = A \cap B \Leftrightarrow \mu_C(x) = \mu_A(x)\mu_B(x). \tag{11.8}$$

An illustration of the operations of complement (11.4), union (11.5) and intersection (11.7) for the exemplary fuzzy sets is given in Figs. 11.2 and 11.3. One should understand that the results of union and intersection will be different when another definition of the operations would be applied.

The union and intersection operations for fuzzy sets (similarly as for crisp sets) are

- cumulative

$$\begin{aligned} A \cup B &= B \cup A \\ A \cap B &= B \cap A. \end{aligned} \tag{11.9}$$

- associative

$$\begin{aligned} A \cup B \cup C &= (A \cup B) \cup C = A \cup (B \cup C) \\ A \cap B \cap C &= (A \cap B) \cap C = A \cap (B \cap C). \end{aligned} \tag{11.10}$$

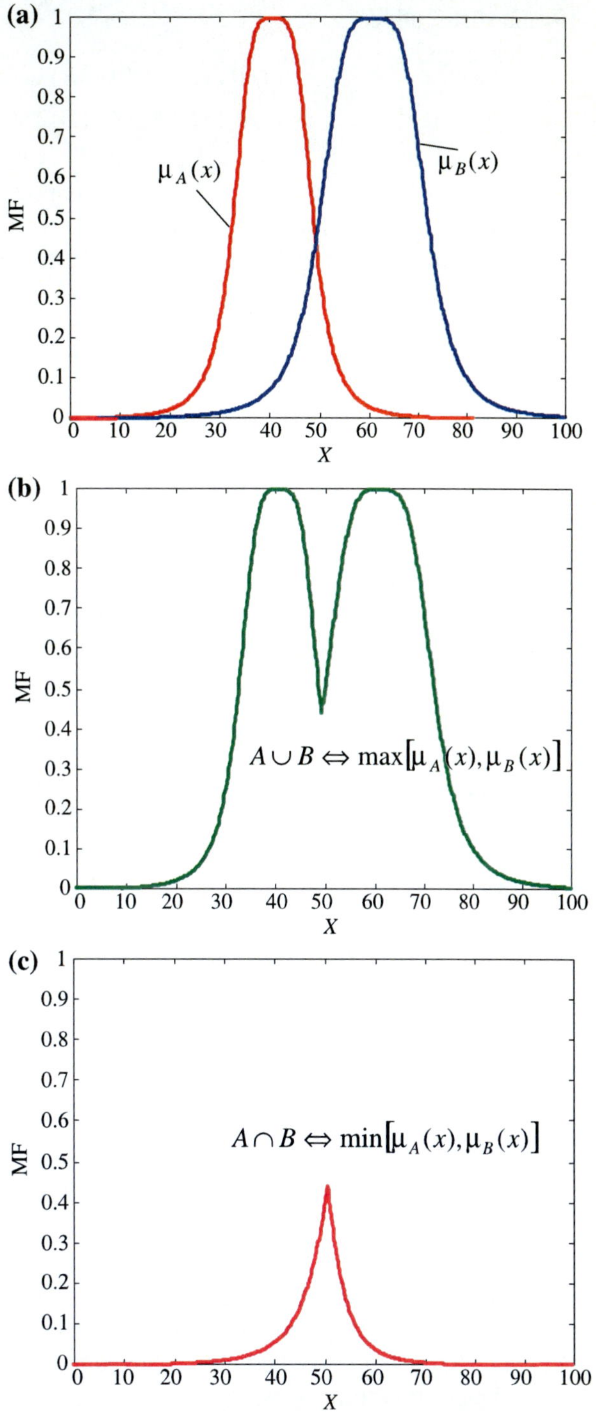

Fig. 11.3 **a** Fuzzy sets and their: **b** union **c** intersection

- and distributive

$$A \cap (B \cup C) = (A \cap B) \cup (A \cap C)$$
$$A \cup (B \cap C) = (A \cup B) \cap (A \cup C). \qquad (11.11)$$

The union and intersection are often expressed in form of the so called *t-conorms* (or *s-norms*) and *t-norms*, respectively. Their definitions may take various forms, it is not a place here to include more; interested readers should refer e.g., to [2, 12].

Unlike for crisp sets the laws of contradiction and excluded middle do not hold for fuzzy sets. One can prove that in general:

$$A \cap \overline{A} \neq \emptyset. \qquad (11.12)$$

and

$$A \cup \overline{A} \neq X. \qquad (11.13)$$

An illustration of the above for an exemplary fuzzy set and its complement is shown in Fig. 11.4. It is seen that neither the intersection of the complementary sets gives an empty set nor the union of them results in a complete set (universe of discourse *X*). This should appeal to the reader's conviction that the rules applicable for traditional (crisp) sets can be transferred to their fuzzy counterparts only with special care and by good understanding of the results.

For further considerations it is also important to recall the definition of a fuzzy number. In protection and control the information on which to base a decision is usually certain measured criterion value. In fuzzy systems this value becomes fuzzified so that instead of crisp real number—a fuzzy representation useful for further fuzzy processing is obtained. The fuzzy number is by definition a fuzzy set A defined over the universe of discourse $\boldsymbol{R}$ being the set of all real numbers:

$$A = \{(x, \mu_A(x)) | x \in \boldsymbol{R}\}, \quad \boldsymbol{R} :\rightarrow [0, 1]. \qquad (11.14)$$

The fuzzy number A in the universe $\boldsymbol{R}$ should be characterized by the MF $\mu_A : \boldsymbol{R} \rightarrow [0, 1]$, that is normalized, convex and at least segmentally continuous. The convexity condition of MF is fulfilled if

$$\underset{x_1, x_2 \in \Re \wedge \lambda \in [0,1]}{\forall} \mu_A(\lambda x_1 + (1 - \lambda)x_2) \geq \min\{\mu_A(x_1), \mu_A(x_2)\}. \qquad (11.15)$$

An example of how the fuzzy criterion signal may look like is shown in Fig. 11.5 [20]. Since the non-fuzzy signal (e.g. magnitude of line current during fault) is dynamic, the corresponding fuzzy signal changes as well. The right-hand side part of Fig. 11.5 shows the fuzzified current magnitude at five chosen time steps, starting from the pre-fault time instant (point A) up to the new steady state of measurement after fault transient disappears (point E). The measured magnitude values have been assigned the fuzzy MFs. The location of center of μ depends on the actual current

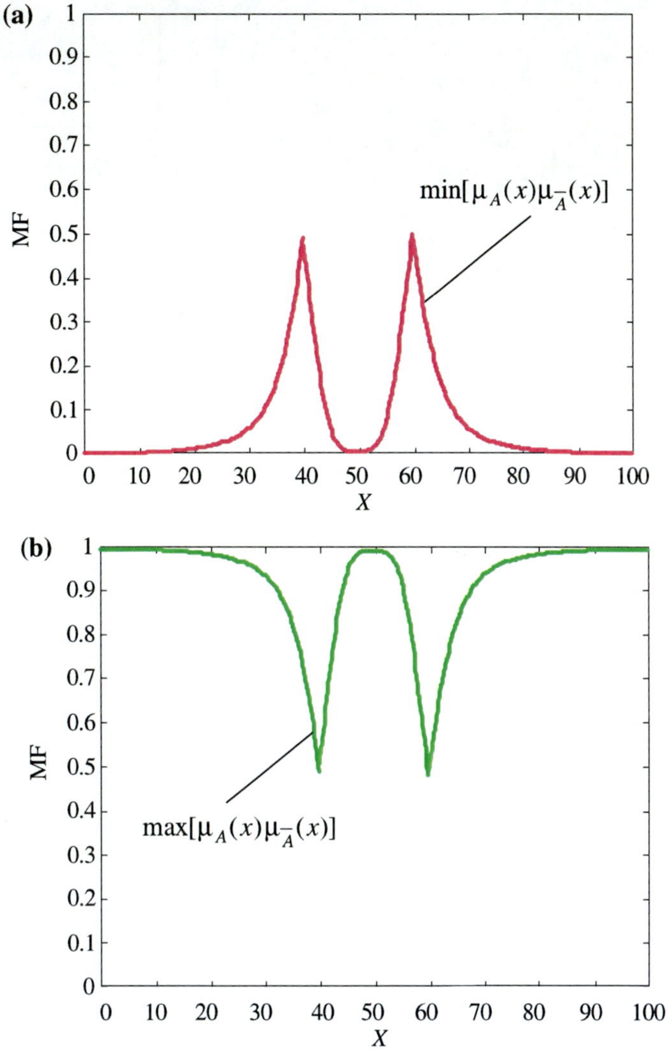

Fig. 11.4 Results of **a** union **b** intersection of the fuzzy set A from Fig. 11.2 and its complement $\bar{A}$

magnitude value, while the degree of fuzziness reflects the degree of confidence to the dynamic measurement results at given time instant. The width of MF is directly related to the rate of change of the criterion signal, which is also a function of applied digital filters and specific measurement algorithms, and in a way encodes the degree of conformity between accurate (but not exactly known) and measured criterion signal values. The amount of information IN in the signal is inversely proportional to the degree of fuzziness, expressed by the area P under the MF μ, and is the highest at the steady state of measurement (points A, E), but the lowest during transient, when the signal magnitude changes dynamically (points B, C and D). Utilization of the fuzzy criteria signals instead of their real counterparts allows for mathematical

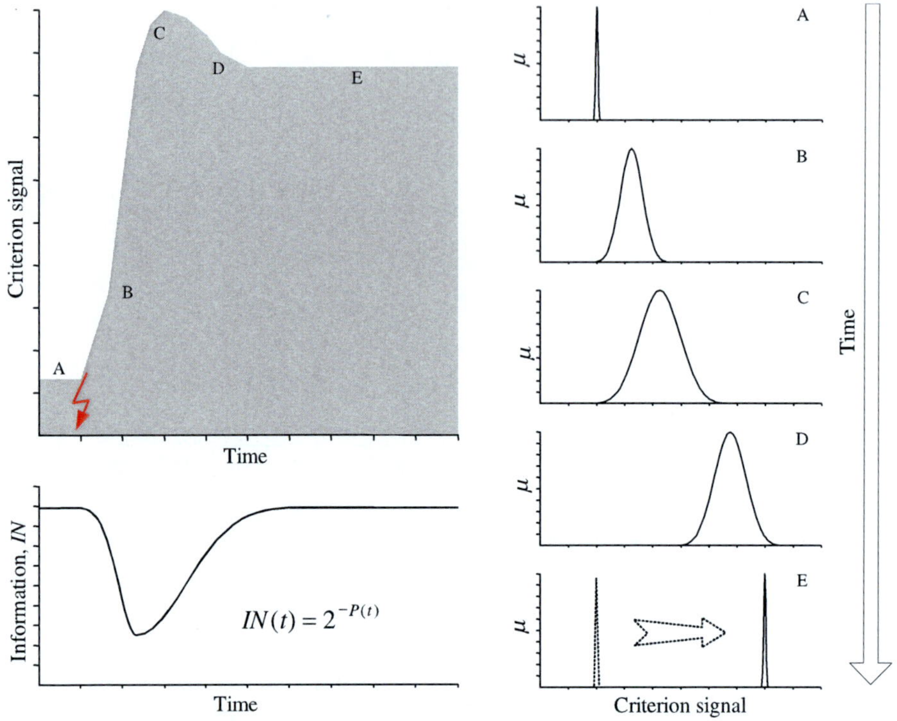

Fig. 11.5 Classical and fuzzy criteria signals—an example

depiction of the measurement uncertainty, especially during transients, that is also just after the fault/disturbance inception.

In Fig. 11.5 a bell-shaped MF was used to represent measurement dynamics. This is not the only option, in engineering applications the following versions may also be applied (the function names used in equations below correspond to the names from MATLAB/Simulink software), to name only a few of them [10, 12]:

- triangle MF

$$\mathrm{trimf}(x; a, b, c) = \max\left(\min\left(\frac{x-a}{b-a}, \frac{c-x}{c-b}\right), 0\right). \tag{11.16}$$

- trapezoidal MF

$$\mathrm{trapmf}(x; a, b, c, d) = \max\left(\min\left(\frac{x-a}{b-a}, 1, \frac{d-x}{d-c}\right), 0\right). \tag{11.17}$$

- Gaussian MF

$$\mathrm{gaussmf}(x; c, \sigma) = \mathrm{e}^{-\frac{1}{2}\left(\frac{x-c}{\sigma}\right)^2}. \tag{11.18}$$

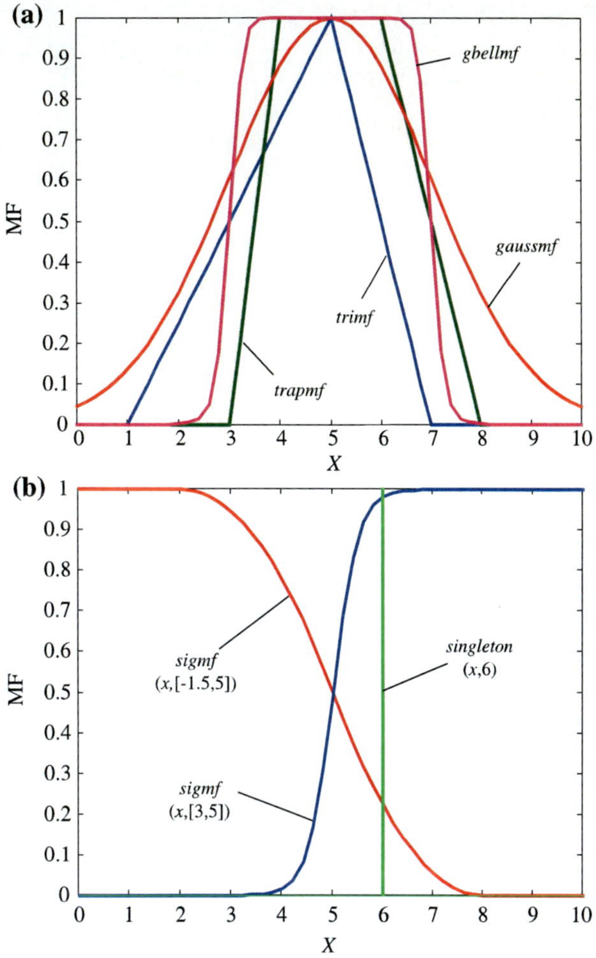

Fig. 11.6 Fuzzy membership function shapes: **a** "band-pass" ones **b** other functions

- generalized bell MF

$$\text{gbellmf}(x; b, c) = \frac{1}{1 + \left|\frac{x-c}{b}\right|^{2b}}. \tag{11.19}$$

- sigmoidal MF

$$\text{sigmf}(x; a, c) = \frac{1}{1 + \mathrm{e}^{-a(x-c)}}. \tag{11.20}$$

- singleton MF

$$\text{singleton}(x; a) = \begin{cases} 1, & x = a \\ 0, & x \neq a\ . \end{cases} \tag{11.21}$$

The first four of the MFs defined above are "band-pass" functions (see Fig. 11.6a), the sigmoidal MF is a single-sided function (either "low-pass" or

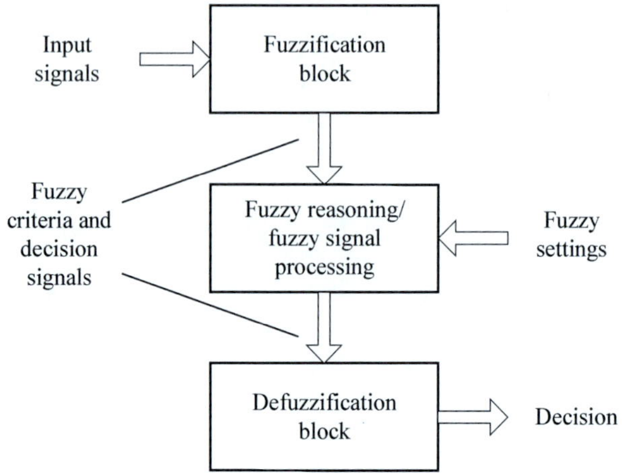

Fig. 11.7 General architecture of a fuzzy protection relay or control unit

"high-pass" depending on the coefficients), while the singleton MF is a point-type function being non-zero at one point only.

Choosing the type of the MFs for given application is a matter of the designer experience and good understanding of the phenomena under consideration. The parameters of MFs are usually set on basis of simulation and/or field-recorded signals and can be further optimized, e.g. with an evolutionary/genetic procedure.

Practical utilisation of the fuzzy set theory as well as the rules of operations on fuzzy numbers lead to the general scheme of the fuzzy protection/control device presented in Fig. 11.7. The following main blocks can be distinguished here [6]:

- fuzzification, where the real input signals are converted into their fuzzy counterparts (fuzzy numbers),
- fuzzy reasoning, where the fuzzy criteria signals are processed and—after comparison with fuzzy settings—some fuzzy decision/output signals are generated,
- defuzzification, which is understood as conversion of the fuzzy outputs into crisp numbers (real output signal or decision).

11.2 Boolean Versus Fuzzy Logic

It is common that most of the relay decisions to be taken in protection devices are of discrete type (0–1). The relay output is either blocking/stabilization that is issued when the protected plant is healthy (value 0) or tripping/alarm, which is a proper decision for all internal fault cases (value 1). There are no intermediate cases, i.e. anything halfway between 0 and 1. In many cases, however, taking proper decision is difficult due to many reasons and sometimes the "don't know" expression would be the best if the premise for given hypothesis is not certain or the measured value is very close to the decision boundary.

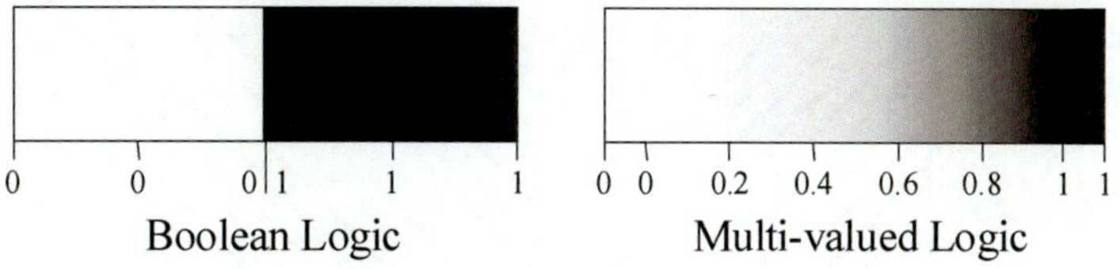

Fig. 11.8 Boolean versus Fuzzy logic—an illustration of the concept

Unlike the two-valued Boolean logic, FL is multi-valued. It deals with degrees of membership and degrees of truth. FL uses the continuum of logical values between 0 (completely false) and 1 (completely true). Instead of just black and white, it employs the spectrum of colors (or color shades, Fig. 11.8), accepting that things can be partly true and partly false at the same time. For instance, one can say that a glass is half-empty or half-full, depending on the context of this statement, which in turn may lead to different implications. By analogy, one may express the fulfillment of given protection criterion as:

- fully met (the criterion signal is far beyond the decision threshold),
- partially met (the criterion signal exceeds the threshold just slightly and/or oscillates around it),
- partially excluded (the measured value is slightly below the threshold, may exceed it for a moment during transient),
- completely excluded (the signal is well below the threshold value).

All the above is closely dependent on the adopted conception of truth. As opposed to the Aristotelian conception of truth, FL allows for a gradual transition between TRUE and FALSE. By comparison to Fig. 14.6, where overlapping of the decision regions were interpreted as superimposing of the conditional probability density functions, here the fuzzy border between two classes of events is defined (Fig. 11.9), where a smooth curve substitutes for a crisp threshold value. One should say that in case (b) no crisp threshold is appropriate for effective decision-making, while with fuzzy approach the degree of setting exceeding can be specified, which can lead to the final protection decision.

Decision-making with FL can be compared to classification of the object to one of two or more sets, where the border between them is not sharp, but is specified with some fuzziness. An element is classified as belonging to given set with some degree between zero and unity. When the degree of membership is "high" (also a fuzzy term) the final decision can be made, otherwise it is either denied or postponed until sound difference between the membership grades for given sets is observed. When

$$|\delta_1 - \delta_2| > \Delta, \tag{11.22}$$

one of the hypotheses (H_1 or H_2) is accepted, the one for which the fuzzy support value δ_1 or δ_2 is the highest, with Δ being certain threshold representing the decision confidence margin.

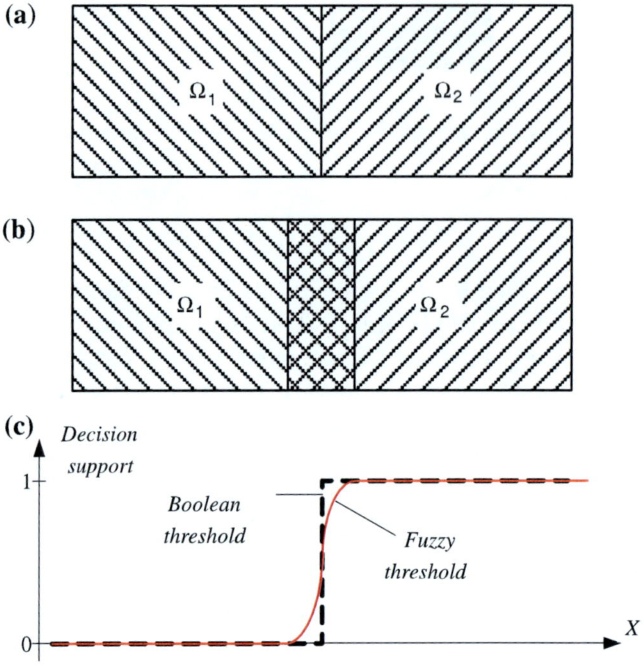

Fig. 11.9 Fuzzy threshold for overlapping hypotheses: **a** ideal decision space division **b** real situation with overlapping of sub-regions **c** traditional and fuzzy thresholds

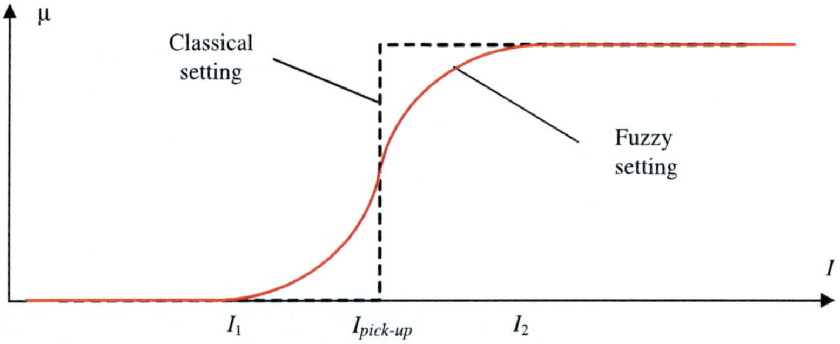

Fig. 11.10 Classical and fuzzy overcurrent decision threshold

As an example of single-criterion protection the fuzzy overcurrent principle can be used (Fig. 11.10). Instead of classical decision threshold $I_{\text{pick-up}}$ one can propose a fuzzy curve, that separates the blocking and tripping regions in a smooth way. Unambiguous decision (with the degree of confidence 1.0) can be taken for current magnitudes lower than I_1 and higher than I_2, whereas for the values within the range (I_1, I_2) appropriate value of the membership μ from the range (0, 1) is assigned, coding the grade of the signal membership to the category of fault cases. The higher the membership value μ, the more clearly (with higher certainty) the

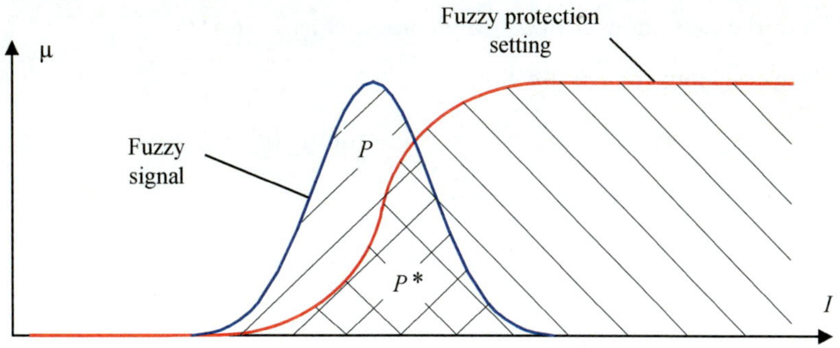

Fig. 11.11 Comparison of fuzzy decision signal with fuzzy threshold

decision to trip the protected plant can be taken. Application of the fuzzy settings creates a remedy for many problems with settings of classical protection relays, being often a peculiar compromise between sensitivity and selectivity.

Figure 11.11 illustrates one of the possible ways of comparison of the fuzzy signal with fuzzy setting. The degree of the threshold exceeding (also a value within [0, 1]) is defined as a ratio of the areas P^* and P, where P^* stands for this part of the area under the fuzzy signal MF that is also situated under the fuzzy setting curve [6, 20], i.e.

$$\nu = \frac{P^*}{P} \tag{11.23}$$

With such a way of comparison, if the signal ν (non-fuzzy number) is interpreted as the grade of meeting given criterion, it is not necessary to perform its defuzzification. It can be directly used for decision-making (through comparison with simple threshold) or working out the final decision in multi-criteria protection schemes, where a number of various criteria signals are analyzed and evaluated in parallel. Introducing the weighting factors w_i for particular criteria values, the resulting support grade for protected plant tripping can be defined as a weighted sum

$$\delta = \sum_{i=1}^{N} w_i \nu_i, \tag{11.24}$$

and the final protection decision is taken after the value of δ exceeds certain non-fuzzy threshold Δ. This approach is called weighting factor aggregation method and can be treated as a simplified version of fuzzy reasoning.

The general reasoning with FL approach is based on processing of a set of fuzzy rules utilizing the implication operation in “IF antecedent THEN consequent” form. The implication $p \rightarrow q$ can be expressed with use of the conjunction (AND), disjunction (OR) and negation (NOT) operands as [22]:

$$p \rightarrow q \Leftrightarrow \; \sim [p \wedge \sim q]. \tag{11.25}$$

Therefore definition of fuzzy implication is as follows:

- with use of min for conjunction

$$\mu_{p\to q}(x,y) = 1 - \mu_{p\cap\bar{q}}(x,y) = 1 - \min[\mu_p(x), 1 - \mu_q(y)]. \tag{11.26}$$

- with use of product for conjunction

$$\mu_{p\to q}(x,y) = 1 - \mu_{p\cap\bar{q}}(x,y) = 1 - \mu_p(x)[1 - \mu_q(y)], \tag{11.27}$$

where $\mu_p(x)$, $\mu_q(y)$ are membership values for the antecedent and consequent, respectively.

In engineering applications, where the cause-effect relationships are usually studied, the implications (11.26) and (11.27) are often applied in simplified forms [8, 9]:

- Mamdani definition

$$\mu_{p\to q}(x,y) = \min[\mu_p(x),\ \mu_q(y)]. \tag{11.28}$$

- Larsen definition

$$\mu_{p\to q}(x,y) = \mu_p(x)\mu_q(y). \tag{11.29}$$

that can be accepted without further proof for all causal engineering systems.

The effects of using (11.28) and (11.29) may be quite different, which is illustrated in Fig. 11.12. The Mamdani definition results in so called MF clipping (Fig. 11.12a), whereas with Larsen definition one may observe MF scaling (Fig. 11.12b). The application of either definition is dependent on the practical problem to be solved.

11.3 Fuzzy Reasoning

A typical Fuzzy Inference System is composed of a number of fuzzy rules (implications) as described above. The rules are usually stored in the so called tables of truth (where all the rules with their inputs and outputs are listed) or in the so called FAMs (Fuzzy Associative Memory) being a kind of relational matrix, as illustrated in Fig. 11.13. In the example shown the acceleration of a car is defined with a set of rules expressed linguistically as functions of distance to trip destination and car velocity. From particular cells of the FAM one can read out the required car acceleration; e.g. for very small distance (VS) and very high car velocity (VH) one can see that the acceleration should be very negative (VN), which means the car should significantly decelerate. The consequence of particular

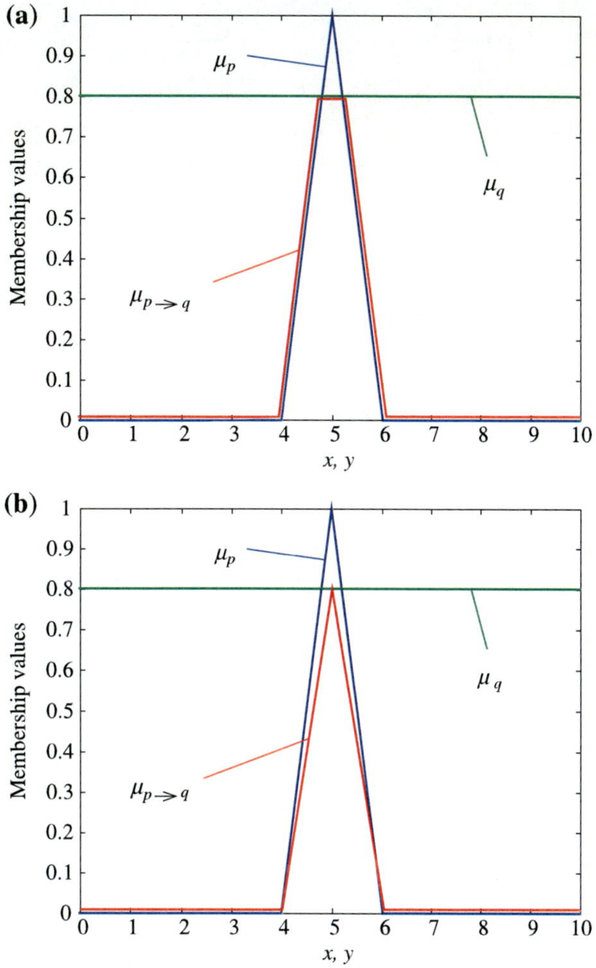

Fig. 11.12 Fuzzy implication according to **a** Mamdani **b** Larsen definitions

Fig. 11.13 FAM example for the car acceleration problem with two-dimensional input space. *VN* very negative, *N* negative, *Z* zero, *SP* small positive, *P* positive, *VP* very positive, *VS* very small, *S* small, *M* medium, *H* high, *VH* very high, *MB* medium big, *B* big, *VB* very big

Velocity \ *Distance*	VS	S	SM	M	MB	B	VB
VH	VN	VN	N	Z	Z	SP	SP
H	N	N	Z	Z	SP	P	P
M	N	Z	Z	Z	SP	P	P
S	Z	Z	Z	SP	SP	P	VP
VS	Z	Z	SP	SP	P	VP	VP

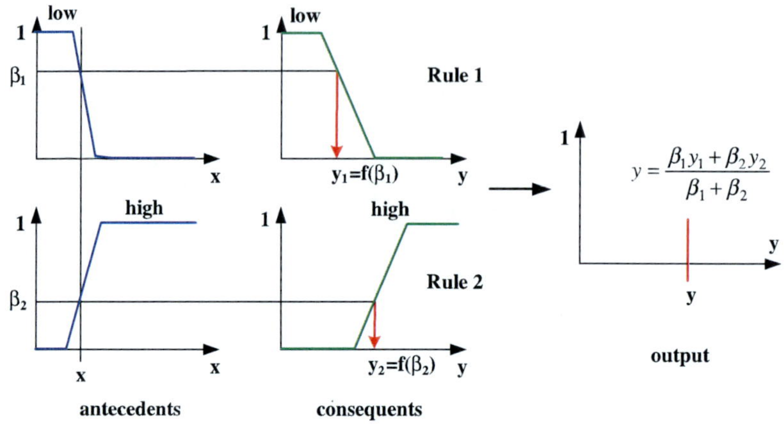

Fig. 11.14 Tsukamoto reasoning scheme with two rules and one input

rules are to be aggregated so that the final fuzzy output is generated and then the output of the scheme ought to be defuzzified to produce equivalent crisp output value.

The process of rules aggregation can be done in various ways. The most popular ones are:

- weighting factor method (applicable for rules with crisp or singleton outputs),
- compositional approach (max–min or max-product schemes, where min/product refers to the way how the implication is performed).

The examples of FIS belonging to the first family are *Tsukamoto* or *Sugeno* schemes, while the *Mamdani* FIS is a representative of the second family of reasoning schemes.

The Tsukamoto reasoning scheme [22] is a monotonic selection scheme, where the value of the rule output (truth membership grade of the rule consequent) can be estimated directly from a corresponding truth membership grade in the antecedent. The illustration of the process is given in Fig. 11.14 for a scheme with two rules. For the input value x the rule antecedents' membership grades are found (β_1, β_2), which is followed by looking for the consequents' output values (y_1, y_2 read out from particular consequents' MFs). The final output of the Tsukamoto FIS is defined as a weighted sum of the rules' outputs, as shown in Fig. 11.14.

Another, quite frequently used example of fuzzy system with weighting factor method aggregation of the rules' outputs, is the Takagi–Sugeno controller [15, 22]. Here, the input part of the controller only is based on fuzzy processing (signal fuzzification), whereas further signal processing is non-fuzzy, determined clearly with some functional relationships.

In Fig. 11.15 the Sugeno reasoning structure is shown for a case of two rules and two inputs. The consequentce of the rules are here defined as singleton values being linear combinations of the input values x_i, where the coefficients of (11.30) are different for particular rule k:

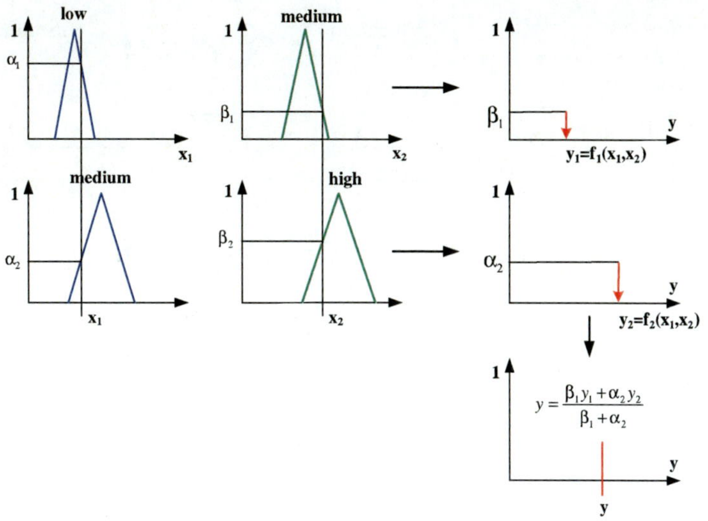

Fig. 11.15 Takagi–Sugeno FIS structure

$$y_k = f_k(X) = \sum_{i=1}^{n} a_{ki}x_i + a_{k0}. \tag{11.30}$$

The weighting coefficients in the final output equation

$$y = \frac{\sum_{k=1}^{n} \delta_k y_k}{\sum_{k=1}^{n} \delta_k}. \tag{11.31}$$

are minimum values of the membership grades for given rule read out from the MFs for all input values, i.e. $\delta_1 = \min(\alpha_1, \beta_1) = \beta_1$ and $\delta_2 = \min(\alpha_2, \beta_2) = \alpha_2$.

Experiences show that Sugeno reasoning schemes are computationally effective and work well with optimization and adaptive techniques, which makes them very attractive in control problems, particularly for dynamic and nonlinear systems. However, their tuning (setting the factors in rule output equations (11.30)) is not straightforward. The coefficients are difficult to be defined intuitively, they are usually set on basis of the results of thorough simulation studies, which can be seen as a serious disadvantage of this scheme.

On the contrast, the Mamdani scheme [9, 22] (illustration in Fig. 11.16) allows to describe the expertise in more intuitive, more human-like manner. However, Mamdani-type fuzzy inference entails a substantial computational burden. Here, as results of rule firing the fuzzy output values are obtained that are represented by areas being parts of output MFs lying under certain membership levels (here, with application of min-type implication—$\min(\alpha_1, \beta_1) = \beta_1$ and $\min(\alpha_2, \beta_2) = \alpha_2$). The schemes' final fuzzy output is reached by aggregation of particular rules' outputs by taking maximum values over all partial output MFs, which gives an area colored in red in Fig. 11.16. The crisp output of the scheme is further obtained by applying one of the available signal defuzzification methods.

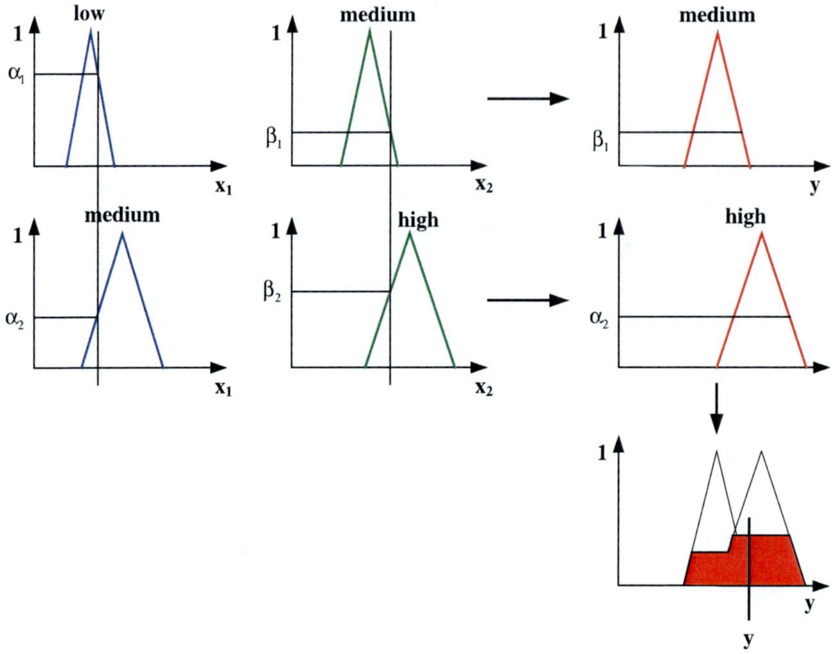

Fig. 11.16 Mamdani FIS as an example of compositional reasoning (min–max)

Fig. 11.17 Defuzzification by searching of MF maximum value

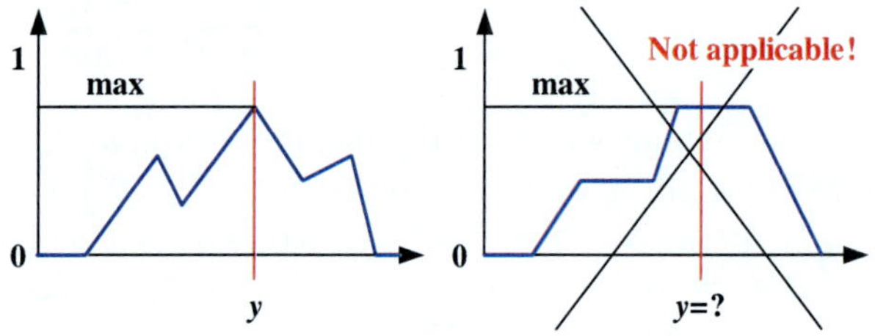

Defuzzification Methods: Back to Crisp Output Value

Since the FIS scheme output should be crisp (non-fuzzy) to be applied as a control/ protection signal, the fuzzy number obtained after rules' aggregation should be converted into crisp value. This process is called defuzzification and can be done, among others, by means of the following methods [12]:

- maximum MAX of the fuzzy number (Fig. 11.17),
- mean value of local maxima MOM (Fig. 11.18) or
- center of gravity COG/weighted sum approaches (Fig. 11.19).

The first two abovementioned methods are not always applicable, which is illustrated in Figs. 11.17 (there is no clear maximum when the MF is partially flat) and 11.18 (the output value may fall outside of the fuzzy number support). In case of the COG method the crisp value is defined according to

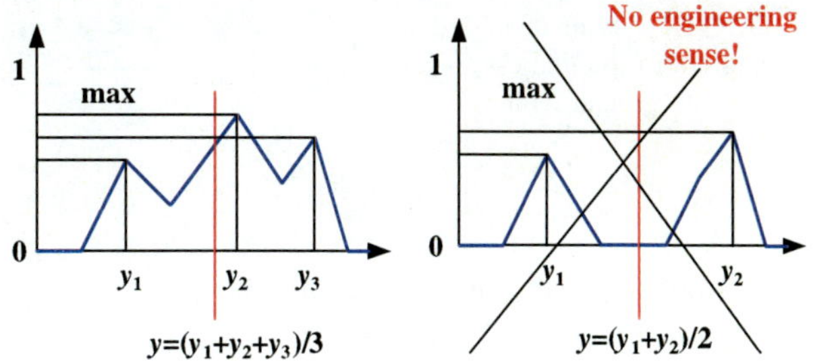

Fig. 11.18 Defuzzification by min-of-maxima method

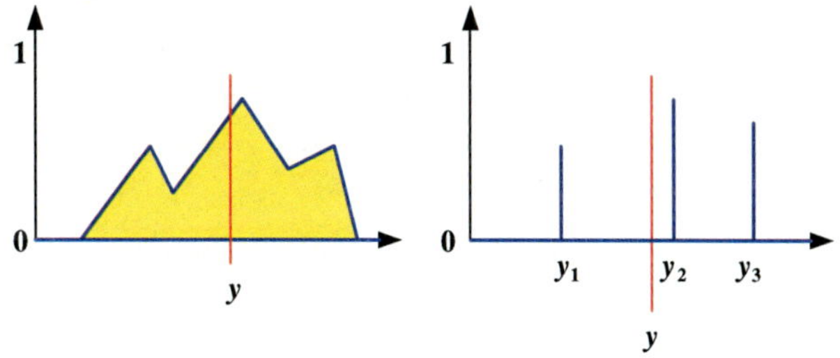

Fig. 11.19 Defuzzification with COG/weighted sum approaches

$$y = \frac{\int_{\mathrm{supp}\,B} \mu_B(y) y \,\mathrm{d}y}{\int_{\mathrm{supp}\,B} \mu_B(y)\,\mathrm{d}y}. \tag{11.32}$$

Instead of the integrals in (11.32) the weighted sum can be used, provided that the output MF is not continuous, as it is with the Takagi–Sugeno scheme. In such a case (11.32) it takes the form,

$$y = \frac{\sum_{\mathrm{supp}\,B} \mu_B(y) y}{\sum_{\mathrm{supp}\,B} \mu_B(y)}. \tag{11.33}$$

11.4 Fuzzy Logic Applications for Protection and Control

At the end of this chapter it is worth to summarize the features of FL systems as well as to point out examples of typical applications of such schemes in power systems. Among the most important virtues of fuzzy systems the following should attract the reader's attention:

- ability of processing of uncertain information, inaccurate and/or corrupted data,
- possibility of expressing of non-sharp relationships and rules in a way close to natural language (linguistic variables, IF … THEN … rules),
- quite easy interpretation of the internal signals of a fuzzy system,
- improvement of efficiency and selectivity by decision-making in protective relays thanks to application of fuzzy settings and fuzzy decision characteristics,
- relatively simple, intuitive setting of the input/output MFs (at least for the first, not-optimized scheme version),

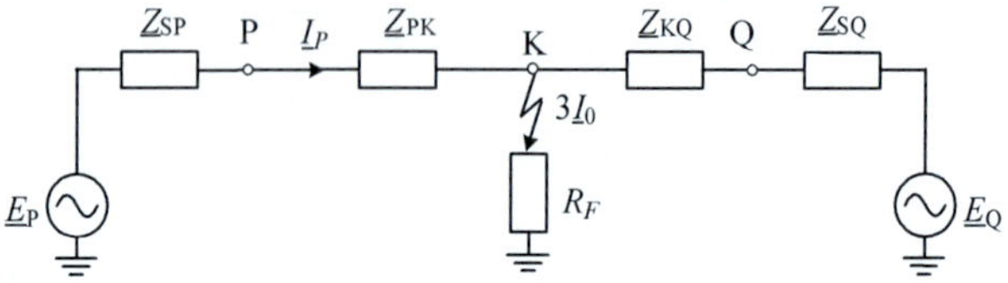

Fig. 11.20 Equivalent diagram of the line P–Q with phase-to-ground fault

- simple description of systems, for which a detailed mathematical model is too complex or is not known.

Among shortcomings of the fuzzy approach the following can be mentioned:

- problems with choice of the effective defuzzification method for given task and its practical implementation in on-line mode,
- problems with obtaining enough data for fuzzy system design, especially for the systems already in operation.

The following (not exhaustive) list of possible applications of fuzzy systems in power systems [1, 3, 4, 7, 11, 13–20] is intended as exemplary only:

- identification of fault type in power system transmission lines,
- fault location in transmission and distribution lines,
- multi-criteria differential protection of power transformers,
- as well as, being examples of non-relaying problems,
- voltage and speed control of synchronous generators,
- diagnostics of power transformers,
- development planning and quality assessment of power networks,
- reactive power flow control,
- load forecasting, etc.

Below an example of fuzzy scheme is described that was intended for improvement of distribution/transmission line distance protection [14, 16]. The scheme incorporates the ideas of signal fuzzification, fuzzy settings and fuzzy decision-making that is based on Mamdani approach and compositional reasoning.

11.4.1 Example of Fuzzy Logic Application for Distance Protection

The distance relaying technique is used as primary protection mainly in transmission lines but in some countries it is also applied for distribution lines, mostly in meshed urban networks. Although the idea of distance protection was introduced a few decades ago, its present-day digital implementations are not free from problems evoked by impedance measurement errors due to both inaccuracy of used algorithms and such system factors as influence of fault resistance, etc. The commonly recognized problem concerns so called reactance effect. Figure 11.20 shows equivalent circuit that can be used to determine the impedance seen by the distance relay installed at the beginning of the line.

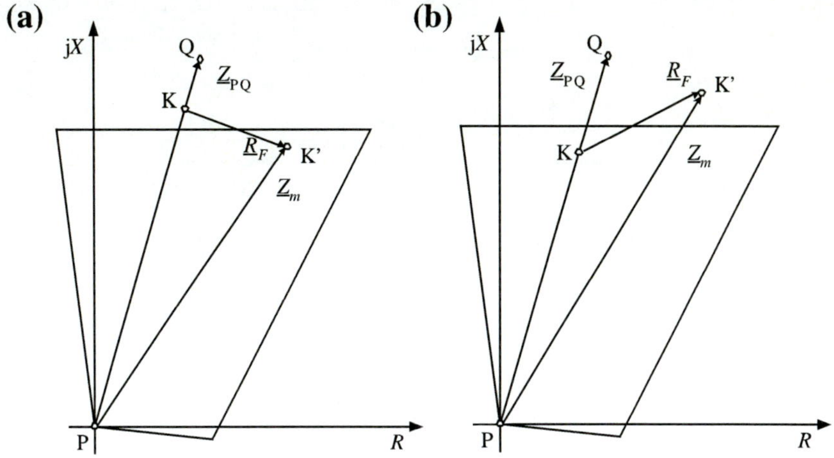

Fig. 11.21 Impedance measured for an intermediate short-circuit: **a** fault out-of-zone, pre-fault current flowing into the line **b** fault in-zone, pre-fault current flowing backwards

It is commonly known that the impedance measured by the relay includes positive sequence impedance of the line section PK and the fault resistance R_F, which is almost always seen as a complex impedance vector $\underline{R}_F$. The reactance part of $\underline{R}_F$ may be negative or positive, depending on the pre-fault current direction. The resulting impedance $\underline{Z}_m$ becomes either decreased or increased and may fall in-zone for some external faults (over-tripping, Fig. 11.21a) or out-of-zone for some faults requiring fast tripping (malfunction, Fig. 11.21b).

As one can see, distance protection relay may face serious problems with discrimination of faulted section of the feeder. It is to be stressed that the majority of contemporary distance relays do not apply any sophisticated algorithms for distance-to-fault determination with elimination of the reactance effect, such as those used in impedance-based fault locators. Common practice is just appropriate shaping of the relay characteristic (expanding the curve in R direction), which is good enough for most cases but cannot eliminate the problem completely. Here the FL technique is proposed to enhance features of traditional distance relays. Below the fuzzy protection scheme developed as well as its testing with the signals obtained from EMTP-ATP simulation of faults are described.

The new protection developed is based on fuzzy concepts as well as adaptive shaping of the relay impedance characteristic for the first zone in dependence on prevailing power system conditions. Figure 11.22 shows the block scheme of the proposed approach.

The scheme from Fig. 11.22 operates as follows. First, using selected algorithms the current and voltage samples are processed (block 1) in order to deliver the criteria values: reactance X (a measure of distance-to-fault) and additional variables $R'_F(n)$, I_L, and I_0 that are used for adaptation of the relay operating characteristic. All the criteria are fuzzified in blocks 2, 3, whereas the fuzzy numbers are obtained in the way illustrated in Fig. 11.23. The left and right borders of the triangle MFs as well as the peak point are calculated according to:

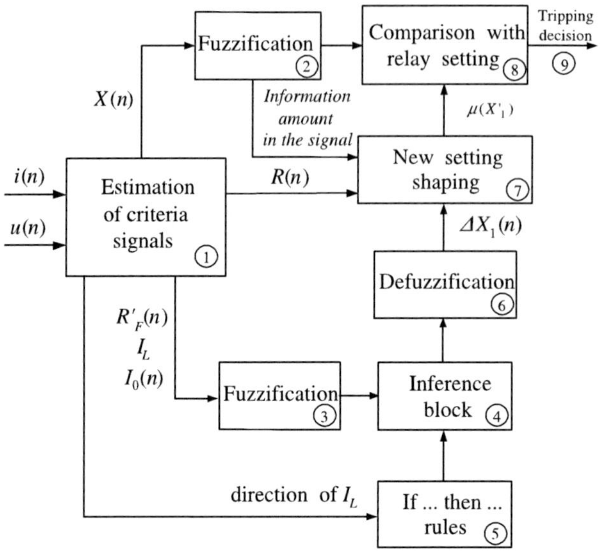

Fig. 11.22 Block scheme of the FL-based adaptive distance protection. *i*(*n*), *u*(*n*) instantaneous current and voltage values, respectively, $R'_F(n)$ estimate of the fault resistance, I_L pre-fault current amplitude (load current), I_0 zero sequence current, *direction of* I_L ("+" or "−") factor indicating direction of I_L in relation to the fault current, which is further used for selection of proper rule base, *X*(*n*) measured reactance, ΔX_1 corrective reactance component, $\mu_S(X)$ membership function of the new fuzzy setting for the reactance, *Information amount in the signal* coefficient dependent on fuzziness degree of membership function of the criterion signal

$$X_m(n) = \frac{1}{N_1/4 + 1} \sum_{k=0}^{N_1/4-1} X(n-k). \tag{11.34}$$

$$X_{\max}(n) = \max_{k=0,\ldots,N_1/4-1} \{X(n-k)\}. \tag{11.35}$$

$$X_{\min}(n) = \min_{k=0,\ldots,N_1/4-1} \{X(n-k)\}. \tag{11.36}$$

which can be interpreted as finding average, maximum and minimum values of the criterion signal *X* over a time period corresponding to a quarter of fundamental frequency cycle, with N_1 being number of samples within 20 ms.

The inference block 4 realizes the fuzzy inference, whereas the reasoning is done by using the product implication method and MAX–MIN aggregation of rule outputs, which can be formalized as follows:

$$\mu_{A^k_{R'_F} \times A^k_{I_L} \times A^k_{I_0} \to B^k}(R'_F, I_L, I_0, X) = \mu_{A^k_{R'_F} \times A^k_{I_L} \times A^k_{I_0}}(R'_F, I_L, I_0) \cdot \mu_{B^k}(X) \tag{11.37}$$

$$\mu_B(X) = \max_{k=1,\ldots,n} \{\mu_{B^k}(X) \min[\mu_{A^k_{R'_F}}(R'_f), \mu_{A^k_{I_L}}(I_L), \mu_{A^k_{I_0}}(I_0)]\}. \tag{11.38}$$

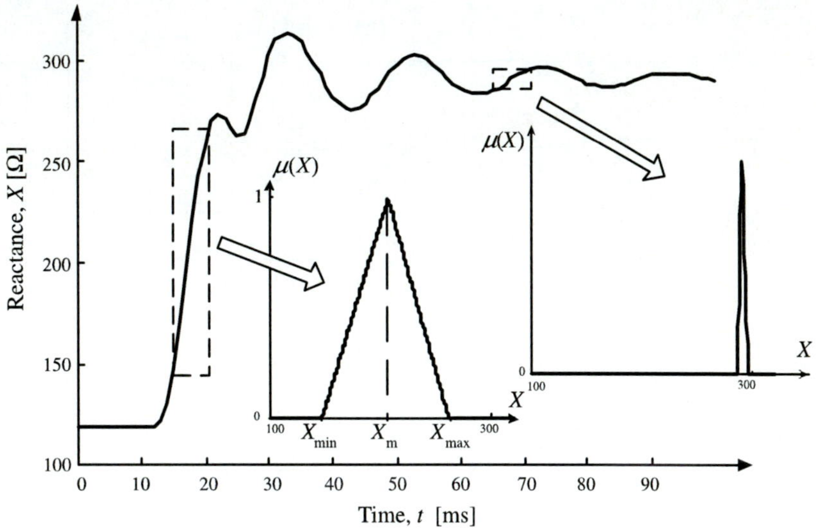

Fig. 11.23 Illustration of signal fuzzification (reactance X)

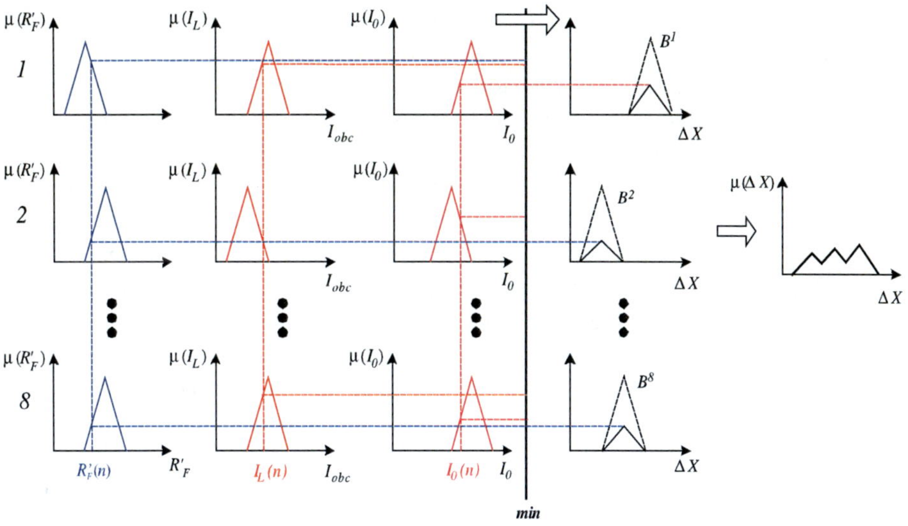

Fig. 11.24 Illustration of the fuzzy inference process

Using the data from EMTP simulations two sets of IF–THEN rules have been obtained. The rules (block 5) have been defined separately for positive and opposite direction of fault current (in relation to the pre-fault flow). The fuzzy reasoning process is shown in Fig. 11.24.

For defuzzification of MF $\mu_B(X)$ the centre of area method is used (block 6 in Fig. 11.22). As a result of this process a non-fuzzy value of the corrective

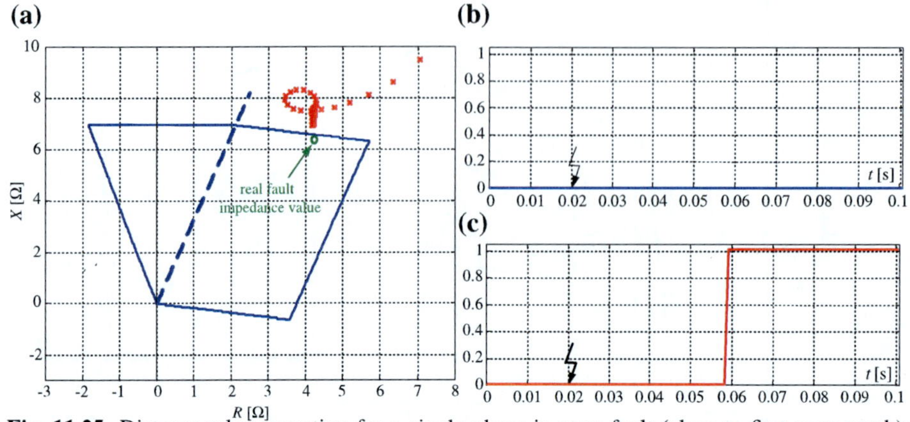

Fig. 11.25 Distance relay operation for a single-phase in-zone fault (close to first-zone reach): **a** trajectory of measured impedance **b** response of classical relay **c** response of FL adaptive relay

reactance component ΔX_1 is obtained that is further used for adaptation of the relay. The new relay 1st zone reach in X direction is defined and then fuzzified taking also into account the estimated information amount in the signal X (block 7). Finally, the fuzzified reactance signal is compared with the new fuzzy setting to generate the final relay decision (block 8). More details about the presented scheme can be found in [14].

The performance of the developed fuzzy adaptive distance relay has been verified with various signals obtained from simulation of in-zone and out-of-zone faults on EMTP-ATP model of a test power system (two sub-systems with a tie line, to be protected). The operation of the new scheme has been compared with distance protection with polygon characteristic as proposed in [2]. Below the relay operation for a selected case of single-phase-to-ground fault is shown (Fig. 11.25). The short-circuit occurred in-zone, at 80% of the line length while the relay reach was set at 85% of the line impedance and thus proper response of the relay was line tripping in shortest possible time (first-zone trip). As seen in Fig. 11.25a the trajectory of measured impedance did not enter the relay first-zone area and the relay without adaptation failed (no response, Fig. 11.25b). The cause for that was the reactance effect that was not compensated for in the classical relay. On the contrary, the developed FL adaptive relay responded properly in the considered case, i.e. classifying the event as first-zone fault (Fig. 11.25c).

The fuzzy distance protection developed has also been tested with the signals from out-of-zone faults. In Fig. 11.26 the case of phase-to-ground fault at 90% of the line length is shown. No malfunctions have been noticed in this case (Fig. 11.26c), contrary to classical non-adaptive relay, where improper operation was observed for some faults due to underestimating of the measured fault-loop impedance (Fig. 11.26b). The FL protection responded also correctly to all unambiguous in-zone faults with the decision time of the same range as for the classical non-fuzzy protection.

One can say thanks to introduction of signal fuzzification and adaptive settings the new relay has gained improved sensitivity and selectivity, especially for faults

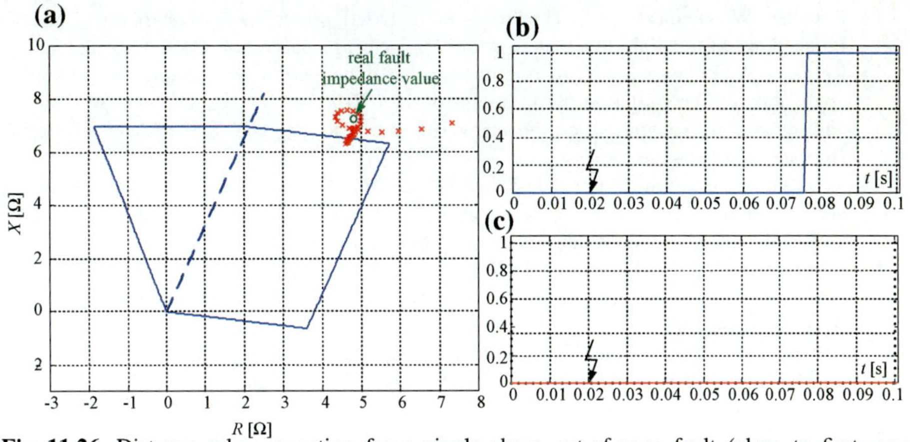

Fig. 11.26 Distance relay operation for a single-phase out-of-zone fault (close to first-zone reach): **a** trajectory of measured impedance **b** response of classical relay **c** response of FL adaptive relay

close to first-zone boundaries. The new relay was able to classify properly all the considered simulation cases, showing more sensitive zone checking than the classical distance relay without application of FL.

References

1. Abdul-Rahman KH, Shahidehpour SM (1994) Reactive power optimization using fuzzy load representation. IEEE Trans Power Syst 9:898–905
2. Cox E (1994) The fuzzy systems handbook. Academic Press, London
3. Dillon TS (Convenor) (1995) Fault diagnosis in electric power systems through AI techniques. Report by TF 38.06.02, Electra 159
4. Ferrero A et al (1995) A fuzzy set approach to fault type identification in digital relaying. IEEE Trans Power Deliv 10:169–175
5. Fuller R (2000) Introduction to neuro-fuzzy systems. Physica-Verlag, Heidelberg
6. Kasztenny B, Rosołowski E, Saha MM, Hillstrom B (1997) Fuzzy sets and logic in power system protection. Eng Int Syst 5:193–203
7. Kezunovic M et al (1998) Practical intelligent system applications to protection, and substation monitoring and control. In: Proceedings of the CIGRE session, Paris, France, paper 34–104
8. Larsen PM (1980) Industrial applications of fuzzy logic control. Int J Man Mach Stud 12:3–10
9. Mamdani EH (1974) Applications of fuzzy algorithms for simple dynamic plant. Proc IEEE 121:1585–1588
10. MATLAB Guide, Fuzzy logic toolbox, the Mathworks Inc., 2009
11. Miranda V, Matos MACC (1989) Distribution system planning with fuzzy models and techniques. In: Proceedings of the 10th international conference on electricity distribution, CIRED, vol 6, pp 472–476
12. Pedrycz W, Gomide F (2007) Fuzzy systems engineering. Toward human-centric computing. Wiley-Interscience, Hoboken

13. Rebizant W, Bejmert D (2005) Fuzzy logic based protection for distribution lines. In: Proceedings of the 40th universities power engineering conference, Cork, Ireland, paper 142
14. Rebizant W, Bejmert D (2006) Improvement of transmission line distance protection with application of fuzzy logic. In: Proceedings of the CRIS workshop on influence of distributed and renewable generation on power system security, O-v-G Universitaet Magdeburg, Germany, pp 115–120
15. Rebizant W, Szafran J, Lee SJ, Kang SH (2003) Adaptive and intelligent systems for generator monitoring and protection purposes. In: Proceedings of the IFAC symposium on power plants and power systems control, Seoul, Korea, paper B01_3_202
16. Rebizant W, Bejmert D, Szafran J (2005) Fuzzy logic based overcurrent protection for MV networks. In: Proceedings of the 15th power systems computation conference, Liege, Belgium, paper 77
17. Rebizant W, Goncerzewicz L, Schiel L (2008) New fuzzy stabilization scheme against transformer inrush currents. In: Proceedings of 8th international IET conference developments in power system protection, Glasgow, UK, pp 38–43
18. Saha MM, Kasztenny B (1998) Application of fuzzy logic in power system protection. In: Proceedings of the international conference on modern trends in the protection schemes of electric power apparatus and systems, New Delhi, India, paper IX-1
19. Wiszniewski A, Kasztenny B (1992) Primary protective relays with elements of expert systems. In: Proceedings of the CIGRE session, Paris, France, paper 34,2, CN
20. Wiszniewski A, Kasztenny B (1995) A multi-criteria differential transformer relay based on fuzzy logic. IEEE Trans Power Deliv 10:1786–1792
21. Zadeh L (1973) A new approach to the analysis of complex systems. IEEE Trans Syst Man Cybern SMC-3:1
22. Zimmermann HJ (1994) Fuzzy set theory. Kluwer Academic Publishers, London

Chapter 12
Application of Artificial Neural Networks

The signal processing methods and algorithms described in preceding chapters were expressed in form of explicit equations, transfer functions and/or logic rules, either in crisp or in fuzzy versions. There are, however, specific tasks and power system operation conditions when, especially for the problems that are complex and difficult to express in terms of traditional means, other solutions should be applied. In such situations, both for signal processing and decision-making Artificial Neural Networks may constitute a good solution.

The Artificial Neural Network ANN is a set of processing units called neurons. The ANN can be used to approximate the relationship between input and output signals of the system (process), which is illustrated schematically in Fig. 12.1. The required approximation abilities of the ANN are acquainted through so called training, with examples of the system (process) operation. After training the ANN is ready to serve as a predictor of the system performance and thus deliver information e.g. for control purposes.

The neural networks are believed to possess a generalization feature, which similarly to human reasoning makes them good tools for identification of patterns, even if not all representative features of the patterns are well defined or when some data is missing. Due to built-in neuron activation functions the ANNs are well suited to represent non-linear problems and bring answers to difficult protection and control questions, some of them being outlined in this chapter.

Below the information on the single neuron features as well as commonly used neural network structures is presented. The training procedures applied for particular ANN structures are described, with special emphasis put to the problems with ANN design. The theoretical part is followed by selected examples of ANN applications for protection and control tasks in power systems.

W. Rebizant et al., *Digital Signal Processing in Power System Protection and Control,* Signals and Communication Technology,
DOI: 10.1007/978-0-85729-802-7_12,

Fig. 12.1 ANN as a process model

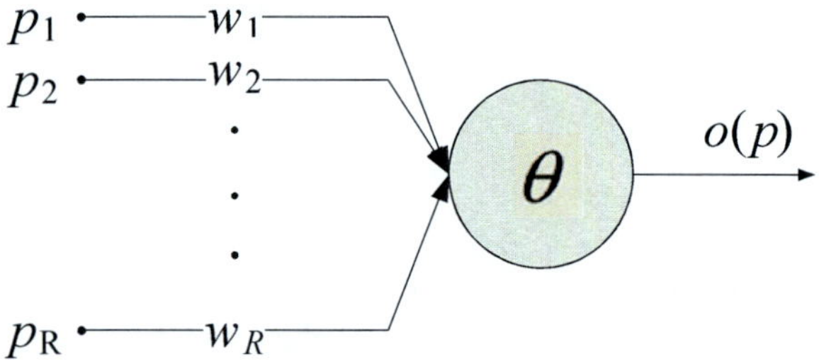

Fig. 12.2 McCulloch and Pitts' neuron model

12.1 Neuron Models and Neural Network Structures

Inspired by the anthropological analogy the ANNs resemble the structure of a human brain, whose smallest component is a single neuron. The neurons are connected with neighbouring elements through their axons (sending information), whereas the incoming signals from other units are received via dendrites synapses. The human brain consists of hundreds of billions of neurons, while signals in brain are noisy spike trains of electrical potential. The important feature of the human neural network is the ability of parallel signal processing as well as distributed memory, which makes our brain robust to noise and failures.

The first concepts related to artificial neural networks were developed by McCulloch and Pitts as early as in 1943 [4]. Their ideas such as threshold and many simple units combined together to give increased computational power are still in use today. The McC&P neuron is depicted in Fig. 12.2, with the corresponding equation (12.1) for the neuron output given below. It is seen that the neuron output is equal to 0 or 1, depending on the calculated weighted sum of the input values $p_1 \ldots p_R$. The inventors assumed further that the inputs p could take only values 1 or 0 (input activated or dead), and the weighting coefficients were either +1 or −1. Therefore the application of such neurons could be very limited.

$$o(p) = \begin{cases} 1 & \text{for } \sum_{i=1}^{R} w_i p_i \geq \theta \\ 0 & \text{for } \sum_{i=1}^{R} w_i p_i < \theta \end{cases} \tag{12.1}$$

A network of several McC&P neurons can be used for realization of basic logic operations, like AND, OR, NAND, NOR, some of which are illustrated in Fig. 12.3.

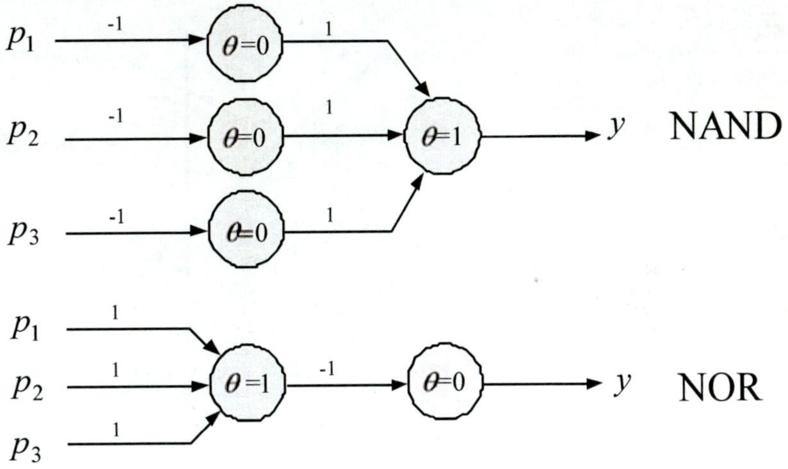

Fig. 12.3 NAND and NOR operands realized with McC&P neurons

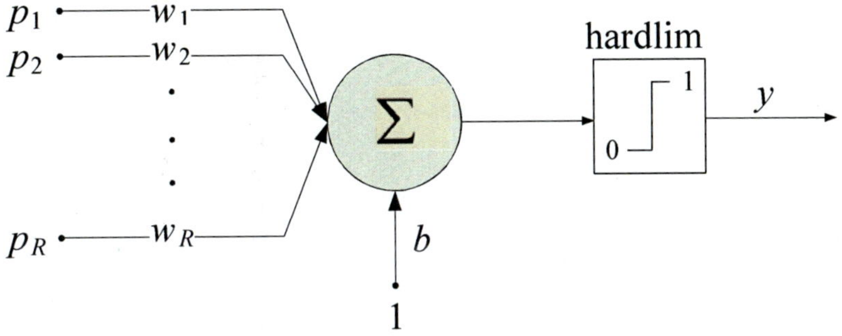

Fig. 12.4 Single-neuron perceptron architecture

The improvement of neuron application possibilities appeared in the 50s and 60s of the last century when many researchers used to work on the so called perceptron theory and related applications. The perceptron (Fig. 12.4) is basically a single neuron structure with adjustable synaptic weights and a hard limiter. Perceptrons are especially suited for simple problems in pattern classification. The hard-limit transfer function gives a perceptron the ability to classify input vectors by dividing the input space into two regions separated by a hyperplane. In the example illustrated in Fig. 12.5 for a two-dimensional problem the separation of the areas corresponding to two classes of events/elements is done with a straight line. Additional "error area" is also marked in the figure, which denotes the fact that for noisy signals the classification may be difficult when the points to be classified are located too close to the decision boundary.

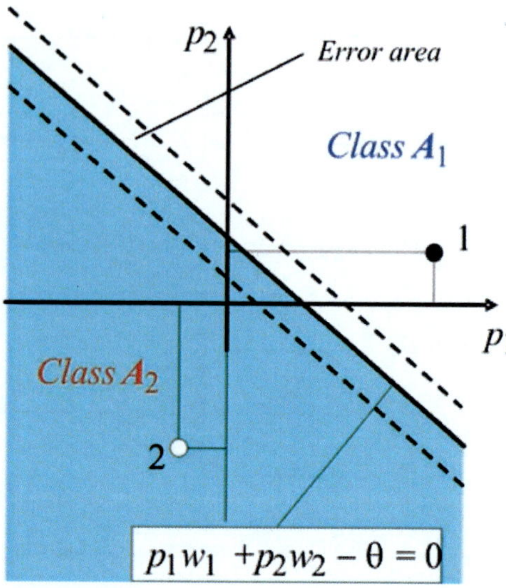

Fig. 12.5 Decision areas for a two-input perceptron used for classification in 2D space

$$p = \begin{bmatrix} p_1 \\ p_2 \\ \vdots \\ p_{R-1} \\ p_R \end{bmatrix} \qquad W = \begin{bmatrix} w_1 & w_2 & \ldots & w_{R-1} & w_R \end{bmatrix} \tag{12.2}$$

$$y = \mathrm{hardlim}(w_p + b) = \begin{cases} 1 & \text{for } Wp + b \geq 0 \\ 0 & \text{for } Wp + b < 0 \end{cases}$$

The two-dimensional perceptron can also be applied for realizing of Boolean logic operations, only through proper setting of the weighing coefficients w and bias b. Unfortunately, there exists no combination of the settings that would enable realizing the XOR function (illustration of the concept and separation problems in Fig. 12.6), not mentioning other non-linear classification problems. For such cases either multilayer perceptron structures (described later) or non-linear neurons and networks are needed.

Another version of a linear artificial neuron is called adaptive linear model (ADALINE), developed by Bernard Widrow (Fig. 12.7) [4]. Instead of hard limit a linear function is introduced here, in simplest version $y = s$, which means that the weighted sum s is forwarded to the neuron output without any change. This type of neuron can be trained to find a linear approximation to a non-linear function. However, a linear network cannot be made to perform non-linear computations.

The neural networks based on linear neurons are characterized by certain limitations:

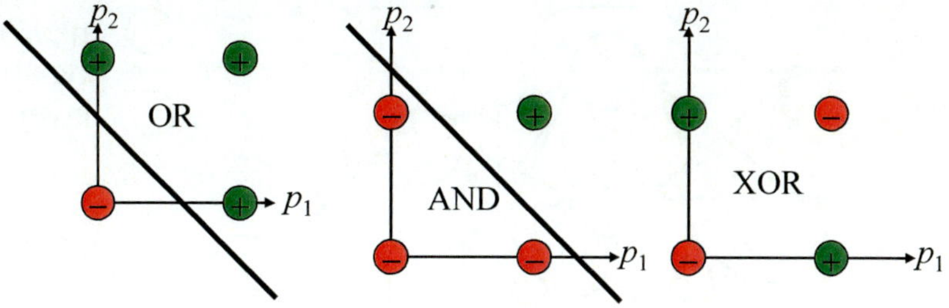

Fig. 12.6 XOR function as an example of non-linearly separable problem

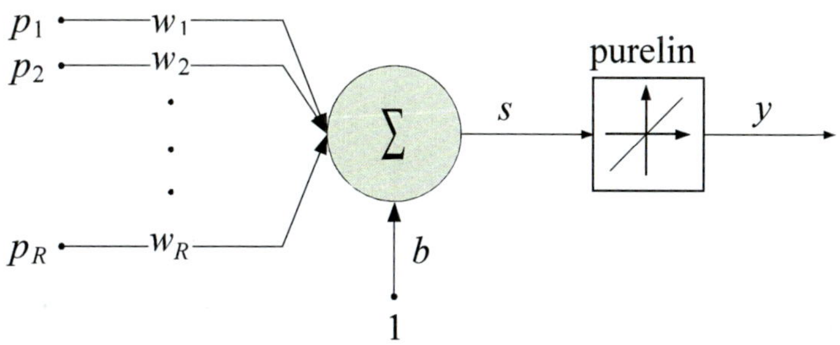

Fig. 12.7 ADALINE neuron model

- They may only learn linear relationships between input and output vectors.
- Even if a perfect solution does not exist, the linear network will minimize the sum of squared errors, if the learning rate is sufficiently small.
- The network will, however, find as close a solution as it is possible given the linear nature of the network's architecture; this property holds because the error surface of a linear network is a multidimensional parabola (since parabolas have only one minimum, a gradient descent algorithm such as the least mean square rule must produce a solution at that minimum).

To solve more complex problems for which single neurons have not sufficient power it is commonly suggested that the neurons are connected together to built networks of various types. The most popular and frequently used structure is the multilayer feed-forward perceptron that is shown in Fig. 12.8. The information (signals) flows in one direction only (feed-forward), there are no feed-back connections. The ANN can be fully connected (all neurons in a given layer are connected to all neurons of the next layer) or can have some connections missing, which would mean that the corresponding weighting coefficients for broken synapses are equal to 0.

The ANNs with linear neurons (ADALINE or hard limited) have limited abilities. Much higher interest is revealed to non-linear nets with the neuron model equipped with so called squashing activation function applied to the sum

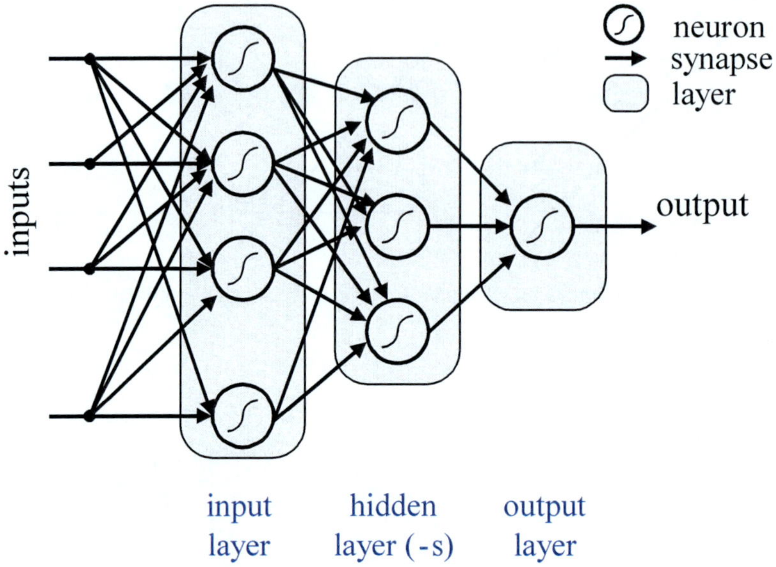

Fig. 12.8 Multilayer perceptron structure

Fig. 12.9 Selected non-linear neuron activation functions

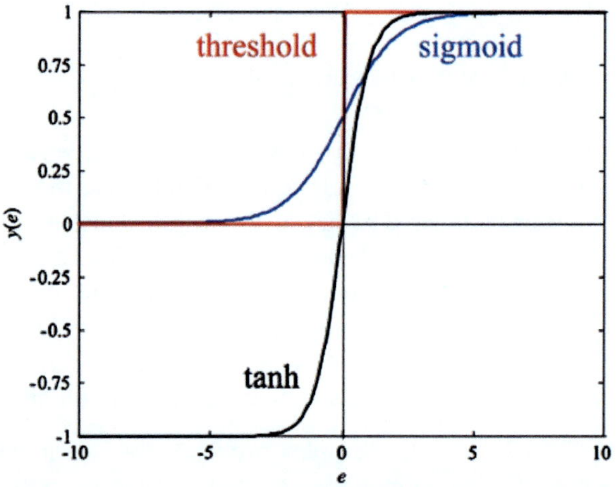

$\boldsymbol{Wp} + b$ at the neuron output. Instead of the hard limit (binary, threshold) a continuous, smooth, differentiable function is introduced, with the output limited to the range [0; 1] or [−1; 1], see Fig. 12.9.

The two most popular differentiable activation functions are [9]:

- Sigmoidal logistic function

$$y = \frac{1}{1 + \exp(-\beta\, e)} \tag{12.3}$$

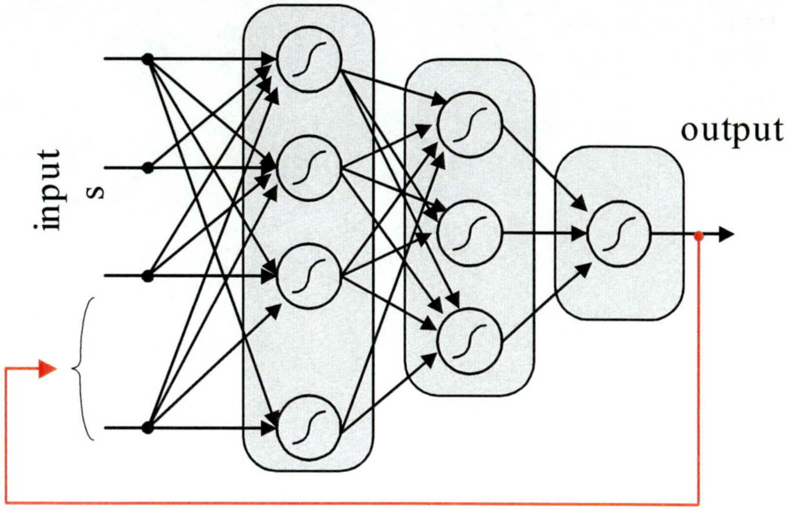

Fig. 12.10 Schematic structure of a recurrent network

- Hyperbolic tangent function

$$y = \tanh(e) = \frac{\exp(\beta e) - \exp(-\beta e)}{\exp(\beta e) + \exp(-\beta e)} \tag{12.4}$$

Applying non-linear activation functions one can expect faster and more consistent ANN training with better convergence, more compact network structures (less neurons) and better generalization of knowledge, which results in higher robustness for noisy inputs.

The other category of networks, in contrast to feed-forward nets, includes recurrent structures that are characterized by feed-back connections (Fig. 12.10). One can imagine that not only the signal/-s from the output layer but also from any other layer can be delivered to the ANN input. Moreover, one can also require that the output values from several time instants (sequence chain of output samples) can be sent to the input or other layers. It is obvious that for ANNs operating on sampled signals the feed-back connection is realized with at least unit delay, which means that the network output is available at the input with a delay amounting to atleast one sampling period.

The recurrent ANNs are usually more efficient and compact (less neurons needed) than their feed-forward counterparts. However, as it results from control theory, the recurrent networks may sometimes be prone to instability, since their equivalent transfer functions may have poles outside of the unity circle. Therefore, after training, when all the coefficients of weights and biases are known, the ANN should be checked for stability, before it is implemented to realize the desired task. Unfortunately, the transfer function as such can be defined for linear structures

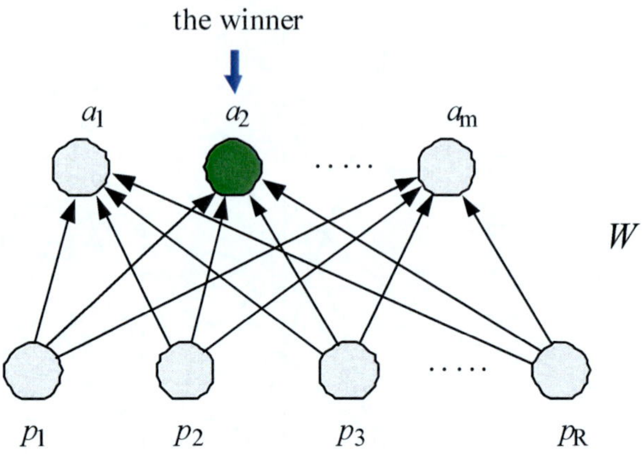

Fig. 12.11 A "winner takes all" network

only, thus especially for non-linear structures their design and further application should be done with care [25].

The neural networks for pattern classification are usually single-output structures. They may, however, also have multiple outputs, each of them assigned to specific purpose. Such networks can be used in multidimensional control schemes, where the outputs of particular neurons of the output layer are sent to appropriate points of the control structure.

A special case of such multi-output ANN—a "winner takes all" network—is depicted in Fig. 12.11. Here, the input x is given to all the network units (neurons) at the same time. The competitive transfer function accepts a net input vector for a layer and returns neuron outputs of 0 for all neurons except for the *winner*, the neuron associated with the most positive element of net input. The winner is the most activated neuron and only its weights are updated during training. The Kohonen rule allows the weights of a neuron to learn an input vector, and because of this it is useful in recognition applications [22].

Such a competitive network learns to categorize the input vectors presented to it. If a neural network only needs to learn to categorize its input vectors, then a competitive network will do. Competitive networks also learn the distribution of inputs by dedicating more neurons to classifying parts of the input space with higher densities of input (see Fig. 12.12). Such networks are also known as self-organizing maps (SOM) that can learn to detect regularities and correlations in their input and adapt their future responses to that input accordingly. SOM allow neurons that are neighbors to the winning neuron to output values. Thus the transition of output vectors is much smoother than that obtained with competitive layers, where only one neuron has an output at a time. The SOM networks are useful in many applications including clustering, visualization and quantization.

The data classification and clustering can also be performed with the so called radial basis function (RBF) networks. The RBF nets are able to approximate an unknown function with a linear combination of non-linear functions (basis functions) that have radial symmetry with respect to a centre. The RBF networks of at

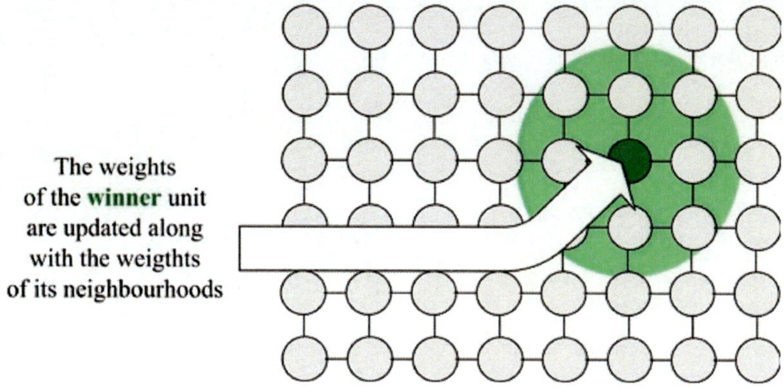

Fig. 12.12 Two-dimensional Kohonen grid network

least three layers are universal approximators, i.e. they can asymptotically approximate any integrable function to a desired degree of precision.

The RBF networks are deeply related to the support vector machines (SVM networks) that are trained according to statistical learning theory incorporating the principle of structural risk minimization. It is claimed that the SVM can guarantee global minimization of error and better generalization than a traditional multilayer perceptron network. Separation of data clusters is done with a hyperplane that maintains a maximum margin in the training set, which leads to high probability of correct predicting and classification of unseen data. The SVMs can be applied for data classification with integrated data visualization with respect to the classification boundaries.

The above introduction to the ANN structures is not and cannot be exhaustive due to space limitation and application orientation of this book. The interested reader should refer to bibliography for more information, e.g. [1, 4, 9].

12.2 ANN Design and Training Issues

The artificial neural networks are characterized by their excellent ability to model the relationships of complex multi-input multi-output processes. Independent of the ANN structure, the neural networks have to be trained on examples before they may be applied for solving any assigned task. The training examples, called patterns, are most frequently originating from computer simulation of power system phenomena, whereas the real world recordings (if available) can be used for final testing of the trained network. Some of the adversaries of the ANN technique stress that because of that the ANNs can not be entrusted fully, as it is in case of mathematically proven explicit algorithms. However, one can say that protection and control devices employing classical methods have to be examined with

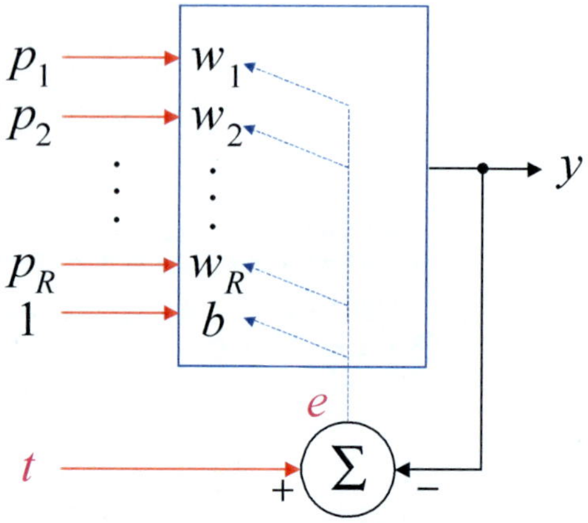

Fig. 12.13 Supervised training of a single neuron

numerous simulation signals and also here a margin of error exists, e.g. by defining of the threshold values for decision making.

This training (learning) of a neural network is to be understood as iterative adjusting of the neurons weights and bias values so that the network output matches the desired value (in case of the so called supervised training) or grouping of the patterns in suitable clusters is achieved for which the desired output is assigned. The training pairs of input–output signals have to be prepared in advance (usually by simulation of power system transients) and then used for training purpose. Some additional patterns are used for network testing, in order to confirm robustness of the trained ANN.

For a clearly defined and not very complex problem the neuron parameters (weights and bias values) can be determined analytically with low effort. However, in most cases—not excluding also the easy ones—the neuron parameters may be adjusted (tuned) by training, either in supervised or unsupervised manner.

The concept of supervised training is illustrated in Fig. 12.13. The neuron weights are being changed iteratively in function of the error value e defined as a difference between the desired neuron output t (called target value) and the neuron actual output y. Initial weights are usually randomly assigned, in the range [−0.5, 0.5], and then updated to obtain the output consistent with the training examples. Neuron tuning is made by small adjustments in the weights, driven by the error value, according to the relationship:

$$\begin{aligned} w_k(i+1) &= w_k(i) + 2\eta \cdot p_k(j) \cdot e(i) \\ b(i+1) &= b(i) + 2\eta \cdot e(i), \end{aligned} \tag{12.5}$$

where i is the iteration number, j—pattern number and η—learning rate (step size).

The algorithm (12.5) represents the Widrow-Hoff training procedure based on an approximate steepest descent procedure, with the aim to minimize the average of the sum of squared errors for all iterations

$$E = \sum_{j=1}^{Q} (t(j) - y(j))^2 \tag{12.6}$$

The main idea of steepest descent is to follow the rule of finding the way downhill within the search space and then taking a step towards minimum, which is repeated for all the training patterns. The training procedure can be run in incremental or batch mode. In incremental mode the weights and biases of the neuron (network) are updated each time an input is presented. In batch training the weights and biases are only updated after all the inputs have been presented. The training procedure is terminated when the convergence conditions are met, that is usually understood as a sufficient decrease of the MSE value (12.6) below a pre-set threshold.

The multilayer ANNs, consisting of many neurons organized in layers, offer much higher pattern recognition and calculation abilities than single neurons. However, their training is much more complex and sophisticated algorithms have to be used to adjust weights and bias values for all neurons. One can observe that the MSE value can be directly calculated for the output layer neuron(-s) only. The errors of the "internal" neurons in hidden and input layers are initially not known since the target values for those neurons are not defined. Thus the output layer errors have to be transferred backwards to calculate the internal error values needed for tuning parameters of the internal neurons. The procedure of training of multilayer perceptron is called backpropagation algorithm (BP). The BP is a recursive algorithm based on a gradient-search optimisation method applied to an error function, whereas the net errors are backpropagated form output layer to hidden layers. The reader interested in more details is referred to the literature [26, 38].

In case of the ANNs with non-linear neurons the shape of their activation function is important, since it may influence computation burden of the training procedure. The form of functions (12.3) and (12.4) is especially advantageous because their derivatives that are calculated at each iteration step during training with gradient descent approach can be expressed with easy formulae being functions of the value y only:

- For the sigmoidal logistic function

$$\frac{dy}{de} = \beta \frac{1}{1 + \exp(-\beta e)} \left[1 - \frac{1}{1 + \exp(-\beta e)} \right] = \beta \cdot y \cdot (1 - y) \tag{12.7}$$

- For the hyperbolic tangent function

$$\frac{dy}{de} = \beta \cdot (1 - y) \cdot (1 + y) \tag{12.8}$$

The above-described procedure of training is called "supervised training". Independent of the ANN complexity the training is organized as follows:

- A set of associated input–output pairs are prepared in advance and then presented to a net which "learns" a model of the process.
- The neurons' weights are iteratively adjusted until the desired accuracy level is achieved.
- A recursive algorithm based on a gradient-search optimization method applied to an error function is executed either in incremental or in batch mode.
- The net errors are backpropagated form output layer to hidden layers (the Backpropagation (BP) procedure is needed for multilayer networks) [38].
- Convergence of the process is checked—the procedure is stopped when the MSE value becomes lower than certain threshold or when a predefined number of iterations (epochs) is exceeded.

The supervised training procedure based on minimization of the network mean square output error may sometimes have problems with convergence to the global minimum within reasonable time or may get stuck in a local minimum. To overcome the problems and assure faster convergence and more accurate results additional supporting techniques have been developed, including variable learning rate and "momentum" (new step is affected by previous step), that are well described in [4]. It is also recommended to randomize the data set before each iteration and to apply cross-validation (checking network response for testing data set during training, learning is stopped when the error for testing data set starts to rise) to get better results.

The unsupervised training (data self-organization concept) is another procedure that is utilized e.g. in case of the SOM networks. The training procedure is organized as follows:

- Given inputs are compared with previously encountered patterns.
- If they are similar to any of the patterns, they will be placed in the same category, otherwise a new category (cluster) will be assigned.
- Category proliferation is controlled by a threshold.
- After the learning (cognition phase) the user defines or labels the clusters according to some criterion.

Detailed ANN training algorithms are related to particular network type and version and are not described here due to space limitation.

For a reader of this book it is good to know what are the problems and issues that should be addressed when an ANN-based solution is to be developed for an application at hand. Generally, one should take into consideration the following points:

- ANN structure type,
- ANN size (number of layers and neurons in particular layers),
- Neuron activation function (may be different in given layers),
- Number and type of input signals,
- Representative set of patterns,

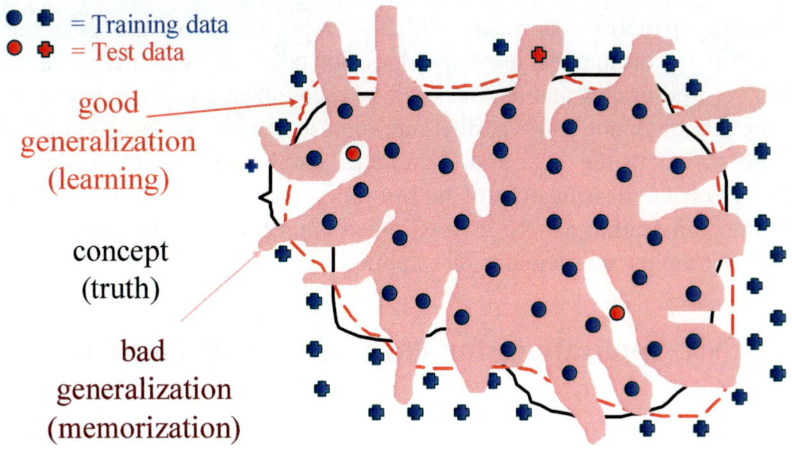

Fig. 12.14 Illustration of good and bad training results

- Initial values of synapse weights and biases (usually random values),
- The training algorithm itself (depends on the ANN type),
- Generalisation versus memorisation dilemma.

Having chosen the ANN type for given task (e.g. multilayer perceptron) one should decide on the size of the neural network. It has been proven that infinitely large neural network with just a single hidden layer is capable of approximating any continuous function [10]. In practice ANNs with one or two hidden layers are in use, with the number of neurons dependent on the number of ANN input signals and the size of training set. The neural network size is sometimes chosen arbitrarily, but it can also be optimized, e.g. with use of a genetic procedure (see Chap. 13).

The number of input signals and their choice is crucial for proper operation of the ANN. It is clear that the signals delivered to the input layer of the neural network should contain the highest amount of information about the events to be classified. Their choice should be done individually for given task, at best by an expert in the field. Both the training and testing sets should be representative in the sense of containing all possible situations that may happen in real life.

The last of the above-mentioned problems (generalization vs. memorization) results from the very fact that the well trained ANN is expected to operate correctly not only for all the patterns from the training data set, but also for all other possibly appearing data that were not presented to the ANN during training. Such a feature means generalization of the acquired knowledge, which is different from focusing on the training cases only (memorization). The difference is illustrated in Fig. 12.14, where the ANN is assumed to be able differentiating the circle points from the cross points. The two categories of points belong to respective parts of the plane that can easily be separated by the black solid line representing the desired division of the plane. When the ANN is "overtrained", it may become focused on the training patterns too much, which is represented by the pink contour in Fig. 12.14. Unfortunately, its shape does not guarantee that the points from the testing set (red circles and crosses) can be recognized correctly. Proper results of training (although not 100% perfect) are represented by the dashed red line, which encompasses both the blue and red circles, excluding all the cross points, thus enabling correct classification of both training and testing cases. The knowledge generalization feature can be assured when appropriate techniques, like randomization of training set elements, are applied.

12.3 ANN Applications for Power System Protection

The artificial neural networks have become popular and have been applied for power system problems for years. This technique arouse interest of researchers in many renowned R&D centers all around the world. Noticeable is also interest within well known organizations (CIGRE, IEEE) and committees (SC38 Power System Analysis and Techniques, SC39 Power System Operation and Control, SC34 Protection) that have established task groups to investigate possibilities of ANN application for power system problems. As a result a number of reports have been prepared, e.g. [6, 19, 28], that summarize latest achievements in the field. Cited references do not exhaust the topic completely, since new proposals are published in journals and conference proceedings every year.

The following (selected) examples of the ANN applications for power system protection and control tasks are quite representative [5, 7, 11, 18, 21, 24, 40–43]:

- Protection of transmission and distribution lines.
  - Fault detection and classification,
  - Fault direction discrimination,
  - Adaptive distance protection,
  - Distance relay for series compensated lines,
  - Autoreclosing and fault location functions,
  - High impedance faults detection,
  - High frequency based relaying.
- Power Transformer
  - Monitoring,
  - Protection functions,
  - Diagnostics.
- Fault location and analysis in substations.
- Other (non-protection) tasks
  - On-line security assessment,
  - Load forecasting,

– Optimisation tasks in power plants,
– Signal analysis,
– Process control and automation.

The authors of this book have also developed a number of ANN-based solutions for protection purposes, among which the following can be named [27, 29–37]:

- Out-of-step protection of synchronous machines,
- Transformer protection with neural inrush detection,
- Detection of high impedance faults,
- Current transformer saturation detection and correction.

For the reader's convenience selected example of ANN application for current transformer saturation detection is described in more detail. This example is representative enough and may serve as an illustration of design and successful application of ANN solution for a protection task [31, 33].

12.3.1 Example of ANN Application for CT Saturation Detection and Compensation

The overall performance of a protection relay is also a function of quality of the analog signal pre-processing path including current and voltage transformers, analog anti-aliasing filters and A/D converters. One of the most seriously deteriorating impacts on protection operation may be observed when traditionally used induction-type current transformers become saturated due to high AC fault current and/or DC current components. It is obvious that protection criterion values calculated on the basis of saturated CT secondary signal may fall quite distantly from their correct values, which might have been determined if the CT primary unsaturated signal was available. Erroneous measurement may in consequence lead to false decisions (e.g. underreaching of overcurrent relays, overestimation of fault loop impedance in distance relays) and protection maloperation [16].

Several approaches may be found in the literature to mitigate or eliminate the impact of CT saturation on protection operation. A large number of papers, which dealt with the CT saturation detection and correction issues, may be divided into four groups:

- No CT current correction is performed, but the information on CT saturation is used for other purpose, e.g. for adaptation of protection settings [37] or performing calculations during CT unsaturated periods only [2].
- Correction performed is based on the information extracted from the secondary current during the saturation interval, which reproduces the primary current waveshape [12–14].

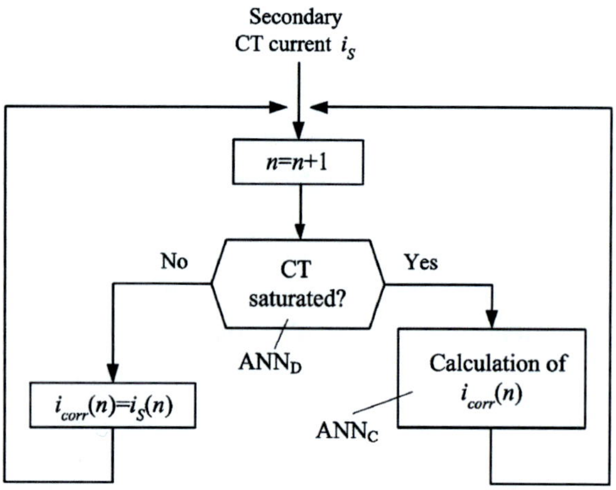

Fig. 12.15 Block scheme of the CT saturation detection and compensation

- Correction is based on information extracted from the secondary current during the non-saturation interval, which aims at determination of the fundamental and DC components of the primary current [23].
- CT saturation detection and/or correction is done with use of the Artificial Neural Networks, trained with the simulation cases of CT transients [3, 17, 39, 44].

Below an approach based on application of artificial neural networks is described. The performance of ANN-based decision units depends to a large degree on the cases used to train the networks. With the variety of possible cases it may be hard ensuring that all possible CT characteristics, residual fluxes, loads and primary current parameters are covered. The solutions described in [17, 39, 44] are quite promising, however, they are based on recurrent networks (prone to instability under some circumstances, [39]) or a set of ANNs designed for sublevels of measured current amplitudes [3], which the authors find as non-optimal.

The basic idea of handling the CT saturation problem adopted here is based on splitting the task into two subtasks, namely saturation detection and correction of the distorted secondary current (Fig. 12.15). When the CT is not saturated (i.e. the detection block has not detected it) the correction block is not activated. Starting from the point of saturation beginning the procedure of secondary current correction is activated. The procedure is operative until the CT goes out of saturation. The process is repeated when the detection block affirms saturation beginning again.

Preparation of the neural networks for both subtasks, choosing proper size of the networks and their input vectors as well as training and testing of the solutions are outlined. The neural CT saturation detection and correction units have been tested with EMTP-ATP-generated current signals. Special attention is also paid to ANN implementation problems. Practical implementation of the neural schemes in real-time with use of a signal processor is also described.

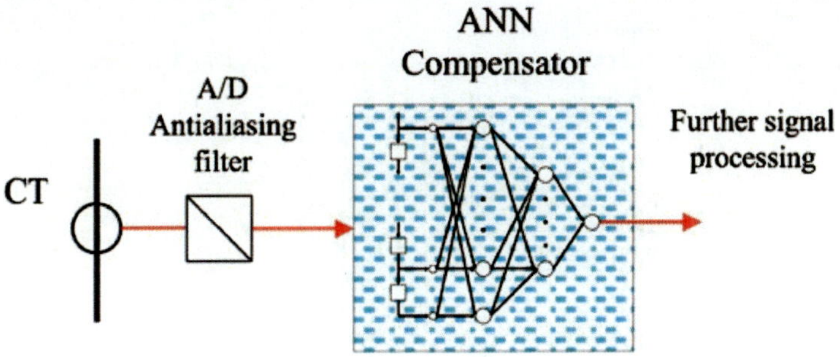

Fig. 12.16 Basic scheme of CT saturation compensation with ANN

Fig. 12.17 Feed-forward ANN structure for CT saturation detection/correction

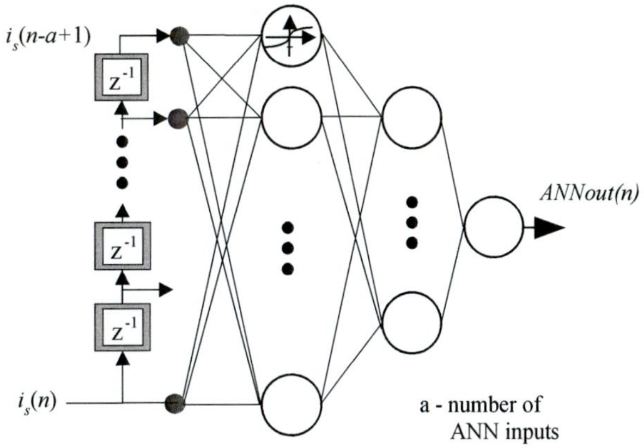

The basic idea of CT saturation compensation with neural networks is depicted in Fig. 12.16. The CT secondary current, after low-pass filtering and converting to digital form, is passed to an ANN, at the output of which the reconstructed primary CT current should appear. The ANN output is further digitally processed to get certain relay criterion values, the ones needed for decision making in the particular relay under consideration.

The structure of ANNs for saturation detection and correction is shown in Fig. 12.17. Two- or three-layer feed-forward ANNs were used with non-linear neurons in input and hidden layers and fully linear output neuron. In case of the saturation detection a single neural network ANN_D (2 layers, structure 13-1) was used for detection of both saturation beginning and ending points. The ANN_D structure was a result of genetic optimization described in [30]. The network was trained to produce output equal 1.0 during the period when the CT was saturated, not only at the time instants when it goes into and out of saturation. Thus the border points of saturation intervals are detected when the ANN_D output changes its state from 0 to 1 and opposite.

A number of options were considered for the ANNs intended for CT saturation correction. The size of the networks ANN_C was fixed (three-layer, 10-7-1 neurons)

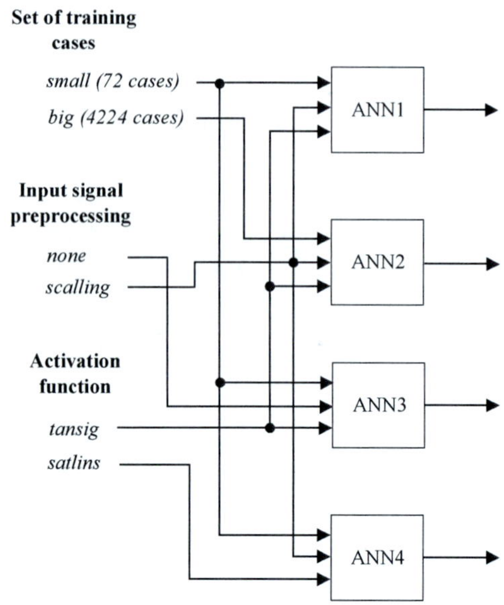

Fig. 12.18 ANNs and training versions considered

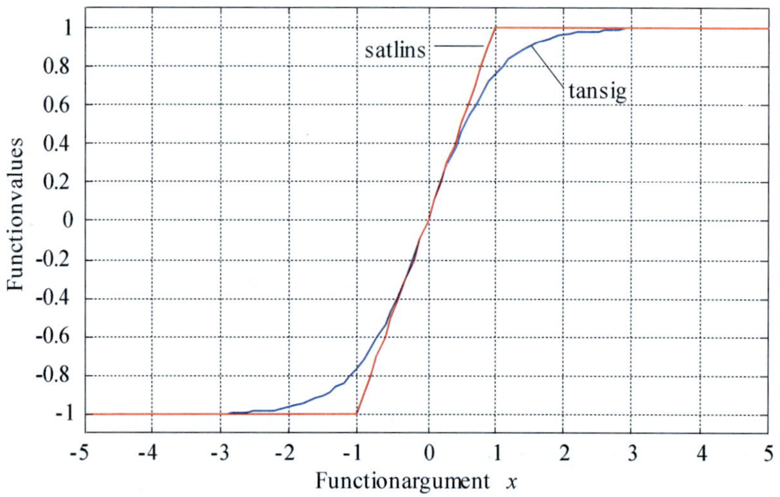

Fig. 12.19 Comparison of activation functions (12.4) and (12.9)

while the other parameters like neuron activation functions, input signal pre-processing (none or scaling) and size of training set were being changed in order to identify the best possible configuration and training options. The parameters that influence the ANN versions are all gathered in Fig. 12.18. In first three ANNs from Fig. 12.18 the non-linear *tansig* activation function (12.4) was used in input and hidden layers. The fourth ANN was built of neurons equipped

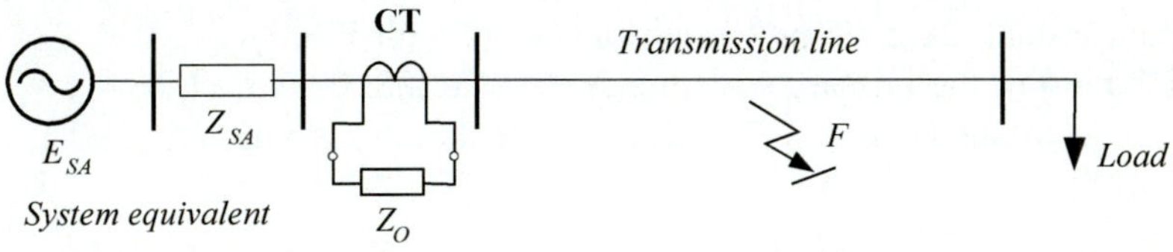

Fig. 12.20 Schematic diagram of the system with CTs under study

with *satlins* activation function. This option was introduced to check how the simplification of activation function influences the performance of ANN and computational burden of the processor, provided the ANN is implemented in hardware. The *satlins* function is defined as

$$\text{satlins}(x) = \begin{cases} -1, & x < -1 \\ x, & |x| \le 1 \\ 1, & x > 1 \end{cases} \tag{12.9}$$

and both activation functions are compared in Fig. 12.19.

The input vector of all ANNs consisted of $a = 15$ most recent samples of the secondary CT current i_s (3/4-cycle sliding window, sampling with 1 kHz). The network ANN3 was trained without any scaling of the input signals, while for three other ANNs their input vector was being scaled (all the samples of the input vector were divided by the maximum over considered data window). Target values adopted during training were primary current samples $i_p(n)$ related to the secondary side that were available from EMTP-ATP simulations. All the ANNs were trained to be used during CT saturation periods only.

The testing of developed units for CT saturation, detection and correction was performed with EMTP-ATP generated signals [8]. Simple 110 kV transmission system fed from one side was modeled, as shown in Fig. 12.20. It comprises the equivalent of feeding system represented by an electromotive source behind an impedance (positive and zero sequence values, $X_0/X_1 = 1.5$) and transmission line transferring power to the end user (load). The simulation model from Fig. 12.20 was equipped with suitable non-linear model of CT, [20]. The CT model developed was an equivalent of standard current transformer 5P20. The last point of the flux-current curve was set at various levels (flux coordinate) to obtain CTs with a number of different inductances and saturation time constants of the secondary circuit. The transformation ratio of CT under study was $\vartheta_{CT} = 500{:}1$ (A/A).

Thorough simulation study of the current transformer operation under various conditions has been performed. Phase-to-ground or phase-to-phase faults in the middle or at the end of transmission line were simulated. To obtain a wide variety of cases the following parameters of the system and transient conditions were being changed in each simulation run:

- Level of the primary current I_{p1} during fault (resulting from the strength of the supplying system): [4.8, 5.2, 5.7, 6.2, 6.9, 7.7, 8.8, 10.2, 12.2, 15.0, 19.7, 28.5] $\times I_n$,

- Fault instant: 20:2:40 ms (various angles of fault inception),
- R/X ratio of the primary system side: [0.07, 0.08, 0.09, 0.1],
- Time constant T_s of the CT secondary side during saturation: [0.3, 0.5, 1, 1.5, 2, 3, 4, 5] ms.

With the above mentioned options a number of 4224 simulation cases have been generated. Further signal processing and testing of the CT detection and correction methods was performed in Matlab environment.

Thorough investigations on efficiency of designed ANN-based CT saturation detector are described in [30]. Its operation was compared with chosen deterministic CT saturation identification scheme based on calculation of the 3rd derivative of CT secondary current [15]. Despite small size of the ANN applied (14 neurons) all starting and ending points of CT saturation intervals were properly detected. In addition, no unnecessary excitations were observed at the beginning of fault, which took place when the non-AI method was applied. The latter feature of the "classic" method was an effect of the fact that the method responds with high peaks in 3rd derivative to any sudden change in current waveform. The method could react properly (with visible impulses) at the beginning of saturation, but had problems with detecting saturation endings, when the CT turns to unsaturated operation mode with much smoother waveform change.

Below the features of designed ANN-based CT saturation compensators are presented with higher attention. The investigations have revealed that the performance of neural compensation units was almost perfect when the network ANN2 (trained with a large number of EMTP cases) was applied. Although only one-third of all patterns were shown to the ANN during training, the robustness and ability to knowledge generalization were more than enough to enable correct operation of the compensator for most cases. The correction errors increase a little bit only for high amplitudes of the primary current for which the internal ANN signals may fall very deep into the regions where the neurons' activation functions are heavily saturated. The network trained on narrow base of patterns (ANN1) was somewhat worse than the ANN2, however the results of operation were not far away from the ideal ones. The network ANN3 (without current scaling) performed badly, which can be seen in Fig. 12.21. The quality of primary CT current reconstruction with use of ANN2 is several times higher than with other ANNs. Definitely the worst performance was observed on average for the ANN3 solution (without scaling), which confirms the significance of properly selected pre-processing of the ANN input signals. It was observed that all the ANNs perform with higher accuracy when the ratio of CT saturation time constant T_s to the sampling period T_S is higher and when the time constant of decaying DC component in CT primary current is higher.

Additional analyses have been done for the network ANN4 (activation function *satlins*) that was not compared in broader sense with other ANNs. The aim of the studies was to check the influence of the simplified activation function on the primary current reconstruction accuracy. As seen in Fig. 12.22, the operation of compensator ANN4 is basically good (the CT primary current waveshape is

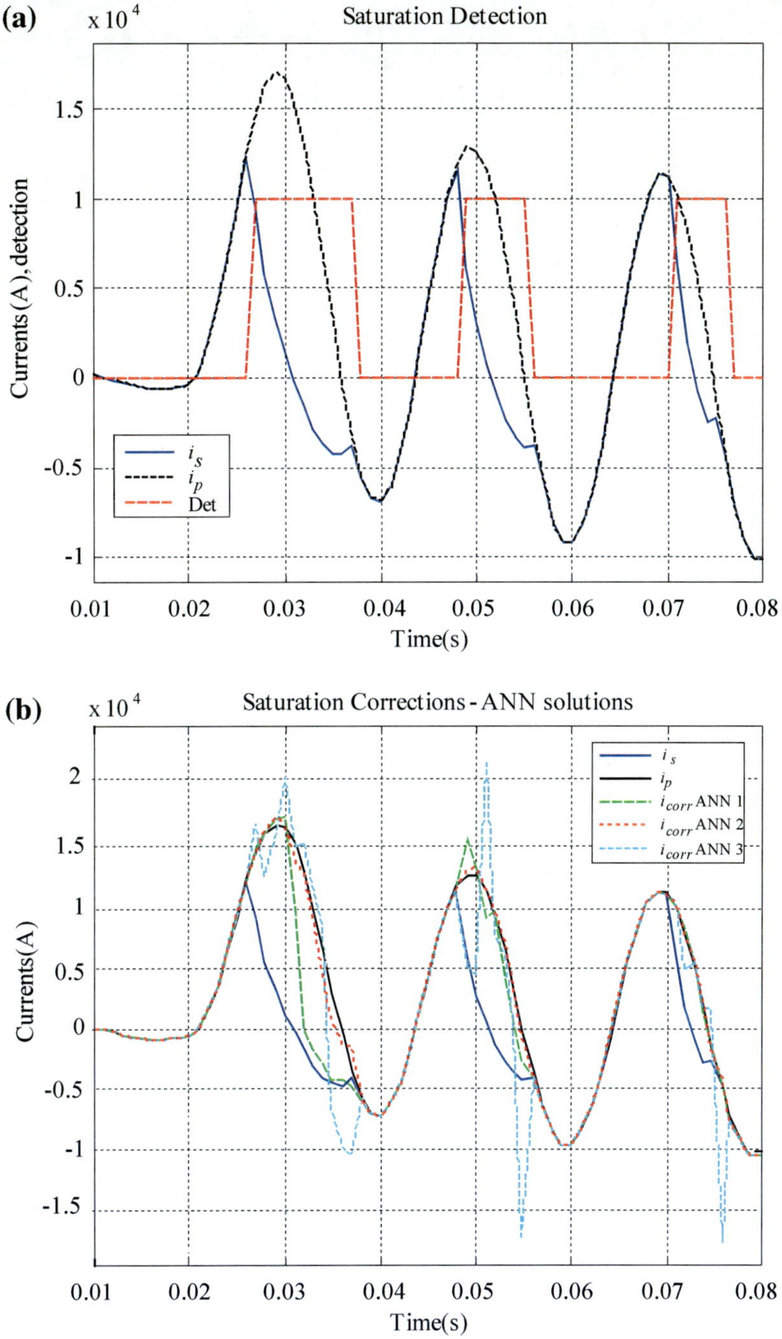

Fig. 12.21 Testing results for the case with $T_s = 1.5$ ms, $I_{p1} = 15I_n$: **a** detection, **b** correction

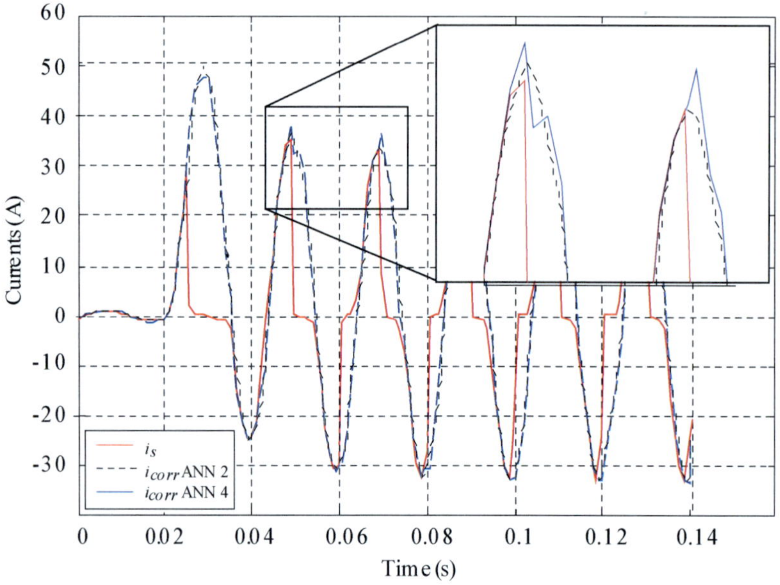

Fig. 12.22 Comparison of accuracy of primary current reconstruction with ANNs equipped with various activation functions (ANN2-tansig, ANN4-satlins)

reconstructed correctly) and lost of accuracy is not very big. The highest errors are observed around the curve maxima where the signal obtained with ANN4 in not as smooth as the one reconstructed by network ANN2. The other aspects related to possibly shorter execution time by on-line implementation of such a network is touched upon below. If the time benefit is high enough and certain loss of reconstruction accuracy is not critical for given application (e.g. in overcurrent relay) then one might say that application of neural network ANN4 is fully recommended.

The technical implementation of the developed ANN-based CT saturation compensators has also been performed. The laboratory stand accomplished comprises the IBM PC computer that was used as a supervisory programming and control unit, the EVM board with TMS320C30 signal processor (TI family, 30 MHz clock, 32-bit, floating-point) and A/D converter (14-bit, with low-pass anti-aliasing filters) that was connected to the ISA PC bus, Fig. 12.23. The testing signals were delivered by the Programmable Function Generator, capable of generating any desired voltage waveform defined by the user with 12-bit resolution. Its programming was done from PC via the RS232C serial interface. The signals to be played in hardware were imported from the files obtained by EMTP-ATP simulations.

In the first step the execution time of two numerical algorithms representing the neural networks ANN2 and ANN4 was determined by measurement. It was found that when the *tansig* function was used (ANN2) the time needed was equal 1150 μs, which is more than the sampling period (1 ms). One of the reasons of

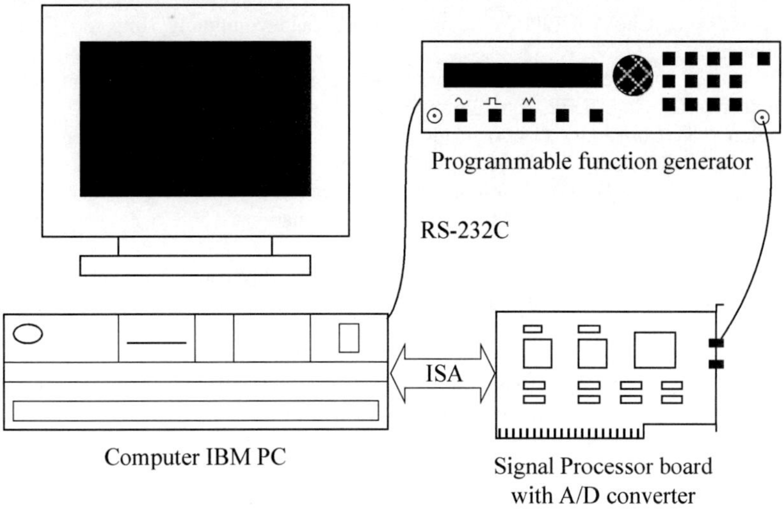

Fig. 12.23 Layout of the laboratory stand with TMS signal processor

such a low calculation speed was the exponent function, realization of which is quite time-consuming in real time. Execution of the algorithm with activation function *satlins* (ANN4) took ca. 960 μs which is a bit less than the time between two consecutive signal samples. Possibilities of time savings were searched by modernizing the algorithm from the numerical point of view. Time-consuming loop instructions were replaced with multiple repeated simple operations. This enabled shortening the execution times ca. by half, i.e. to 650 μs for the network ANN2 and to 420 μs for the network ANN4. It is to be stressed that the real-time tests were done with relatively slow signal processor and even by such a hardware arrangement the required time for execution of the algorithms of the neural CT compensators within one time-step were close to 0.4 ms. Deeper entry in assembler could bring about considerable shortening of the algorithm execution time. With faster signal processors (available at present above 200 MHz) and the entire program code written in the machine language even 10-time higher sampling rates could be used. That means that practical implementation of the ANN-based CT saturation compensators is utterly attainable.

References

1. Bothe HH (1998) Neuro-Fuzzy-Methoden. Einfuehrung in Theorie und Anwendungen. Springer-Verlag GmbH, Berlin Heidelberg, ISBN: 978-3-540-57966-4
2. Bunyagul T, Crossley P, Galac P (2001) Overcurrent protection using signals derived from saturated measurement CTs. In: Proceedings of PES summer meeting, vol 1. Vancouver, pp 103–108
3. Chen KW, Glad ST (1991) Estimation of the primary current in a saturated transformer. In: Proceedings of the 30th conference on decision and control, vol 3. Brighton, England, pp 2363–2365

4. Chow TWS, Cho SY (2007) Neural networks and computing. Imperial College Press, London
5. Dalstain T, Kulicke B (1995) Neural network approach to fault classification for high speed protective relaying. IEEE Trans Power Deliv 10:1002–1011
6. Dillon TS (Convenor) (1995) Fault diagnosis in electric power systems through AI techniques. Report by TF 38.06.02, Electra 159
7. El-Sharkawi MA (1995) Neural network application to high performance electric drives systems. In: Proceedings of the IEEE IECON international conference, vol 1, pp 44–49
8. EMTP-ATP Manuals (2001), EEUG
9. Fuller R (2000) Introduction to neuro-fuzzy systems. Physica-Verlag, Springer, Heidelberg
10. Funabashi K (1989) On the approximate realization of continuous mappings by neural networks. Neural Networks 2:183–192
11. Halinka A, Winkler W, Witek B (1995) Fault detection and recognition in generator-transformer units by neural network based adaptive protection. In: Proceedings of the 30th UPEC conference, vol 1. London, UK, pp 82–84
12. Kang YC, Lim UJ, Kang SH (2004) Compensating algorithm for the secondary current for use with measuring type current transformers. In: Proceedings of the international conference on advanced power system automation and protection, Jeju, Korea, pp 3–8
13. Kang YC, Park JK, Kang SH, Johns AT, Aggarwal RK (1996) Development and hardware implementation of a compensating algorithm for the secondary current of current transformers. IEE Proc Electr Power Appl 243:41–49
14. Kang YC, Park JK, Kang SH, Johns AT, Aggarwal RK (1997) An algorithm for compensating secondary current of current transformer. IEEE Trans Power Deliv 12:116–124
15. Kang Y, Kang S, Crossley P (2003) An algorithm for detecting CT saturation using the secondary current third-derivative function. In Proceedings of the IEEE Bologna powertech conference, pp 320–326
16. Kasztenny B, Mazereeuw J, Jones K (2001) CT Saturation in industrial applications—analysis and application guidelines. GE Multilin, Canada, Publ. GET-8501
17. Kasztenny B, Rosołowski E, Łukowicz M, Iżykowski J (1997) Current related relaying algorithm immune to saturation of current transformers. In: Proceedings of the developments in power system protection, Conference publication No. 434, pp 365–368
18. Kezunovic M et al (1994) Neural network applications to real-time and off-line fault analysis. In Proceedings of the conference intelligent system application to power systems, Montpelier, France, pp 665–671
19. Kezunovic M et al (1998) Practical intelligent system applications to protection, and substation monitoring and control. In: Proceedings of the CIGRE Session, Paris, France, Paper 34–104
20. Kezunovic M, Fromen CW, Phillips F (1994) Experimental evaluation of EMTP-based current transformer models for protective relay transient study. IEEE Trans Power Deliv 9:405–413
21. Koglin HJ, Kostyla P, Lobos T, Waclawek Z (1988) Voltage waveforms analysis for arcing faults detection on transmission lines. In: Proceedings of the 11th international conference on power system protection, Bled, Slovenia, pp 147–152
22. Kohonen T (1984) Self-organization and associative memory. Springer, New York
23. Li F, Li Y, Aggarwal RK (2002) Combined wavelet transform and regression technique for secondary current compensation of current transformer. IEE Proc Gener Transm Distrib 149:497–503
24. Lu CN, Wu HT, Vemuri S (1992) Neural networks based short term load forecasting. In proceedings of the IEEE power engineering society winter meeting, WM 12-55 PWRS
25. Łukowicz M, Rosołowski E (1998) Artificial neural network based dynamic compensation of current transformer errors. In: Proceedings of the 8th international symposium on short circuit currents in power systems, Brussels, Belgium, pp 19–24
26. MATLAB - Neural Network Toolbox

27. Michalik M, Lukowicz M, Rebizant W, Lee SJ, Kang SH (2008) New ANN-Based algorithms for detecting HIFs in multigrounded MV networks. IEEE Trans Power Deliv 23:58–66
28. Neibur D (Convenor) (1995) Artificial neural networks for power systems. Report by TF 38.06.06, Electra 159
29. Rebizant W (2000) ANN based detection of OS conditions in power system. In: Proceedings of the 12th international conference on power system protection PSP2000, Bled, Slovenia, pp 51–56
30. Rebizant W, Bejmert D (2005) Current transformer saturation detection with genetically optimized neural networks. In: Proceedings of the IEEE powertech conference, St. Petersburg, Russia, paper 220
31. Rebizant W, Bejmert D (2007) Current-transformer saturation detection with genetically optimized neural networks. IEEE Trans Power Deliv 22:820–827
32. Rebizant W, Bejmert D, Schiel L (2007) Transformer differential protection with neural network based inrush stabilization. In: Proceedings of the IEEE powertech conference, Ecole Polytechnique Federale de Lausanne, Switzerland, paper 607
33. Rebizant W, Bejmert D, Staszewski J, Schiel L (2007) CT saturation detection and correction with artificial neural networks. In: Proceedings of the 2nd international conference on advanced power system automation and protection, Jeju, Korea, paper 504
34. Rebizant W, Szafran J, Feser K, Oechsle F (2001) Evolutionary improvement of neural classifiers for generator out-of-step protection. In: Proceedings of the IEEE porto powertech conference, Porto, Portugal, vol 4, paper PRL1-223
35. Rebizant W, Szafran J, Feser K, Oechsle F (2002) Evolutionaere Optimierung neuronaler Klassifikatoren fuer den Generatorschutz. ELEKTRIE, Berlin 56:51–56
36. Rebizant W, Szafran J, Oechsle F (2001) Out-of-step protection with genetically optimized neural networks. In: Proceedings of the 10th international conference on present-day problems of power engineering, vol. 2. Gdansk-Jurata, Poland, pp 39–46
37. Rebizant W, Hayder T, Schiel L (2004) Prediction of CT saturation period for differential relay adaptation purposes. In: Proceedings of the international conference on advanced power system automation and protection, Jeju, Korea, pp 17–22
38. Rumelhart DE, MccLelland (1986) Parallel distributed processing: exploration in the microstructure of cognition. MIT Press, Cambridge
39. Saha MM, Iżykowski J, Łukowicz M, Rosołowski E (2001) Application of ANN method for instrument transformer correction in transmission line protection. In: Proceedings of the IEE development in power system protection conference, Publication No. 479, pp 303–306
40. Santoso NI, Tan OT (1990) A neural network based real-time control of capacitors installed on distribution systems. IEEE Trans Power Deliv 5:266–272
41. Uhrig RE (1991) Potential applications of neural networks to nuclear power plants. In Proc Am Power Conf 53(2):946–951
42. Wong KP, Phuoc TN, Attikiouzel Y (1990) Transient stability assessment for single machine power systems using neural networks. In: Proceedings of the IEEE conference on computer and communication systems, pp 32–36
43. Wu QH, Hog BW, Irwin GW (1992) A neural network regulator for turbo generators. IEEE Trans Neural Networks 3:95–100
44. Yu DC, Cummins JC, Wang Z, Yoon HJ, Kojovic LA (2001) Correction of current transformer distorted secondary current due to saturation using artificial neural networks. IEEE Trans Power Deliv 16:189–194

Chapter 13
Genetic and Evolutionary Algorithms for PSP

In this chapter the procedures originating from biological evolution are presented that can be applied e.g. for intelligent searches and optimization in power systems. The procedures belong to the so-called "biological programming" family, which is not limited to the genetic algorithms (GA) only. In wider sense the neural networks described in Chap. 12, being an analogy of human brain, are also good examples of this family.

The procedures presented below follow from the Darwin theory of evolution of species. Many examples prove that Darwinian mechanism generates a kind of optimization process. Simplifying, one can say that in real world only the best individuals from given population have a chance to survive, and this is forwarded to the next and further generations of offsprings, which become better and better adjusted to the surrounding world. As such the mechanism should deliver the best solutions to given problem (here—survival), which is now tried to be transferred to some other domains, including physics, economy and natural and technical sciences.

It can be stated that in many domains one has to deal with the classical problem of optimization. Economy particularly has become a specialist of that field. Generally speaking, a large part of mathematical development during the XVIII century dealt with that topic (remember those always repeated problems where you had to obtain the derivative of a function to find its extremes). One can surely admit that purely-analytical methods widely proved their efficiency; nevertheless, they suffer from an insurmountable weakness: reality rarely obeys to those wonderful differentiable functions that you try to use to describe the world around.

John Holland, from the University of Michigan began his work on (GA) at the beginning of the 1960s. A first achievement was the publication of *Adaptation in Natural and Artificial System* in 1975 [12]. Both genetic and evolutionary techniques deliver their solutions performing statistical search within a pre-defined population. It is important to understand that the functioning of such an algorithm does not guarantee success. We are in a stochastic system and a genetic pool may be too far from the solution, or for example, a too fast convergence may halt the

W. Rebizant et al., *Digital Signal Processing in Power System Protection and Control,* Signals and Communication Technology,
DOI: 10.1007/978-0-85729-802-7_13, © Springer-Verlag London Limited 2011

process of evolution. These algorithms are nevertheless extremely efficient, and are used in fields as diverse as stock exchange, production scheduling or programing of assembly robots in the automotive industry.

Below the selected aspects of genetic and evolutionary algorithms are outlined. Special attention is paid to their application for protection and control in power systems. The theory is then followed by selected examples of genetic optimization of ANNs as well as genetic search for optimal settings of over-current relays.

13.1 Basics of Evolution and Genetics for Technical Problems

Evolutionary and genetic algorithms are a result of emulation (numerical implementation) of the evolution principles observed in the nature, with particular concern given to natural selection appearing in the population of living beings. Evolution is a natural process leading to the maintenance of a population's ability to survive and reproduce in a specific environment. This ability (quality) is called evolutionary fitness that can also be viewed as a measure of the organism's ability to anticipate changes in its environment. Emulation of natural evolution is an iterative process involving creation a population of individuals, evaluating their fitness, generating a new population through genetic changes and repeating the steps a number of times.

Generally speaking, among the stochastic search methods that mimic the metaphor of natural biological evolution one can distinguish:

- Genetic Algorithms (GAs), the most general optimization approach with binary coding and full range of genetic operators) [11].
- Evolutionary strategies (ESs, use natural problem-dependent representations, and primarily mutation and selection as search operators) [4].
- Evolutionary programing (EPs, the solutions are in the form of computer programs, and their fitness is determined by their ability to solve a computational problem) [8, 16].

In this chapter the GAs are only described; however, the examples provided later belong both to GA and ES groups of methods.

The GAs are mainly applied for solving various optimization problems. It can be said that genetic optimization iteratively improves the quality of solutions until an optimal, or at least feasible, solution is found. From the other traditional optimization approaches the GAs are singled out with the following features:

- The GA does not process the problem parameters directly but in a coded form.
- Searching for an optimum is performed commencing not from a single starting point but from a certain population of initial guesses.
- The genetic optimization process is controlled by suitably defined goal function.
- Probabilistic selection rules are applied.

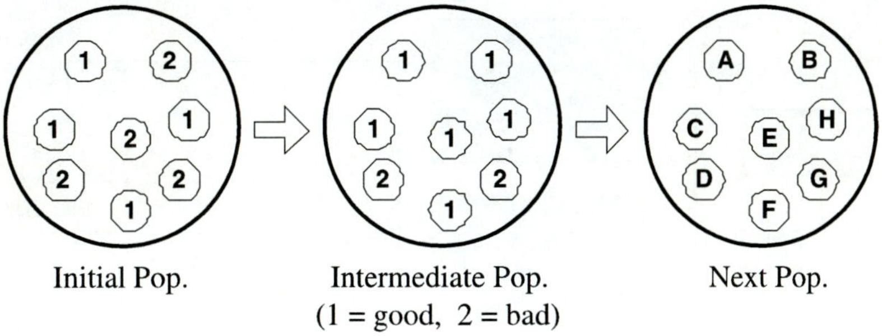

Fig. 13.1 Basic scheme of the genetic-optimization process

In the field of genetic optimization the concepts and definitions borrowed from genetics are exploited, such as

- *Population*—a set of individuals of selected number (living in the same environment and competing with each other).
- *Individuals*—representatives of the prospective problem solutions given in a coded form (treated also as points in the search universe).
- *Chromosomes* (coding chains or sequences)—ordered sequences (vectors) of *genes*, encoding all the information about an individual.
- *Gene*—single element of the *genotype*, and in particular of the chromosome.
- *Genotype* (i.e., words-structure)—a set of chromosomes of given individual.
- *Phenotype*—a set of values corresponding to given genotype, i.e., its decoded structure (the "outward, physical manifestation" of the organism, the organism itself).

Natural principles according to the theory of Darwin, such as heredity, crossover, mutation, and selection, are used over several generations to develop and improve the characteristics of the individuals. An important role in GA plays the idea of *adaptation function*, which is also called *matching function* or *goal function*. The adaptation function allows assessing the adaptation grade of each individual in the population, which forms the base for selection of the best individuals, according to the general rule that "only the strongest can survive".

The general scheme of the basic genetic procedure is presented in Fig. 13.1. The algorithm starts from creating of the initial population, which can be understood as a random choice of individuals that are represented by vectors of their encoded parameters (chromosomes). In each iteration of GA the adaptation grade of each individual is assessed, basing on which the intermediate population is established with the best individuals from previous stage being represented more frequently. The members of intermediate population are then subjected to further genetic modifications and as a result the next population of individuals is created.

Consecutive iterations of GA are called *generations*, while the new individuals can be treated as *offspring* (descendants) of elder population. The GA usually runs in a closed loop with the interruption conditions defined according to the specific requirements of the problem to be solved. In optimization analyses the algorithm

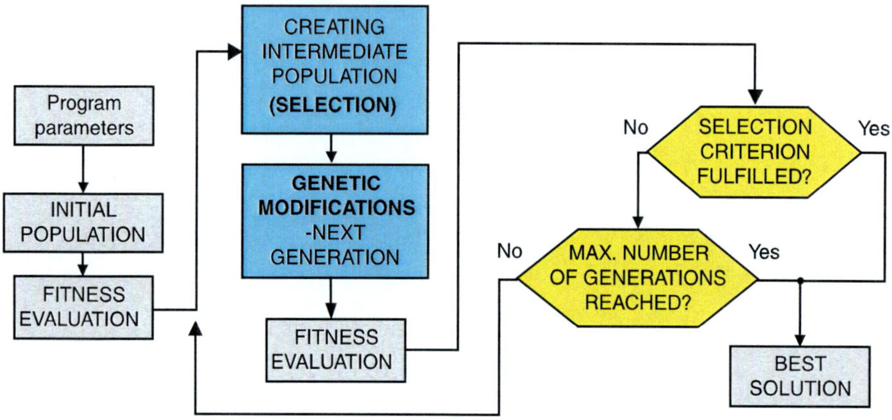

Fig. 13.2 GA optimization loop

interruption can occur when preset level of the goal function is reached, if need be—with given accuracy. The GA procedure can also be interrupted when its further operation does not improve the optimization outcomes (no better results can be expected). If the GA interruption conditions are satisfied the best chromosome is accepted as final solution. Otherwise the algorithm operation is continued until certain time has passed or the prescribed number of iterations is reached (Fig. 13.2).

13.1.1 Selection Versions

After the individuals of the population have been assessed the best of them are selected for further processing (genetic changes). The intermediate population is formed from the individuals selected according to one of the following methods (not exclusive list):

- Wheel of roulette (individuals are ordered according to their fitness; the better their chromosomes are, the more chances to be selected they have), Fig. 13.3.
- Stochastic universal sampling (a number of *NP* parents are selected from the ranked set of individuals), Fig. 13.4.
- Truncation selection (first x% of individuals with the highest fitness are selected for further breeding).
- Tournament selection (randomly selected pairs of individuals compete, the ones with higher fitness win).

The idea of the wheel of roulette method is illustrated in Fig. 13.3. In this example the population consists of ten individuals represented by their chromosomes (Chr1–Chr10). The sizes of wheel sections correspond to the quality of individuals, thus one may expect that the individuals with higher fitness would

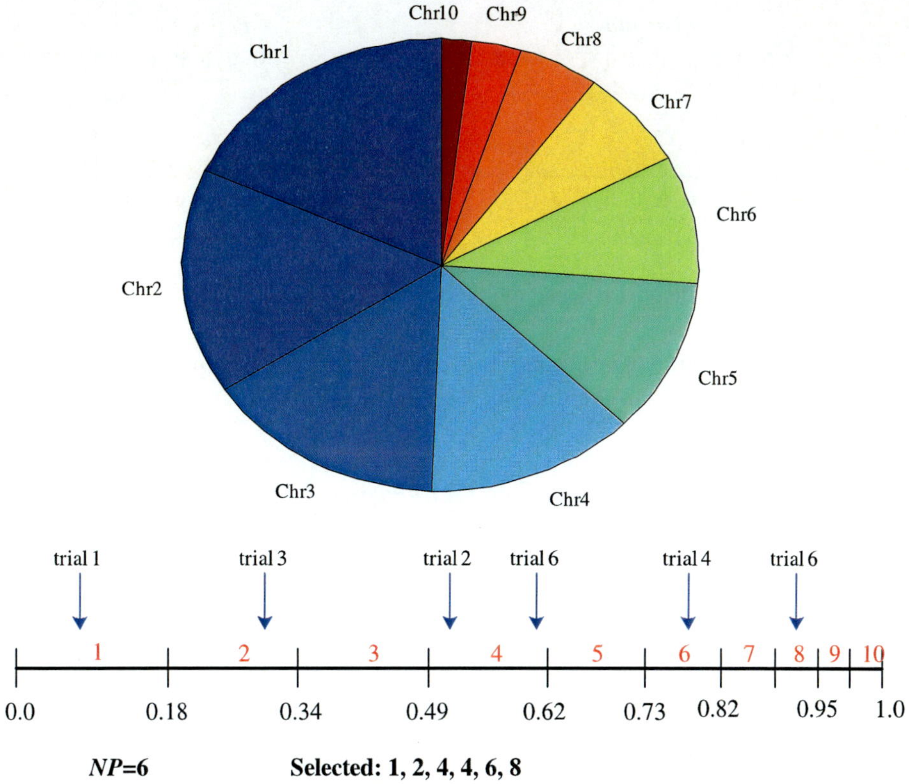

Fig. 13.3 Selection with the wheel of roulette method

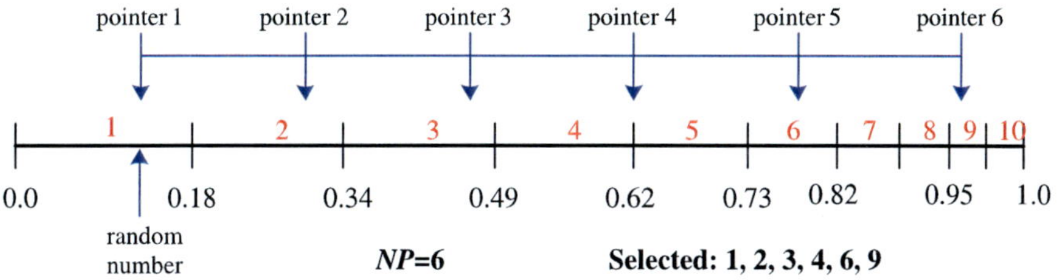

Fig. 13.4 Selection with the stochastic universal sampling method

have higher chance to proceed to the next generation. One can see that both "good" and "poor" individuals can be selected; moreover, some of them can be drawn many times. Here, after NP = 6 trials (wheel revolutions) the individuals 1, 2, 4, 4, 6 and 8 were selected.

The procedure of stochastic universal sampling is shown in Fig. 13.4. Here, the comb of pointers separated one from another by the same distance is used to indicate the selected set of individuals. Its left-hand side position is determined by drawing a random number. With this method the probability of repeated selection

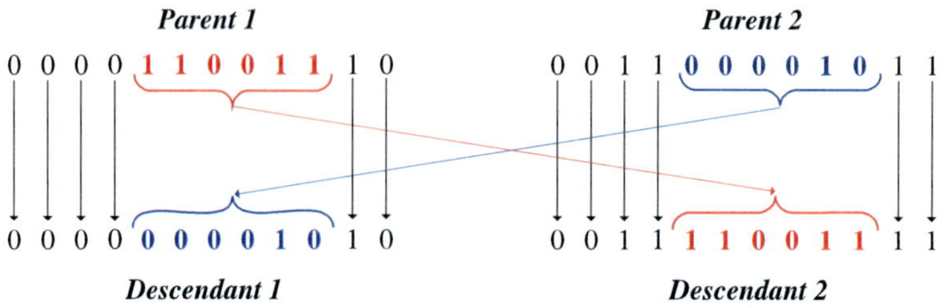

Fig. 13.5 An example of two-point crossover

of the same individual is limited and can happen only when the section for an individual is wider than the distance between the pointers.

13.1.2 Emerging of the Next Generation

The bunch of selected individuals is further used to produce the population of the next generation. The descendants/offsprings are created from their parents in the process of crossover, mutation and recombination, whereas the chromosomes of descendants are a result of chromosome material changes of two (or just one) parent individuals. The genetic operations mentioned should be understood as

- Crossover—exchange of some parts of chromosome information of two selected parent individuals, Fig. 13.5.
- Mutation—small change of the chromosome information (usually at one or two positions of the chromosome string), Fig. 13.6.
- Recombination—creating of new individuals by finding their "location" along a line or at line crossings (discrete or real-valued recombination), Fig. 13.7.

The crossover on two strings of genes can be single-point (when the chromosomes are cut at single point, the same for both individuals), multi-point (with multiple cutting points, in Fig. 13.5—two-point cutting) or uniform operation (when bits are randomly copied from the first or from the second parent).

It should be mentioned that the individuals to take part in given genetic operation are selected from the intermediate population randomly. The probabilities of genetic operations should be defined at the beginning of optimization procedure, being constant numbers or values changing with the generation number (with time). It can happen that not all individuals are subjected to genetic operations, some of them can pass to the next generation without any change. It is usual practice that the population of next generation contains the same number of individuals as the preceding one, which simplifies automation of the GA procedure.

Summarizing, one can enumerate the following steps of GA implementation:

- Specify the problem, define constraints and optimality criteria,
- Chose the way of encoding (binary, permutation, value, tree ...),

Parent	*Offspring*
0 0 0 0 1 1 0 0 1 1 1 0	0 0 0 0 1 1 0 1 1 1 1 0

Fig. 13.6 Mutation example

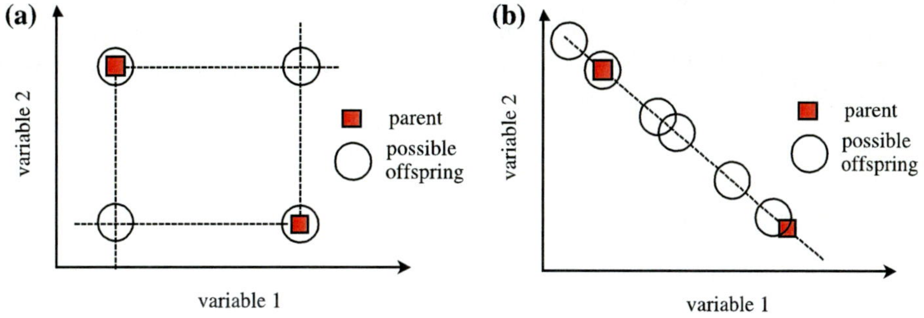

Fig. 13.7 Recombination examples: **a** 2-D grid locations, **b** locations along a line

- Represent the problem domain as a chromosome,
- Define a fitness (goal) function to evaluate the chromosome performance,
- Construct the genetic operators (way of implementation for given encoding),
- Define probabilities for genetic operations (constant or changing),
- Generate initial population of individuals,
- Run the GA and tune its parameters.

13.2 Application Examples

The genetic and evolutionary algorithms have been widely used for many technical problems, including optimization in nonlinear and dynamical systems, designing neural networks, strategy planning, function minima finding, etc. They are also used for machine learning and for evolving simple programs.

The applications of GA and ES techniques for power-system problems are also numerous, including:

- Power-system planning (optimal location of FACTS devices) [9],
- Economic load dispatch [15],
- Optimal power flow [17],
- Unit commitment problem solving [3].

and specifically for protection and control issues:

- Supply restoration and optimal load shedding [18],
- Power-system stabilizer design [1],

- Power-system stability margin determination [7],
- Frequency relaying based on GA using FPGAs [6],
- ANN optimization for the generator out-of-step protection [22, 23],
- Optimization of the ANNs for CT saturation detection task [20, 21],
- Optimization of generator-voltage regulator (AVR) settings [19].

Below two examples of GA application for optimization of neural network structure as well as for finding optimal setting of the generator-voltage regulator (AVR) are presented that are representative enough to illustrate all the issues mentioned in the above introductory part.

13.2.1 Optimization of the ANN Structure

The GA principles described have been successfully used by the authors in ANN optimization for the generator out-of-step protection decision module [22, 23]. Here, the procedure as applied for optimization of the ANNs for CT saturation detection task is presented [20, 21]. The general idea of using neural networks for current transformer saturation detection and compensation was outlined in Chap. 12. With application of GA procedure the neural network structure was being sought that was required to be optimal from the viewpoint of both operational efficiency and small number of neurons. The latter requirement is to be observed when the ANNs should be implemented on-line, i.e., when the execution time of its code is of importance.

In Fig. 13.8 block scheme of the genetic optimization procedure as applied for ANN structure optimization is given. At the beginning an initial population of neural networks is randomly created. While the number of neurons in the input and output layer is fixed according to the classification problem, the number of hidden layers and the number of neurons in these layers are randomly selected. The neural networks are trained with selected typical patterns and validated with all available patterns. After the adaptation function of each individual was determined (quality index Q calculated), an intermediate population is created where successful individuals are reproduced more likely. This may be realized with one of following selection algorithms (Fig. 13.9):

- "Wheel of roulette" method (a segment of circle proportional to ANN quality index is assigned to each individual, ANN selection for intermediate population is done by drawing, Fig. 13.9a),
- "Tournament" method (the ANNs are joined in pairs at random, the individuals compete with each other to determine which of the two can introduce its two replicas into the intermediate population,and whether deterministic as well as probabilistic selection is possible, Fig. 13.9b),
- "Ranking place" method (a ranking list according to ANN quality indices is created, intermediate population is composed of the first half of individuals (top of the ranking list) and their replicas, Fig. 13.9c).

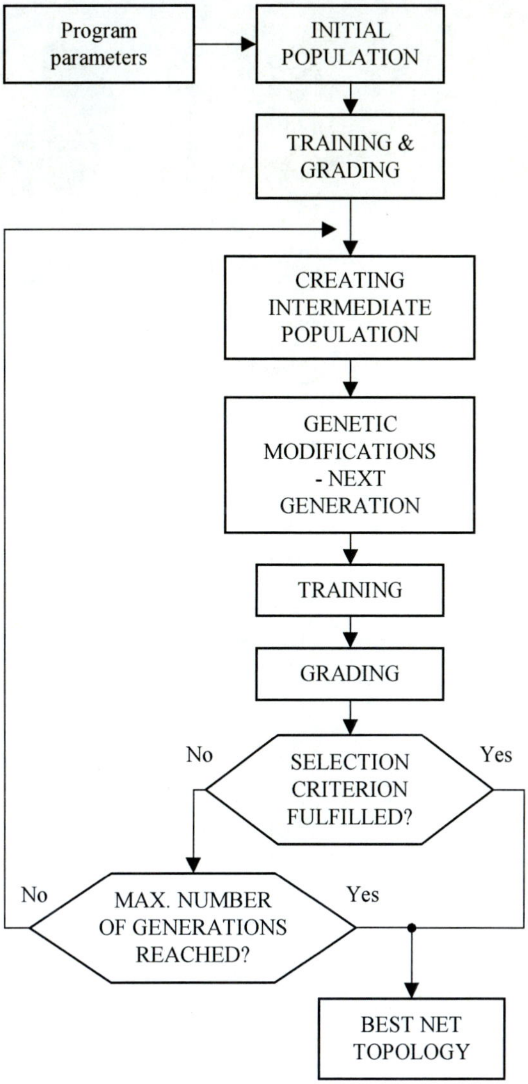

Fig. 13.8 Flow chart of the GA procedure

After selection of ANNs for the intermediate population is done, several genetic operations can be applied to create individuals of the next population. The most important one is the crossover of two parent individuals (Fig. 13.10) to produce new descendants. A crossover can be an exchange of single neurons, groups of neurons or whole layers between two ANNs. Crossovers are very important and frequent at the beginning of a GA so that a wide variety of different individuals can be produced. With the increasing number of generations the probability of crossovers reduces since later in the process the displacement of weaker individuals by the successful one becomes more important than the creation of various new individuals. Furthermore, mutations can take place which change the network topology randomly by adding or removing neurons. Mutations are used to avoid

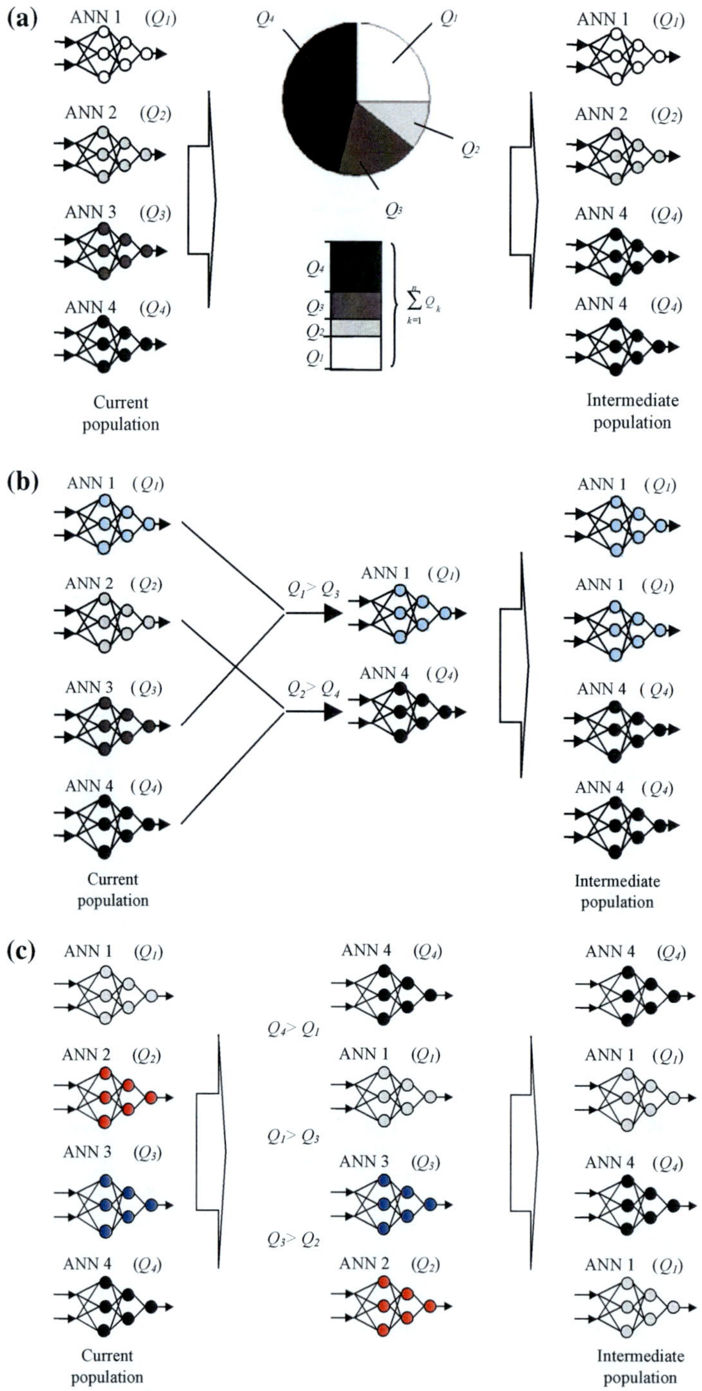

Fig. 13.9 Illustration of sample selection procedures: **a** the "wheel of roulette" method **b** the "tournament" method and **c** the "ranking place" method

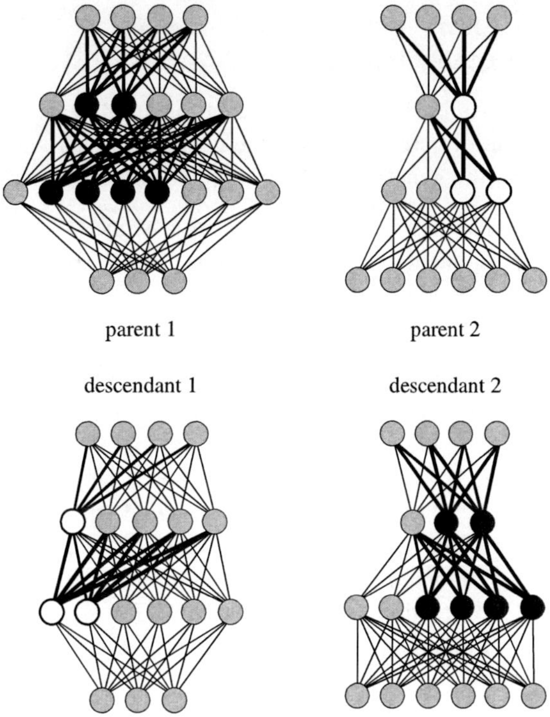

Fig. 13.10 Crossover of groups of neurons to create new descendant individuals

the optimization that is done around a local minimum. Four kinds of mutations can be done: delete neuron, add neuron, delete layer or duplicate layer. The probability of mutations to occur during the evolution is constant. Another important genetic operation is recombination, which stands for the creation of a descendant being an identical copy of a parent. Of course, this operation is used more and more frequently in the later evolution to make the most adapted individual successful.

The newly created population is trained again and then subjected to quality determination (grading). The consecutive populations of ANNs are created, trained and graded until the optimization criterion is fulfilled. The evolutionary process described should end up in an optimum that represents the most appropriate neural network topology.

Below the genetic procedure is described with more details and several implementation aspects (crucial for final results) are discussed.

Application of the GA principles for ANN structure optimization is an example of natural coding (not binary, here—real valued). The coding of particular individuals was very compact—all the ANNs of given population were represented by the set (record) of the following parameters (genes):

- ANN consecutive number k,
- Number of layers l_k,
- Vector of layer sizes (number of neurones in layers) $\boldsymbol{N}_k$,
- Matrix of the connection weights $\boldsymbol{W}_k$,

- Matrix of the neurones' biases $\boldsymbol{B}_k$,
- Quality index Q_k.

At each consecutive optimization step (for successive population of nets) appropriate changes in the genome of each individual were carried out, depending on the genetic operation applied to the net (-s):

- The new layers sizes $\boldsymbol{N}_k$ are assigned.
- Elements of the matrices $\boldsymbol{W}_k$ and $\boldsymbol{B}_k$ are cut away, newly established or interchanged between the ANNs.

Optimization of the neural CT saturation detector was performed by going out of a population of neural networks that consisted of 20 individuals. The ANNs were being trained with the first half and tested with the second half of pattern signals originating from EMTP-ATP simulation of current transformer operation. The ANN input vector consisted of 10 most recent samples of the CT secondary current, which may be seen as signal windowing with 10 ms long data window (by sampling frequency $f_s = 1$ kHz). The Levenberg–Marquardt training algorithm was adopted with the desired output of the ANN set to 1.0 for the periods of linear CT operation and 0.0 when the CT was saturated.

Implementation of the GA procedure brought about results, which, among the others, depend on the definition of the ANN quality index (adaptation function). Two versions of the index were analyzed, i.e. efficiency index Q_{eff} and efficiency/size index $Q_{\mathrm{eff/size}}$, according to the formulae

$$Q_{\mathrm{eff}} = \frac{\text{number of correct decisions}}{\text{number of all testing cases}} \tag{13.1}$$

$$Q_{\mathrm{eff/size}} = \frac{1}{(1 - Q_{\mathrm{eff}}) \cdot 2 \cdot n_{\mathrm{ANN}}} \tag{13.2}$$

where n_{ANN} stands for ANN size (total number of neurons).

With the quality index (13.1) the best neural nets from the population considered were assigned values close to 1.0, while the worst ones were graded with values approaching 0.0. One has to understand that the assessment of ANN with use of the efficiency index Q_{eff} is done with respect to the ANN performance quality only (in terms of percentage of correctly classified cases) without taking into consideration the ANN size. Such an approach can sometimes lead to quite big neural networks, implementation of which may create problems if they are going to be applied in on-line operating protection or control systems (high-computational burden, proportional to the ANN size). In order to drive the optimization process in both efficiency and ANN size directions the quality index (13.2) was proposed. The values of $Q_{\mathrm{eff/size}}$ (not limited to 1.0) are inversely proportional to the total number of neurons of ANNs being assessed, thus giving a chance of obtaining efficient yet reasonably small (and implementable) neural networks. In first attempt the simple ratio of Q_{eff}–n_{ANN} was considered; however, such a quality index could sometimes lead to very small ANNs but of poor classification abilities. After further investigations the $Q_{\mathrm{eff/size}}$ index in form of Eq. (13.2) was adopted. The analyses confirmed that such an approach could provide much better optimization results in both ANN efficiency and size aspects.

The best ANNs obtained for respective quality indices were:

- For Q_{eff} index—ANN having 14 neurons (13–1), classification efficiency equal to 0.978, and
- for $Q_{\mathrm{eff/size}}$ index—ANN having 6 neurons (3–2–1), classification efficiency of 0.954.

As one can see, optimization with combined index (13.2) ceased with twice smaller neural network by only slight decrease of ANN classification abilities.

The operation of described genetic optimization algorithm was thoroughly tested taking also into account various ways of selection procedure realization (Fig. 13.9) as well as various probability values for occurrence of particular genetic operations. The genetic operations mentioned above were realized with regard to neural networks in the following way:

- Crossover operation (with probability p_c) consisted in exchange of some neurons between two ANN individuals; the crossover could concern single neurons (with probability p_{sn}), groups of neurons (p_{gn}) or whole layers (p_l).
- Mutation operation (with probability p_m) was realized in one of the following versions: delete a neuron, duplicate a neuron, delete a layer of neurons or duplicate a layer of neurons.

In Table 13.1 the optimization results of the neural CT-saturation detectors are gathered for various selection methods applied and probability values of the genetic operands adopted. Here, the ANNs were being assessed with the efficiency index Q_{eff}. It is seen that the resulting neural networks are characterized by similar effectiveness of CT-state identification. The best ANN ($Q_{\mathrm{eff}} = 0.9783$) was obtained for the algorithm Rn4, i.e. for selection with ranking place method, probability of crossover set low ($p_c = 0.3$) and probability of mutation set high ($p_m = 0.6$). Apart from high-classification abilities, its important advantage is also quite small size (the considered ANN consisted only of 14 neurons laid out in two layers). Such a compact neural structure can easily be implemented even in on-line operating protection and control systems. From Table 13.1 one can also draw some conclusions with regard to the influence of the selection method applied on the results of GA operation. Independent of the values of probabilities of genetic operations, the selection method based on "ranking place" approach brought about the ANNs with the highest values of efficiency index Q_{eff}. With this respect the worst one turned out to be the "tournament" selection algorithm with random choice of individuals.

The analysis of results gathered in Table 13.1 with respect to influence of the probabilities of genetic operations reveals the fact that the best optimization results are obtained for the algorithms with probability of mutation p_m set higher than probability of crossover p_c. It is also advantageous when the probability of

Table 13.1 Topologies and quality indices of the ANN classifiers obtained with the GA for various combinations of selection method and probabilities of genetic operations

Probability values	Optimization results (best ANN size and quality) for selection process realized with the method of			
	Wheel of roulette	Ranking place	Tournament with deterministic choice	Tournament with random choice
$p_m = 0.25$ $p_c = 0.65$ ($p_l = 0.75$ $p_{gn} = 0.20$)	ANN-R1: (11–1) $Q_{eff} = 0.9637$	ANN-Rn1: (14–1) $Q_{eff} = 0.9742$	ANN-Td1: (5–9–3–1) $Q_{eff} = 0.9720$	ANN-Tr1: (6–10–1) $Q_{eff} = 0.9563$
$p_m = 0.25$ $p_c = 0.65$ ($p_l = 0.40$ $p_{gn} = 0.55$)	ANN-R2: (4–11–1) $Q_{eff} = 0.9707$	ANN-Rn2: (6–15–1) $Q_{eff} = 0.9720$	ANN-Td2: (11–8–8–1) $Q_{eff} = 0.9670$	ANN-Tr2: (11–11–4–4–1) $Q_{eff} = 0.9683$
$p_m = 0.60$ $p_c = 0.30$ ($p_l = 0.75$ $p_{gn} = 0.20$)	ANN-R3: (6–7–7–1) $Q_{eff} = 0.9720$	ANN-Rn3: (7–11–12–1) $Q_{eff} = 0.9733$	ANN-Td3: (9–6–1) $Q_{eff} = 0.9650$	ANN-Tr3: (9–11–4–1) $Q_{eff} = 0.9574$
$p_m = 0.60$ $p_c = 0.30$ ($p_l = 0.40$ $p_{gn} = 0.55$)	ANN-R4: (4–8–7–7–1) $Q_{eff} = 0.9653$	ANN-Rn4: (13–1) $Q_{eff} = 0.9783$	ANN-Td4: (15–3–11–5–1) $Q_{eff} = 0.9723$	ANN-Tr4: (6–2–1) $Q_{eff} = 0.9687$

crossover for group of neurons p_{gn} is higher than for the whole layers p_l. Not going into detail it is worth to mention that the probability values do not have to be constant during the run of GA. Usually the value of p_c is higher at the beginning of the optimization process and decreases with time to the advantage of probability p_m, which becomes higher for later generations.

In Fig. 13.11 the course of quality index Q_{eff} during the run of GA algorithm (version with "wheel of roulette" selection method R2) is shown. Looking at the curve of quality index for the best ANN in current population (Q_{eff_max}) one can find that the process convergence is the fastest for the first few iterations. After several generations the pace of optimization decreases and from ca. 25th generation no significant improvement of the neural classifiers is observed. The average quality index (Q_{eff_av}) calculated for the entire population of ANNs is not a monotonic function, however, it increases slowly with time, which means that consecutive populations of nets consist of individuals on better average and better adapted for the task they are trained to.

The research performed also included an attempt of ANN structure design with heuristic "trial and error" approach. A number of 1,000 neural networks were randomly created, trained and then tested with EMTP-ATP simulation signals. It allowed getting the neural classifiers with 94% efficiency at the most (the best out of 1,000 created ANNs was characterized by efficiency Q_{eff} index equal to 0.94 and size (9–7–1), i.e. consisting of 17 neurones), which is less than obtained with the genetic optimization approach, independent of the selection method and probabilities of genetic operations.

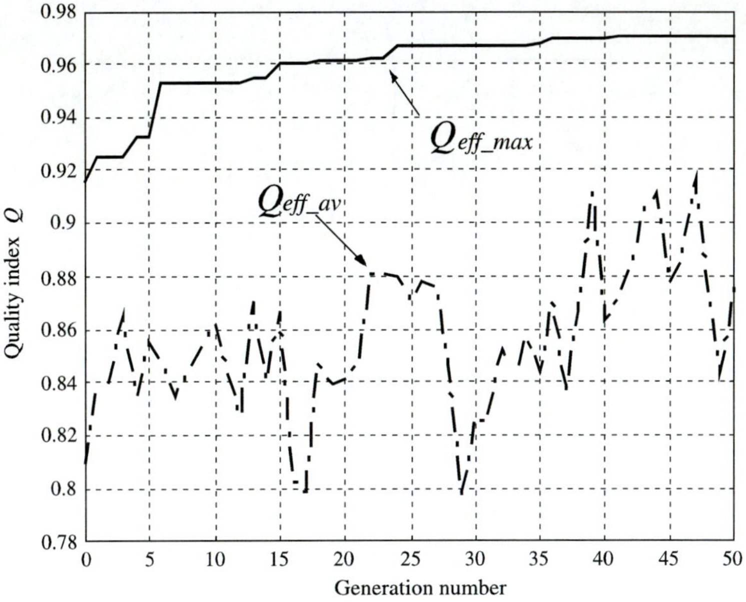

Fig. 13.11 Optimization process for the algorithm R2 (see Table 13.1)

Fig. 13.12 Synchronous machine with local control

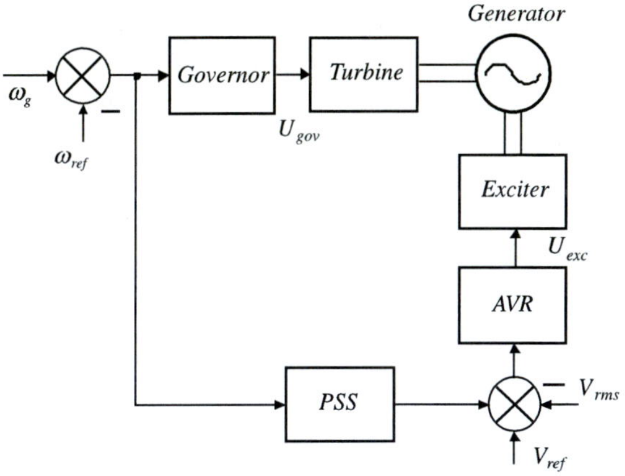

13.2.2 Optimal Selection of the Generator AVR Settings

The basic control of a synchronous generator includes voltage regulator and turbine-governor units (Fig. 13.12). The automatic voltage regulator (AVR) adjusts the power output of the generator exciter in order to control the magnitude of generator's terminal voltage. The turbine-governor, on the other hand, adjusts the steam-valve position to control the mechanical power output of the turbine [2, 10]. The main attention is paid here to voltage regulator units that are intended to be

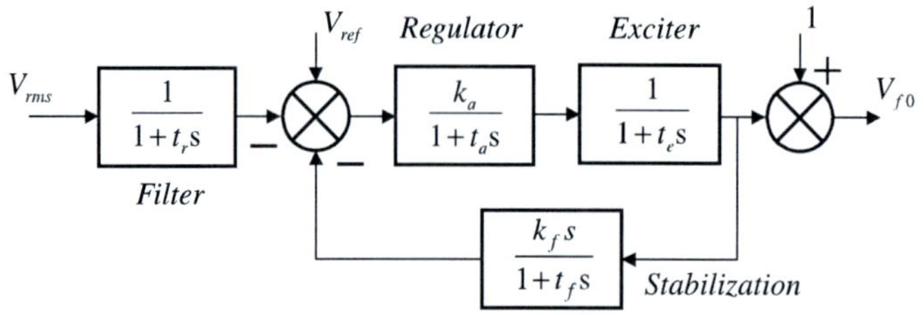

Fig. 13.13 Arrangement of the studied AVR unit

optimized to achieve possibly best transient and steady-state response in terms of dynamic features of the system and control accuracy [19].

Generator–voltage control as well as turbine-governor units are appropriately designed taking into account generator ratings, exciter type, prime mover features and many other factors. Both the control units are usually tuned separately to respond properly to either generator voltage or speed deviations (control-decoupled design concept). Quite rare are attempts of coordination of both controls, although it can be proved that such an approach may bring significant improvement of generator behavior under transient conditions (e.g. [24, 25]). The parameters of control units are also seldom optimized in a systematic way for broad range of generator operating conditions. More attention is delivered to tuning of PSS units that can be done e.g. with a GA or a population-based incremental learning (PBIL) [5]. Here the genetic approach is applied to optimization of generator AVR or AVR in coordination with turbine-governor unit. The optimized AVR units are tested with simulation signals and the results of generator operation are discussed.

Depending on the generator and exciter type the appropriate regulating circuits may be of various structure. Block diagrams of several standard types of generator AVR systems have been developed by the IEEE Working Group on Exciters [13]. The same refers to the models of generator speed control systems [14]. The diagrams may be used in power-system transient studies (e.g. stability problems), if their operation is important from the point of view of the phenomena to be analyzed.

The basic scheme of the AVR studied is shown in Fig. 13.13. The diagram presented refers to the system with continuously acting regulator and rotating exciter that is often classified as the AVR type I [13]. The AVR adopted contains the input filter and the feedback loop control part. The exciter itself is represented by an inertial block, while the regulator part is splitted in two parts, the amplification and stabilization ones. The input signal of AVR unit is the r.m.s. value of generator output voltage. The stabilizing compensator in the feedback loop is used to improve the dynamic response of the voltage regulator by reducing excessive overshoot. The compensator provides damping signal (filtered first derivative of the exciter output) that is subtracted from the voltage deviation signal. Nonlinearities due to exciter saturation and limits on exciter output are also taken into account but are not shown in Fig. 13.13. One should understand that the maximum and minimum limits of the regulator are imposed so that large input error signals cannot produce regulator output that exceeds practical limits.

The AVR parameters can be divided among two groups:

- Non-optimizable (hardware-related):
  - V_{ref} Reference voltage
  - U_{exc} Excitation voltage (AVR output)
  - t_r Time constant of the input filter
  - t_e Exciter time constant
- Parameters to be optimized:
  - k_a Regulator gain
  - t_a Regulator time constant
  - k_f Feedback connection (stabilization) gain
  - t_f Feedback connection time constant

The values of original (non-optimized) AVR settings proposed in [10] were as follows: $k_a = 195.6$; $t_a = 1.3$; $k_f = 0.5$; $t_f = 1$. The values were taken as a starting point for the optimization (initial population AVRs were assigned settings close to the set mentioned) and as a reference for testing of the optimized regulators and comparison of their behavior with the non-optimized one.

The above described genetic procedure was applied for AVR-settings optimization. The population of AVRs consisted of 20 individuals, i.e. 20 sets of parameters to be optimized. The AVR parameters were assumed to be changed within the following ranges: $k_a = (50\text{–}300)$, $t_a = (0.2\text{–}3.0)$, $k_f = (0.01\text{–}2.0)$, $t_f = (0.1\text{–}2.0)$. The so-called initial population of individuals was randomly created. The individual was understood as a chromosome of genes coding all the features of the structure to be optimized. Here the chromosome contained the parameters of AVR controller. The controllers with parameters as coded in genomes of the individuals were further applied in EMTP-ATP, where certain situations of power system operation were simulated. After the quality of each individual AVR was determined (grading, details further) an intermediate population was created, where successful individuals were reproduced more likely. On this intermediate population several genetic operations were applied (crossover, mutation) to create a new population.

Maximum number of generations was equal 20, while the goal fitness level of the sought solution was set to 1,200. The genetic operations were realized with the following assumptions:

- Crossover probability: 100% (i.e. each of the individuals was subjected to this operation),
- Mutation probability within given population: 35%,
- Mutation probability for particular genes: 50%,
- Reproduction of individuals for further generations: with the tournament method with deterministic selection.

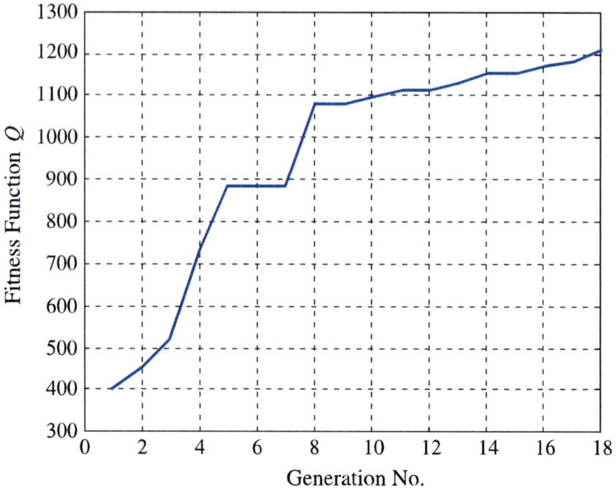

Fig. 13.14 Optimization process for the AVR assessed with fitness function Q according to Eq. 13.3

The quality of each individual is assessed with appropriately defined fitness function. An optimal AVR unit should provide possibly fastest response to any changes of generator's terminal voltage and minimal (or zero) for steady-state regulation error. With such a specified optimization goal the fitness function may be defined in various ways, taking into account specific parameters of the machine output signal such as overshoot and time-to-steady-state. Here the fitness function Q_e was defined as

$$Q = \sum_{c=1}^{F} \left(\frac{1}{N} \sum_{n=1}^{N} \left(V_{\mathrm{rms}}(n) - V_{\mathrm{ref}})^2 \right) \right)^{-1} \tag{13.3}$$

where F—number of considered EMTP-ATP simulation cases, N—number of samples over which the AVR operation is assessed and n—time index (sample number).

From the Eq. 13.3 it follows that the fitness function adopted is a digital version of the integral of regulation error squared (ISE index), averaged for all considered simulation cases.

The genetic optimization procedure stops when either the prescribed number of iterations (generations) is reached or the best individual fitness value exceeds certain threshold.

The convergence of the optimization process can be observed in Fig. 13.14. It is seen that the genetic run ends up after <20 generations, when the preset fitness level equal to 1,200 was reached. The most significant increase of the average fitness function of the AVRs is visible within first eight generations. Further run of the procedure does not change the optimal solution a lot, however, the population of voltage controllers becomes slowly dominated by the final solution, i.e. the best AVR.

The quality index (fitness value) of the initial non-optimized AVR set as mentioned above was equal $Q = 203.36$. The best individual AVR from the last

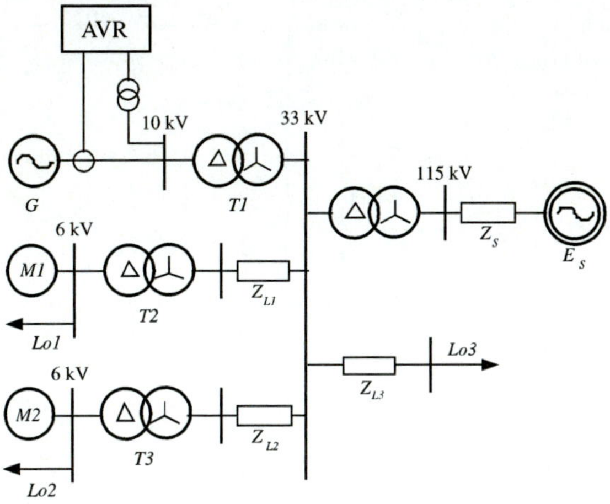

Short-circuit power of the 110 kV system: S_{zw}=1500MVA
The system transformer TS: S_n = 63 MVA; 110kV/33kV; Yd11;
Generator G: Type TB-60-2; S_n=75 MVA; P_n=60 MW; U_n=10.5kV; T'_{d0}=11.7s, T_{q0}'=0.6s;
Power station transformer T1: S_n=63MVA; 33kV/10.5kV; Yd11;
Load transformers (Yd11): (T2): S_n=20MVA; 33kV/6kV; U_z=8.1%; P_{Cu}=123kW; (T3): S_n=40MVA; 33kV/6kV;
Lines: (L1): R_1=0.095Ω; X_1=0.061Ω; (L2): R_2=0.067Ω; X_2= 0.082Ω; (L3): R_3= 4.34Ω ; X_3= 3.83Ω;
Non-rotating loads: Lo1: S=(5+j5)MVA; Lo2: S=(10+j8)MVA; Lo3: S=(25+j20)MVA;
Induction motors M1, M2: P_n =630kW ring type; U_n=6 kV; η=0.94; cosφ=0.85; n_n=1500rev/min, s_n=0.01; H=0.16s.

Fig. 13.15 Single-line diagram and basic parameters of the test power system

population after optimization process was characterized by the fitness value $Q = 1210.7$, which is approximately six times higher than for the AVR without optimization. The resulting optimal AVR obtained the following settings: $k_a = 245.14$; $t_a = 0.322$; $k_f = 0.018$; $t_f = 1.956$.

The signals for AVR assessment were obtained from EMTP-ATP simulations of the test power system. A typical industrial network shown in Fig. 13.15 was considered. The local power station with generator G of nominal voltage 10.5 kV supplies the main 33 kV busbar (GPZ) via its step-up transformer T1. The main busbar is also supplied from the 110 kV system via transformer TS. The non-rotating load connected directly to the main busbar is represented by an impedance. The rotating loads used in the system operate at the voltage level of 6 kV so two major separated 6 kV busbars are considered that are supplied via lowering transformers T2 and T3. The 6 kV loads are divided into non-rotating (Lo1, Lo2, represented by impedances) and rotating ones (groups of induction motors M1 and M2). The network parameters are given under Fig. 13.15.

A number of several various cases of the system normal and abnormal operating conditions have been prepared. The simulation scenarios considered included:

- Normal operation at rated load,
- Switching on and off certain load parts,

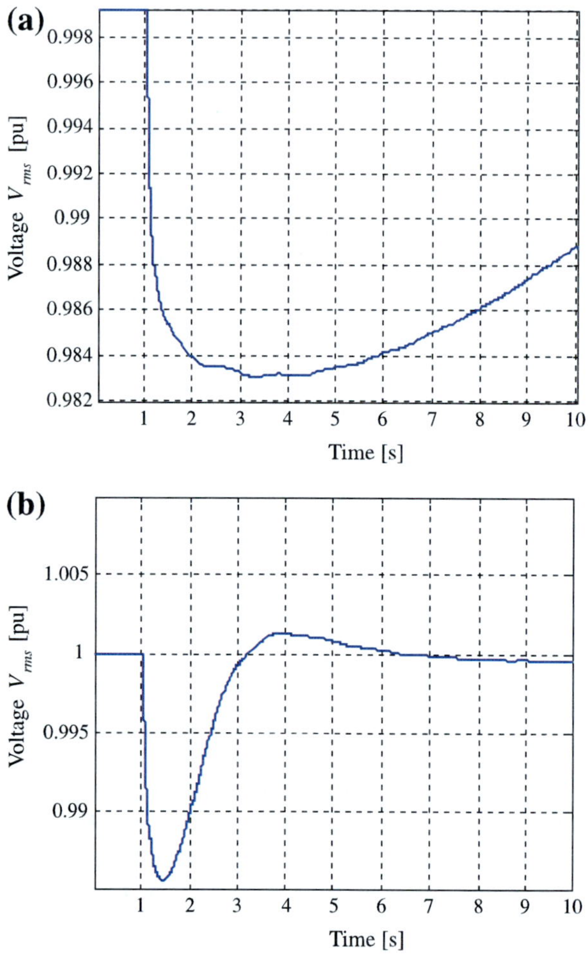

Fig. 13.16 Generator terminal voltage for a case of switching on the load Lo3: **a** original AVR, **b** optimized AVR

- Faults at various places in the system (at the generator terminals, behind transformer T1, at the 6 kV side of load transformers, etc.),
- Loss-of-mains (islanding operation).

Within the loop of GA the simulation runs were being repeated for all AVR units from the population of individuals to be optimized. After the performance of all AVRs was assessed and the new population of individuals was created, the EMTP-ATP simulations were run again with the new settings of the AVRs.

Selected example of the optimized AVR performance is shown in Fig. 13.16 for the case of load switching (Lo3 on). Terminal voltage of the generator *G* with the optimized AVR can be compared with the signals obtained with the non-optimized controller. One can notice the superior behavior of the AVR after optimization for the presented case. The optimized AVR helps keeping maximal voltage deviation smaller and enables faster voltage recovery after sudden change of generator operating conditions. The oscillations of the generator rms voltage are effectively damped and the steady-state is reached much faster than when the non-optimized AVR is employed.

References

1. Abdel-Magid YL, Abido MA (2003) Optimal multiobjective design of robust power system stabilizers using genetic algorithms. IEEE Trans Power Syst 18:1125–1132
2. Anderson PH, Fouad AA (1977) Power system control and stability. The Iowa State University Press, Ames
3. Arroyo JM, Conejo AJ (2002) A parallel repair genetic algorithm to solve the unit commitment problem. IEEE Trans Power Syst 17:1216–1224
4. Beyer HG (2001) The theory of evolution strategies. Springer, Berlin
5. Chen L, Petroianu A (1999) Application of evolutionary algorithms to the optimization of sub-optimal H_∞ PSS tuning. In: Proceedings of the 13th PSCC conference, Trondheim, pp 170–176
6. Coury DV, Oleskovicz M, Delbem ACB, Simões EV, Silva TV, De Carvalho JR, Barbosa D (2009) Frequency relaying based on genetic algorithm using FPGAs. In: Proceedings of the 15th international conference on intelligent systems application to power systems, Curitiba, Brazil, paper 250
7. Dong ZY, Makarov YV, Hill DJ (1998) Analysis of small signal stability margins using genetic optimization. Electr Power Syst Res 46:195–204
8. Eiben AE, Smith JE (2003) Introduction to evolutionary computing. Springer, Berlin
9. Gerbex S, Cherkaoui R, Germond A (2001) Optimal location of multi-type FACTS devices in a power system by means of genetic algorithm. IEEE Trans Power Syst 16:537–544
10. Glover JD, Sarma M (1994) Power systems analysis and design. PWS Publishing Company, Boston
11. Goldberg DE (1989) Genetic algorithms in search optimization and machine learning. Kluwer Academic Publishers, Boston
12. Holland JH (1975) Adaptation in natural and artificial systems. The University of Michigan Press, Ann Arbor
13. IEEE Committee Report: Computer Representation of Excitation Systems (1968) IEEE Trans PAS 6:1459−1463
14. IEEE Committee Report: Dynamic Models for Steam and Hydro Turbines in Power System Studies (1973) IEEE Trans PAS 6:1904−1913
15. King WDJ, Özveren CS, Bradley DA (2007) Economic load dispatch optimization of renewable energy in power system using genetic algorithm. In: Proceedings of the powertech conference, Lausanne, Switzerland, Paper 531
16. Koza J (1992) Genetic programming: on the programming of computers by means of natural selection. MIT Press, Cambridge
17. Lai LL, Ma JT, Yokoma R, Zhao M (1997) Improved genetic algorithms for optimal power flow under both normal and contingent operation states. Electr Power Energy Syst 19:287–292
18. Luan WP, Irving MR, Daniel JS (2002) Genetic algorithm for supply restoration and optimal load shedding in power system distribution networks. Proc IEE Gener Transm Distrib 149:145–151
19. Rebizant W, Bejmert D (2006) Genetic optimization procedure for design of synchronous generator voltage and speed control systems. In: Proceedings of the international symposium on modern electric power systems, Wroclaw, Poland, pp 210–215
20. Rebizant W, Bejmert D (2007) Current-transformer saturation detection with genetically optimized neural networks. IEEE Trans Power Deliv 22:820–827

21. Rebizant W, Bejmert D (2007) Genetic optimization of ANN-based decision units for protection purposes. Przegląd Elektrotechniczny 4:129–133
22. Rebizant W, Szafran J, Feser K, Oechsle F (2001) Evolutionary improvement of neural classifiers for generator out-of-step protection. In: Proceedings of the IEEE porto powertech conference, vol. 4. Porto, Portugal, Paper PRI1223
23. Rebizant W, Szafran J, Feser K, Oechsle F (2002) Evolutionäre Optimierung neuronaler Klassifikatoren für den Generatorschutz. ELEKTRIE, Berlin 56:51–56
24. Sharaf AM, Lie TT (1997) An artificial neural network coordinated excitation/governor controller for synchronous generators. Electr Mach Power Syst 25:1–14
25. Yong T, Lasseter RH, Cui W (1998) Coordination of excitation and governing control based on fuzzy logic. IEEE publication no. 0-7803-4403-0/98, pp 737–742

Chapter 14
Expert Systems

Expert System (ES) is a kind of software that simulates the problem-solving behavior of a human expert of given domain. ES can be used to solve a complex problem or give an advice, mainly in cases when the amount of data to be processed is very high. There is no sense to apply ESs for simple decision problems, they should be used wherever a conventional data processing (DP) solution is not possible or may appear troublesome. Referring to Fig. 14.1 one can say that instead of traditional DP by algorithmic, well-defined methods, expert systems perform processing of knowledge and make intelligent searches.

Why use Expert Systems? When a situation or data to handle is very complex and the solution of the problem is neither visible at the first sight nor can be easily found with available explicit formulae, an expert of given domain is usually called to perform some reasoning and give an advice. Since human experts are not always available, and even if they are, their advice is costly, may not always be reliable or consistent, not to mention problems with organizing meetings with them every time when it is needed. Therefore, building an artificial ES that would replace a human expert becomes a reasonable and cost-effective alternative for many technical problems.

It is obvious that an ES is as good as the experts that shared their domain-specific knowledge and the manner they solve problems. Important is also the way the knowledge and reasoning mechanisms are stored and organized in a formalized way by a knowledge engineer and programmers. The structure of an ES, reasoning mechanisms and other aspects related to their application are all touched in the following sections of this chapter.

It is good to remember that expert systems are not panacea for all conceivable decision problems. Their usage is limited to the domain and problems they have been designed for. The ESs are not always up to date, they do not learn, they also do not offer any "common sense" (general view), and are in most cases inflexible. And, of course, human experts are still needed to setup and maintain the system.

W. Rebizant et al., *Digital Signal Processing in Power System Protection and Control,* Signals and Communication Technology,
DOI: 10.1007/978-0-85729-802-7_14,

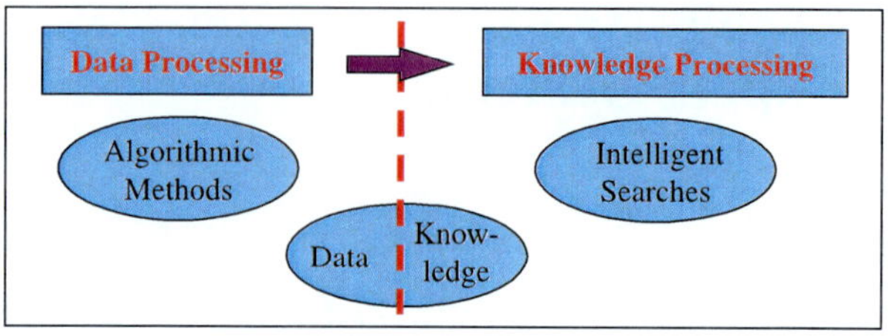

Fig. 14.1 Data versus knowledge processing

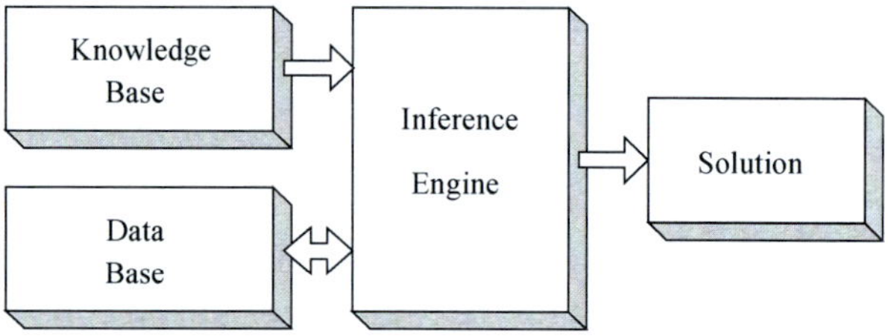

Fig. 14.2 Structure of an expert system

Nevertheless, they have been found useful and are quite often applied, also for power system problems, including protection and control issues.

14.1 Components of an Expert System

An ES is usually organized as it is schematically shown in Fig. 14.2. The components of ES are:

- Knowledge base—containing the knowledge about the system, its functioning, rules of problem solving, etc,
- Data base—including the facts, which generally describe the domain and the state of the problem to be solved,
- Inference Engine—with the reasoning principles and conflict resolution strategies.

The block connections in Fig. 14.2 suggest that there exist unidirectional (read-only) information flow from the KB to the Inference Engine and bidirectional exchange of data with the DB. This is because during data and knowledge processing new facts are created that should be stored in the Data Base for further usage.

The most important blocks of the ES structure, i.e., Knowledge Base and Inference Engine contain the information provided by human experts. Potential experts are those who possess knowledge and strong practical experience in particular domain. They are often skillful persons who can do things that other people cannot. The experts should, however, be capable of expressing their knowledge in the form of rules for problem solving, which is not always simple since some conclusions are sometimes drawn without deep consideration, on the basis the basis of experience and "rules of thumb" that cannot be easily explained.

Knowledge can be understood as a theoretical or practical understanding of a subject or a domain, the sum of what is currently known. For the usage in an ES the knowledge can be represented in the form of:

- Mathematical logic (well suited for objects with numerical values),
- Production rules (condition–action),
- Meta-rules (based on meta-knowledge–knowledge about knowledge, how to use and control knowledge),
- Semantic networks, consisting of nodes (concepts or meanings) and links (relations),
- Frames (information is grouped in records, with multiple levels),
- or in mixed form of the above, when required.

The most common way of knowledge representation is the use of production rules. Simple rules are a realization of the implication operation $p \rightarrow q$ in the form "IF antecedent THEN consequent". The antecedent of a rule incorporates two parts: an object (linguistic object) and its value. The object and its value are linked by an operator, where the operator identifies the object and assigns the value. Operators such as "IS", "ARE", "IS NOT" and "ARE NOT" are used to assign a symbolic value to a linguistic object. Expert systems can also use mathematical operators to define an object as numerical and assign it to the numerical value, e.g., "IF 'age of the customer' <18 AND 'alcohol content' >35% THEN 'it is not allowed to sell the drink'".

The rules can be fuzzified if needed, as described in Chap. 11, then leading to a Fuzzy Expert System.

The rules used in an ES may have different meaning:

- Relation (IF 'fuel tank' is 'empty' THEN 'car' is 'dead'),
- Recommendation (IF 'season' is 'autumn' AND 'sky' is 'cloudy' AND 'forecast' is 'drizzle' THEN 'advice' is 'take an umbrella'),
- Directive (IF 'car' is 'dead' AND 'fuel tank' is 'empty' THEN 'action' is 'refuel the car'),
- Strategy (IF 'car' is 'dead' THEN 'action' is 'check the fuel tank'; step1 is complete, IF 'step1 is complete' AND 'fuel tank' is 'full' THEN 'action' is 'check the battery'; step2 is complete), or
- Heuristic (IF 'spill' is 'liquid' AND the 'spill pH' < 6 AND 'spill smell' is 'vinegar' THEN the 'spill material' is 'acetic acid').

The generic Inference Engine is the heart of an ES. Its purpose is to:

- Simulate the problem-solving strategy of an expert,
- Control the actions taken by the system,
- Execute chosen rules and/or activates new rules,
- Link the rules given in the knowledge base with the facts provided in the database,

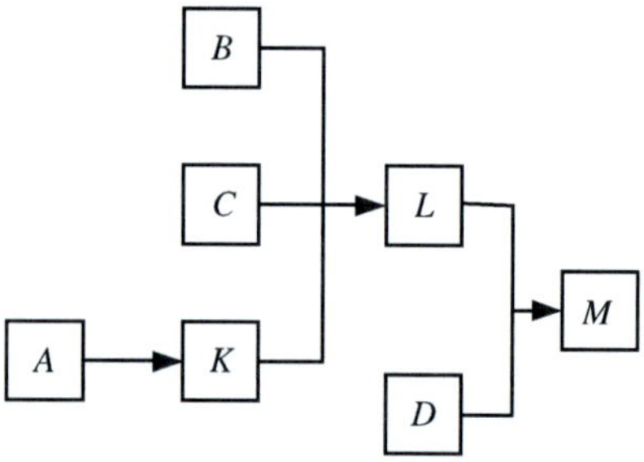

Fig. 14.3 An example of chain of rules

- Identify which rules could be applied,
- Employ a conflict resolution strategy, if needed (the appropriate rules are selected).

The above issues are explained in more detail in next section.

14.2 Knowledge Processing Methods

The rule-based expert systems contain a number of IF … THEN … rules. The rules stored in the knowledge base are compared with facts contained in the database. When the IF (condition) part of the rule matches a fact, the rule is fired and its THEN (action) part is executed (match-fire procedure). The matching of the rule IF parts to the facts produces inference chains. The inference chain indicates how an ES applies the rules to reach a conclusion.

An example of a rule chain is presented in Fig. 14.3, with the rules defined as follows:

Rule 1: IF D is true AND L is true THEN M is true

Rule 2: IF B is true AND C is true AND K is true THEN L is true

Rule 3: IF A is true THEN K is true

In case of vast and long-rule chains arriving at the conclusions is not trivial and may create a lot of technical problems. Examining a vast set of rule chains can be performed either with so-called *forward chaining* (data-driven inferencing) or with *backward chaining* (goal-driven inferencing) (Fig. 14.4). The techniques can be characterized as follows:

- Forward chaining
  - A technique for gathering information and then inferring from it whatever can be inferred,
  - Many rules may be executed that have nothing to do with the established goal,
  - If our goal is to infer only one particular fact, the forward chaining inference technique would not be efficient,
  - Any rule can be executed only once,
  - The match-fire cycle stops when no further rules can be fired;

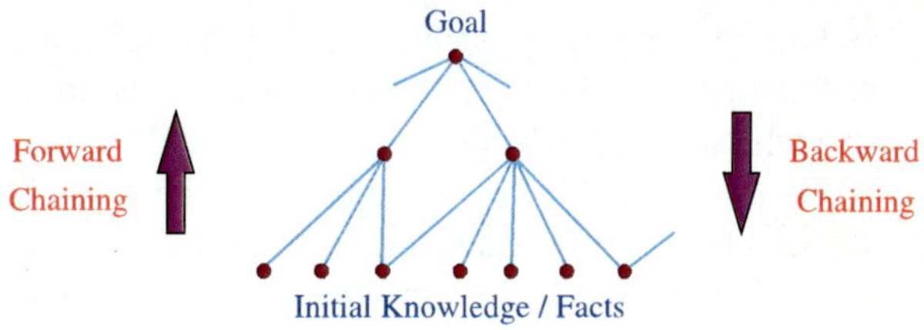

Fig. 14.4 Two methods of inferencing

- Backward chaining
  - The goal (a hypothetical solution) is assumed and the inference engine attempts to find the evidence to prove it.
  - First, the knowledge base is searched to find rules that might lead to the desired solution.
  - If such a rule is found and its IF (condition) part matches data in the database, then the rule is fired and the goal is proved.

Depending on the way of reasoning by the human expert, forward or backward chaining is appropriate in given situation. If an expert first needs to gather some information and then tries to infer from it whatever can be inferred, then the forward chaining inference should be used. However, if an expert begins with a hypothetical solution and then attempts to find facts to prove it, then the backward chaining inference engine is the right one.

It can sometimes happen that two or more rules may provide conflicting conclusions. In such a case special procedure called conflict resolution strategy should be used. The rule selection strategies may be as follows:

- Fire the rule with the highest priority (order of rules in the knowledge base is to be defined).
- Fire the most specific rule (a specific rule processes more information than a general one).
- Fire the rule that uses the data most recently entered in the database (time tags should be attached to each fact).
- Select the rule whose facts can be determined with the least amount of risk (or are already known).
- Select the rule that reduces search space the most.
- Select the rule that has strongest influence on the hypothesis, which is currently the highest rated.

14.3 Designing of an Expert System

As stated above, ESs incorporate the factual and empirical knowledge of experts in particular subject area. Modern ESs utilize great amount of knowledge in the form of simple if–then-else rules but also as more complex structures (objects, frames and algorithms). In order to identify the useful rules and to structure and store the knowledge is appropriate way, the rule acquisition requires the co-operation of two specialists—a knowledge engineer and a domain expert.

Design of an ES starts from defining the problems to be solved and determining all conditions, constraints and requirements. Then one needs to identify experts in given domain, capable of solving the problem and give clear explanation on how they come to the final solution. Basing on their expertise one has to now construct ES components, implement results in an algorithmic way (databases, links, control), after which the system should be extensively tested, reviewed and optimized, if necessary.

The parties involved in designing and implementing an ES are:

- Human expert (can solve problems; we desire to solve the problems without him/her),
- Knowledge engineer (can communicate with human expert to obtain and model the knowledge that we need in the system),
- Programmer (builds and maintains all the necessary computer programs),
- User (wants to use expertise to solve problems).

Good co-operation of the four sides can contribute to creating an ES that is efficient and fulfills all requirements of the user.

While designing an ES it is good to check the following aspects related to acquired knowledge representation:

- Expressive adequacy (is particular knowledge representation sufficiently powerful?),
- Reasoning efficiency (time of reasoning in case of ES containing lots of knowledge),
- Meta-representation (how is the knowledge structured?),
- Incompleteness (how the inference is performed over incomplete knowledge?),
- Real-world knowledge (how can we deal with attitudes, e.g., beliefs; how can paradoxes be avoided?).

An ES can perform well when all the above questions have thoroughly been analyzed and the answers have been implemented in the ES software. In case of troubles or when the advice is wrong it is often difficult to judge who is responsible for it. Even a brilliant expert is only a human and thus can make mistakes, so can ES also do. The errors can also arise from inadequate definition of the requirements (the user), inappropriate structuring of the human expert knowledge (the knowledge engineer) or due to false programming (the programmer of the ES shell).

Summarizing this part it is good to understand the similarities and differences between the way of problem solving and other aspects of inferencing performed by human experts, related expert systems and conventional programs. The reader is referred to Table 14.1 below for details. More theory related to ES technique can be found e.g., in the three selected books [4, 9, 11].

Table 14.1 Aspects of problem solving as performed by human experts, expert systems and conventional programs

Human experts	Expert systems	Conventional programs
Use knowledge in the form of rules of thumb or heuristics to solve problems in a narrow domain	Process knowledge expressed in the form of rules and use symbolic reasoning to solve problems in a narrow domain	Process data and use algorithms, a series of well-defined operations, to solve general numerical problems
In a human brain, knowledge exists in a compiled form	Provide a clear separation of knowledge from its processing	Do not separate knowledge from the control structure to process this knowledge
Capable of explaining a line of reasoning and providing the details	Trace the rules fired during a problem-solving session	Do not explain how a particular result was obtained and why input data was needed
Use inexact reasoning and can deal with incomplete, uncertain and fuzzy information	Permit inexact reasoning and can deal with incomplete, uncertain and fuzzy data	Work only on problems where data is complete and exact
Can make mistakes when information is incomplete or fuzzy	Can make mistakes when data is incomplete or fuzzy	Provide no solution at all when data is incomplete or fuzzy
Enhance the quality of problem solving via years of learning and practical training. This process is slow, inefficient and expensive	Enhance the quality of problem solving by adding new rules or adjusting old ones. When new knowledge is acquired, changes are easy to accomplish	Enhance the quality of problem solving by changing the program code, which affects both the knowledge and its processing, changes difficult

14.4 Expert System Applications

Capabilities of ES are well recognized, they include: strategic goal setting, planning, design, decision-making, quality control and monitoring, diagnosis that may be exploited in almost any conceivable branch of industry.

One has to underline that, unfortunately, there are only a few applications of ES to power system protection reported (off-line tasks solved):

- Protection settings co-ordination [7].
- Alarm processing [10].
- Designing the protection system of a power transformer [13].
- Post-fault analysis [16].
- Fault diagnosis [18].

As yet there is no application reported of the ES technique employed as a decision-making tool in an on-line operating protective relay, mainly due to complexity of such systems that might not be able to respond suitably fast and in real-time regime. Therefore, unlike in the other chapters on artificial intelligence techniques, no example of ES application is provided in this chapter.

Among the other power system (non-relaying) ES applications the following are worth being cited:

- Reactive power management [1],
- Power plant events diagnosis [2],
- Load forecasting [3],
- Power quality monitoring [5],
- Maintenance scheduling [6],
- Distribution system control [8],
- Classifying power quality disturbances [12],
- Power system reliability analysis [14],
- Substation switching [15],
- Network planning [17], and
- Power system restoration [19].

References

1. Ananthapadmanabha T, Kulkarni AD, Rao ASG, Rao KR, Parthasarathy K (1996) Knowledge-based expert system for optimal reactive power control in distribution system. Int J Electr Power Energy Syst 18:27–31
2. Arroyo-Figueroa G, Alvarez Y, Sucar LE (2000) SEDRET—an intelligent system for the diagnosis and prediction of events in power plants. Expert Syst Appl 18:75–86
3. Chiu CC, Kao LJ, Cook DF (1997) Combining a neural network with a rule-based expert system approach for short-term power load forecasting in Taiwan. Expert Syst Appl 13:299–305
4. Darlington K (2000) The essence of expert systems. Pearson Education, Harlow
5. Dash PK, Jena RK, Salama MMA (1999) Power quality monitoring using an integrated Fourier linear combiner and fuzzy expert system. Int J Electr Power Energy Syst 21:497–506
6. Dillon TS, Podbury C (1990) A dynamic frame based maintenance scheduler. In: Dillon TS, Laughton MA (eds) Expert system applications in power system. Prentice Hall, New York, pp. 153–179
7. Ganjavi MR, Krebs R, Styczynski Z (2006) Design of a pilot knowledge-based expert system for providing coordinated setting values for power system protection devices. In: Proceedings of MEPS'06 conference, Wroclaw, Poland, pp. 354–360
8. Fujii Y, Miura A, Tsukamoto J, Youssef MG, Noguchi Y (1992) On-line expert system for power distribution system control. Int J Electr Power Energy Syst 14:45–53
9. Giarratano JC, Riley GD (1998) Expert systems: principles and programming. PWS Publishing, Boston
10. Hasan K, Ramsay B, Ranade S, Ozveren CS (1994) An object-oriented expert system for power system alarm processing and fault identification. In: Proceedings of the 7th mediterranean electrotechnical conference, vol. 3. pp. 909–912
11. Jackson P (1999) Introduction to expert systems. Pearson Education Ltd, Harlow
12. Liao Y, Lee JB (2004) A fuzzy-expert system for classifying power quality disturbances. Int J Electr Power Energy Syst 26:199–205
13. Lifeng L, Zhongde G, Qixun Y, Zhongmin B (1995) An expert system for designing the protection system of a power transformer. Electr Power Syst Res 35:59–64
14. Matijevics I, Józsa L (1995) An expert-system-assisted reliability analysis of electric power networks. Eng Appl Artif Intell 8:449–460

15. Nacsa J, Kovács G, Kopácsi S (1999) Intelligent application AT the 400/120 kV substation of the paks nuclear power plant. In: Proceedings of the powertech conference, Budapest, Hungary, paper BPT99-109-23
16. Styvaktakis E, Bollen MHJ, Gu IYH (2002) Expert system for classification and analysis of power system events. IEEE Trans Power Deliv 17:423–428
17. Teive RCG, Silva EL, Fonseca LG (1998) A cooperative expert system for transmission expansion planning of electrical power systems. IEEE Trans Power Syst 13:636–642
18. Yang C, Okamoto H, Yokoyama A, Sekine Y (1992) Expert system for fault section estimation of power systems using time-sequence information. Int J Electr Power Energy Syst 14:225–232
19. Yongli Z, Hogg BW, Zhang WQ, Gao S, Yang YH (1994) Hybrid expert system for aiding dispatchers on bulk power systems restoration. Int J Electr Power Energy Syst 16:259–268

Chapter 15
Artificial Intelligence: Summary and Hybrid Schemes

The artificial intelligence methods presented in Chaps. 11–14 require a lot of computational power but, in return, provide flexibility and possibility of handling imprecise or missing data. Despite of their differences, they all offer soft signal processing skills, thus one can say that they all form a family of soft computing that can be defined as follows:

Soft Computing is a collection of methodologies (working synergistically, not competitively) which, in one form or another, reflects its guiding principle: exploit the tolerance for imprecision, uncertainty, approximate reasoning and partial truth to achieve tractability, robustness and close resemblance with human-like decision making [2].

Some of the users claim that all the problems can be solved with neural networks, being universal approximators. Some prefer to use fuzzy reasoning schemes or expert systems to get soft distinguishing of events and perfect decision-making. With genetic approach one can achieve optimization, which supplements the family of soft computing.

In the following sections advantages and disadvantages of the AI methods are compared. Possibilities of hybrid combinations of various approaches are outlined and final recommendations for power system applications are given.

15.1 Comparison, Advantages and Disadvantages of AI Techniques

The artificial intelligence techniques have gained a lot of attention due to their undeniable *advantages* that are enumerated below.

The strengths of fuzzy logic (FL) schemes result from:

- Ability of representing linguistically phrased input features for processing,
- Representing multi-class membership of ambiguous patterns,

W. Rebizant et al., *Digital Signal Processing in Power System Protection and Control,* Signals and Communication Technology,
DOI: 10.1007/978-0-85729-802-7_15, © Springer-Verlag London Limited 2011

- Generating rules and inferences in linguistic form,
- Extracting ill-defined image regions, primitives and properties and describing relations among them as fuzzy subsets.

The ANNs provide natural classifiers having:

- Resistance to noise,
- Tolerance to distorted patterns/images (ability to generalize),
- Superior ability to recognize overlapping pattern classes or classes with highly nonlinear boundaries or partially occluded or degraded images,
- Potential of parallel processing.

The genetic/evolutionary algorithms are characterized as efficient, adaptive and robust search processes, producing near-optimal solutions and have a large amount of implicit parallelism.

The expert systems are capable of making decisions in the face of many arguments, coming to the solution on the base of knowledge and reasoning rules defined by a human expert as well as explaining a line of reasoning and providing the details when necessary.

As usual, each technique is not free from *drawbacks* that may prevent them from being applied for some technical problems. The most important difficulties arise from the fact that:

- The FL and ES schemes are basically not trainable,
- Interpreting of ANN internal signals is difficult, if not impossible,
- Robustness of the ANN scheme is difficult to ensure,
- Improving or extending operational range of an ANN scheme is possible usually with renewed training of the neural network with new patterns,
- Troubleshooting in case of an ES requires careful examination of all rules (and adding new rules, if necessary),
- Computational burden by implementation of vast expert systems or big ANN structures may exceed technical capabilities of available hardware.

15.2 Hybrid Solutions

Application of a single AI technique (with its strengths, capabilities and assumptions) to solve a real-world power system problem may not bring satisfactory results. Integration of two or more techniques may be required in some cases. As mentioned in previous section, a combination of various AI techniques in one scheme may contribute to mutual reinforcement of advantages and elimination or at least significant reduction of their weaknesses. Numerous books (e.g. [2, 8, 13]) and papers (cited later) are available on that topic.

A combination of various AI techniques are aimed at providing flexible information processing capability for representation and evaluation of various real-life ambiguous and uncertain situations. The hybrid systems:

- Combine at least two intelligent techniques,
- Are highly capable of reasoning and learning in an uncertain and imprecise environment,
- Allow for gaining greater tractability, robustness and lower the cost of solutions.

Applying in one system such techniques as FL, ANN, GA and ES is meant as complementary rather than competitive. The virtues of particular techniques:

- Dealing with imprecision and uncertainty (FL),
- Possibility of learning and curve fitting (ANN),
- Search and optimization (GA),
- Decision-making in face of many often different in nature input signals (ES) are merged together to provide an optimal solution for given problem.

Any combination of AI techniques is conceivable; the examples may include fuzzy-expert systems, neural-expert hybrids, fuzzy-neural networks, fuzzy controlled genetic algorithms, etc. The literature study confirms that neuro-fuzzy hybridization is the most visible integration realized so far.

A hybrid intelligent system can be good or bad—it depends on components which constitute the hybrid. Metaphorically one can say that a good hybrid would be "British Police, German Mechanics, French Cuisine, Swiss Banking, Italian Love and Polish Hospitality", but "British Cuisine, German Love, French Mechanics, Italian Banking, Polish Police and Swiss Hospitality" would be a bad one. Therefore careful analyses are needed to select the right components for building a good hybrid system for the application at hand.

Generally, the hybrid schemes proposed are either fused (melted) schemes or cooperative hybrids. The two classes of schemes are outlined below with some more detail.

15.2.1 Fused Hybrid Schemes

The first category of hybrid schemes include those where one of the schemes is melted into another one to realize certain partial function or where given AI technique is represented by another one in order to exploit some virtues of this technique (e.g. training of a fuzzy scheme).

In the paper [19] a fuzzy version of the neural network model called Fuzzy Self-Organizing Map was introduced. The neurons of the original ANN model were replaced by fuzzy rules, which are composed of fuzzy sets. The fuzzy sets defined an area in the input space, where each fuzzy rule fires. The output of each rule was a singleton. The outputs were aggregated by a weighted average, with the firing strengths of the fuzzy rules acting as the weights. Thus the Fuzzy Self-Organizing Map performed a mapping from a n-dimensional input space to one-dimensional output space. The learning capability of the Fuzzy Self-Organizing Map enabled it to model a continuous valued function to an arbitrary accuracy. The learning was first done by self-organizing the centers of the fuzzy sets according to Kohonen's Self-Organizing Map learning laws. After that, the fuzzy sets and the outputs of the fuzzy rules were initialized. Finally, in the last phase, the fuzzy sets were tuned by an algorithm similar to Kohonen's Learning Vector Quantization. Simulation results show good accuracy and fast convergence of the fuzzy-neural scheme.

Another example of the fused hybrid scheme is the so called Adaptive Network Fuzzy Inference System introduced in [4]. It is idea based on the fact that hybrid neuro-fuzzy systems are homogeneous and usually resemble neural networks. Here, the rule base of a fuzzy system is interpreted as a special kind of neural network. Fuzzy sets can be regarded as weights whereas the input and output variables and the rules are modeled as neurons. One can say that the neurons of the network represent the fuzzy knowledge base. The advantage of such hybrid NFS is its architecture since both fuzzy system and neural network do not have to communicate any more with each other. They are one fully fused entity and thus can also learn, both online and offline.

One should note that the ANFIS model [4] implements a Sugeno-like fuzzy system (Fig. 15.1a, described also in Chap. 11) in a network-like structure. The possibility of parameters adjusting via training (similar as for neural network schemes) is an important feature of the ANFIS structure. The paradigm of ANN-like training is applied here to a FIS system.

The ANFIS hybrid system has been applied e.g. to realize an efficient protection scheme against out-of-step (OS) conditions of synchronous machines [15]. The structure of an ANFIS with three inputs and one output is shown in Fig. 15.1b. The ANFIS consists of three layers of nodes performing different operations on incoming signals. The nodes in particular layers are responsible for determination of membership grades for each linguistic term, executing the rules and generating the weighted output. Additional factor is introduced to normalize the firing strengths of the rules with respect to the sum of all firing strengths. A hybrid training algorithm being a combination of the least squares method and back-propagation gradient descent method was used to prepare the FIS for the OS classification task.

The FL-based OS protection scheme developed has been thoroughly optimized and tested with ATP-generated case signals. The scheme designed displays almost perfect efficiency and high speed of OS detection. With the scheme designed the OS cases are identified much earlier comparing to standard impedance-based protection schemes. Wide robustness features of the ANFIS-based scheme have also been achieved.

A fuzzy-neuro hybrid can also be obtained with introduction of fuzzified neurons. A classical neuron represented by equation

$$y = \varphi\left(\sum_{k=1}^{n} w_k x_k\right) \tag{15.1}$$

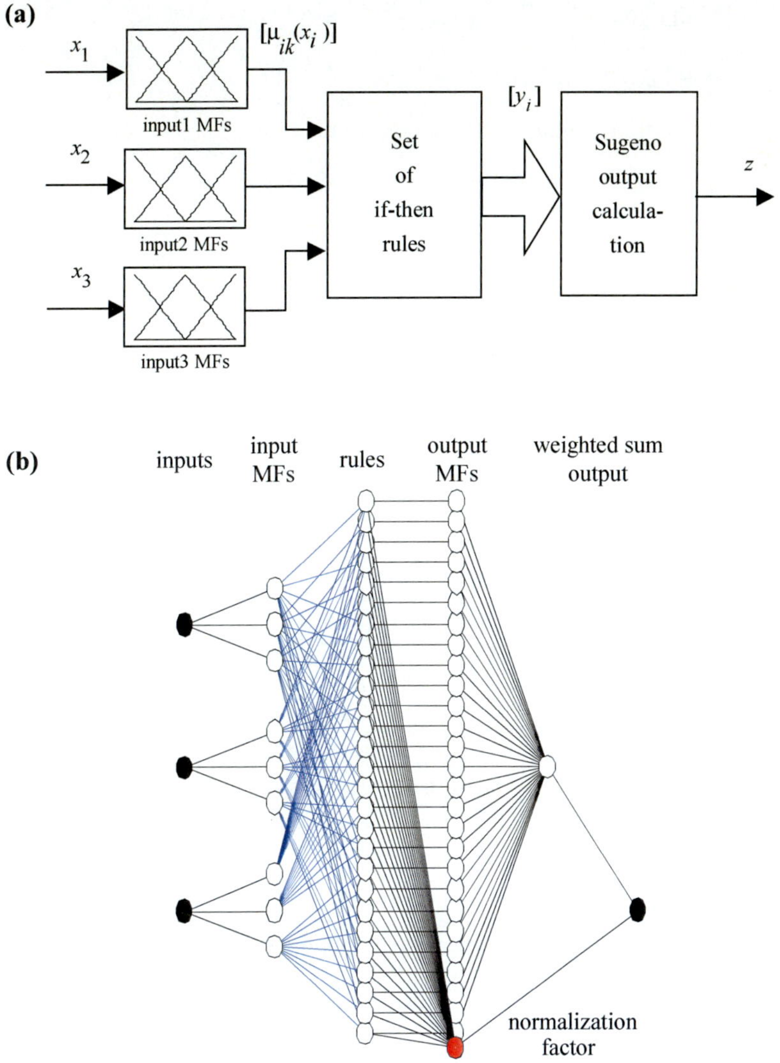

Fig. 15.1 FIS studied: **a** general structure, **b** adaptive network representation

becomes fuzzified when

- Any T-norm or S-norm is introduced to combine the input data with weighting factors,
- Aggregation is performed with any T-norm or S-norm,
- φ is any continuous function.

The following are just few examples of fuzzy neurons that can be defined in such a way [2, 10]:

- Fuzzy AND neuron

$$y = AND(p_1,\ldots,p_n) = T(p_1,\ldots,p_n) = T(S(w_1,x_1),\ldots,S(w_n,x_n)) \qquad (15.2)$$

with $T = \min$, $S = \max$ (min–max composition):

$$y = \min(w_1 \vee x_1,\ldots,w_n \vee x_n) \qquad (15.3)$$

- Fuzzy OR neuron

$$y = \mathrm{OR}(p_1,\ldots,p_n) = S(p_1,\ldots,p_n) = S(T(w_1,x_1),\ldots,T(w_n,x_n)) \qquad (15.4)$$

with $T = \min$, $S = \max$ (max–min composition)

$$y = \max(w_1 \wedge x_1,\ldots,w_n \wedge x_n) \qquad (15.5)$$

- implication-OR fuzzy neuron

$$y = S(I(w_1,x_1),\ldots,I(w_n,x_n)) \qquad (15.6)$$

- Kwan and Cai's max fuzzy neuron

$$e = \varphi(\max(w_1x_1,\ldots,w_nx_n) - \Theta) \qquad (15.7a)$$

$$y = \mu(e) \qquad (15.7b)$$

- Kwan and Cai's min fuzzy neuron

$$e = \varphi(\min(w_1x_1,\ldots,w_nx_n) - \Theta) \qquad (15.8a)$$

$$y = \mu(e) \qquad (15.8b)$$

With last two neurons fuzzification is obtained with introduction of the fuzzy membership function μ.

15.2.2 Cooperative Hybrid Schemes

In the case of cooperative hybrid systems, any of the AI technique employed is assigned to work independently from the others. Their inputs/outputs can be exchanged, which usually results in cascaded connections of sub-schemes, rarely some other connection versions can be met. The AI techniques applied cooperate, but do not permeate each other.

The most popular for applications in protection and control are the schemes employing neural networks and /fuzzy-logic set theory. Numerous examples can

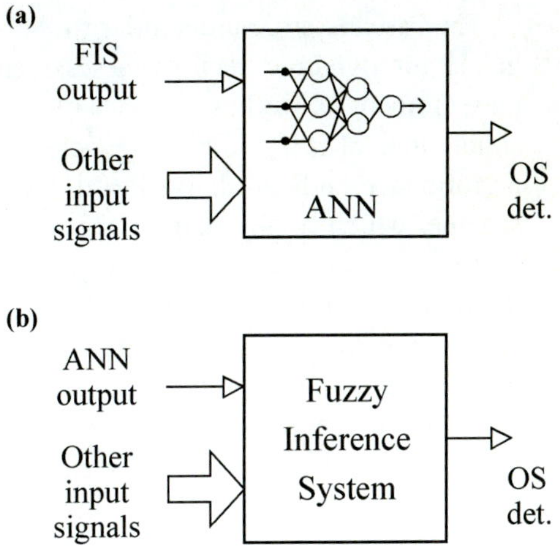

Fig. 15.2 Hybrid schemes for OS detection: **a** fuzzy-neuro, **b** neuro-fuzzy

be found in the literature [2, 8, 13]. The ANN may learn the parameters from the fuzzy system. This can be either performed offline or online while the fuzzy system is applied. The neural network can also provide fuzzy sets or even fuzzy rules for a fuzzy system by training on examples [8]. Rule weights can also be determined by a neural network [13].

The investigation on application of various AI techniques to out-of-step detection/prediction of synchronous machine has been presented in [16]. The basic protection scheme was employing a single neural network. Relatively compact structures (ANN with 9-9-1 neurones) were achieved which implies that they may be easily implemented on-line on traditional signal processors. The developed neural OS detectors proved to be robust against changing power system conditions (different fault types) and may be used for synchronous machines of various ratings.

Further optimization of the protection scheme has been done by introducing some concepts of FL, thus making the scheme an AI hybrid. Two cascade-type neuro-fuzzy configurations have been examined. The fuzzy inference system (FIS) module was used as a pre-processor of natural power system signals (the FIS output was further processed by an ANN) or as an interpreter of the ANN outputs (Fig. 15.2). Apart from the output signal of the pre-processing unit (FIS or ANN, respectively) some additional inputs may also be added. Their character, number as well as time position are to be examined in order to become the most optimal operation of the scheme under investigation.

The EMTP simulations confirmed superior performance of the hybrid FIS-ANN scheme. Additional processing of FIS output by an ANN brought about significant narrowing of the uncertainty area around tripping threshold. The ANN outputs are closer to 0 and 1 and thus more reliable decision could be issued. As a result the hybrid system became more robust than the other single-technique schemes developed. The developed hybrid OS detectors were able to recognize coming OS conditions 100–1,000 ms before they actually took place (a prediction action) thus providing a chance to undertake proper preventive operations to maintain system stability and to avoid presumable stresses to the protected machine.

The cascade structure similar to the one from Fig. 15.2 has also been proposed in [1] for dynamic security assessment of a power network. Signals from the measurement as well as the ones obtained from state estimation and time-domain simulations of selected contingencies are first processed by a neural network. The performance indices delivered at the ANN output go further through a fuzzy scheme, which finally outputs the predicted system state.

Another neuro-fuzzy scheme presented in [22] was proposed with the aim to improve transmission line fault detection and classification. Here, the ART neural network is applied to set the prototypes of trained clusters that are further used in the process of fault detection and classification. The inputs of the scheme are the data from power system simulation, data from substation historical database as well as signal samples from real system delivered by CVTs and CTs. The authors claim that the hybrid system developed is much more efficient that the other ones,also including single-technique protection versions.

The other examples of hybrid schemes as applied for power system problems include:

- Optimization of the generation companies' bidding strategy (fuzzy simulation and neural network combined with GA) [3],
- New solution for economic power dispatch (genetic algorithm combined with interior point methods harmony search) [14],
- Power quality analysis in distribution networks with a fuzzy-expert system (combining the usefulness of fuzzy logic in interpreting the fuzzy inputs and the expert system shell) [7],
- Power system fault diagnosis with fuzzy-expert system (COFES—incorporating fuzzy symbol classification through an enhanced knowledge-base system) [12],
- Peak load forecasting with use of a fuzzy expert system (incorporating linguistic fuzzy IF–THEN rules and expert's opinions) [5],
- On-line fault diagnosis on a transmission network with use of a fuzzy-expert system (information received includes the open/closed states of circuit breakers and the operational response of protection relays in conjunction with the topology of the transmission network) [18],
- Alarm interpretation and fault diagnosis (generic neuro-expert system architecture that can overcome difficulties faced by stand-alone ES and ANN schemes) [6],
- Transmission expansion planning using neuro-GA hybrid (neuro-computing is hybridized with genetic algorithms) [21],
- Wind-solar generation power control (a genetic-based self-adaptive hierarchical fuzzy controller is developed) [20],
- Digital power metering (fuzzy-based adaptive approach employing a genetic algorithm) [9],
- Fuzzy power system stabilizer design (optimization with genetic procedure) [11],
- Monitoring and diagnostics of large power transformers (an adaptive neuro-fuzzy system identification is applied) [17].

The above list is by no means exhaustive; the reader can find a lot more in many scientific journals and conference proceedings.

References

1. Assis TML, Nonara AA, Valentini TM (2009) Power system dynamic security assessment through a neuro-fuzzy scheme. In: Proceedings of the 15th international conference on intelligent system applications to power systems, Curitiba, Brazil, Paper 48
2. Fuller R (2000) Introduction to neuro-fuzzy systems. Physica-Verlag, Springer, Heidelberg
3. Huang DW, Han XS (2009) Study on generation companies' bidding strategy based on hybrid intelligent method. In: Proceedings of the 9th international conference on hybrid intelligent systems, Shenyang, China, Paper C14-06
4. Jang JSR (1993) ANFIS: adaptive-network-based fuzzy inference systems. IEEE Trans Syst Man Cybern 23:665–685
5. Kiartzis SJ, Bakirtzis AG, Theocharis JB, Tsagas G (2000) A fuzzy expert system for peak load forecasting. Application to the Greek power system. In: Proceedings of the 10th mediterranean electrotechnical conference, vol 3. pp 1097–1100
6. Khosla R, Dillon T (1992) A neuro-expert system architecture with application to alarm processing in a power system control centre. In: Proceedings of the 4th international conference on tools with artificial intelligence, pp 471–472
7. Kochukuttan H, Chandrasekaran A (1997) Development of a fuzzy expert system for power quality applications. In: Proceedings of the 29th southeastern symposium on system theory, pp 239–243
8. Kosko B (1992) Neural networks and fuzzy systems. A dynamical systems approach to machine intelligence. Prentice-Hall, Englewood Cliffs
9. Kung CH, Devaney MJ, Huang CM, Kung CM (1997) Fuzzy-based adaptive digital power metering using a genetic algorithm. In: Proceedings of the IEEE instrumentation and measurement technology conference, vol 1. pp 218–221
10. Kwan HK, Cai T (1994) A fuzzy neural network and its application to pattern recognition. IEEE Trans Fuzzy Syst 2:185–193
11. Menniti D, Burgio A, Pinnarelli A, Sorrentino N, Costanzo A (2007) Fuzzy stabilizers tuned by genetic algorithm in order to improve power system oscillation damping. In: Proceedings of the international conference on power engineering, energy and electrical drives, pp 291–299
12. Monsef H, Ranjbar AM, Jadid S (1997) Fuzzy rule-based expert system for power system fault diagnosis. IEE Proc Gener Transm Distrib 144:186–192
13. Nauck D, Klawonn F, Kruse R (1997) Foundations of neuro-fuzzy systems. Wiley, Chichester
14. Pandi VR, Panigrahi BK, Mallick MK, Abraham A, Das S (2009) Improved harmony search for economic power dispatch. In: Proceedigs of the 9th international conference on hybrid intelligent systems, Shenyang, China, Paper C14-05
15. Rebizant W, Feser K (2001) Fuzzy logic application to out-of-step protection of generators. In: Proceedings of the IEEE PES summer meeting, Vancouver, Canada, Paper 01SM061
16. Rebizant W, Feser K (2001) Out-of-step protection with AI methods. In: Proceedings of the 7th international IEE conference developments in power system protection, pp 295–298
17. Roizman O, Davydov V (1999) Neuro-fuzzy computing for large power transformers monitoring and diagnostics. In: Proceedings of the 18th international conference of the North American Fuzzy Information Processing Society, pp 248–252
18. Tan JC, Crossley PA, Mclaren PG (2001) Fuzzy expert system for on-line fault diagnosis on a transmission network. In: Proceedings of the IEEE power engineering society winter meeting, Columbus, OH, vol 2. pp 775–780

19. Vuorimaa P (1994) Fuzzy self-organizing map. Fuzzy Sets Syst 66:223–231
20. Wang B, Wu J, Yang J, Zhao S (2004) A distributed hybrid wind-solar power control system based on genetic algorithm and hierarchical fuzzy control. In: Proceedings of the 5th world congress on intelligent control and automation, vol 6. pp 5185–5188
21. Yoshimoto K, Yasuda K, Yokoyama R (1995) Transmission expansion planning using neuro-computing hybridized with genetic algorithm. In: Proceedings of the IEEE international conference on evolutionary computation, Perth, WA, doi:10.1109/ICEC.1995.489129
22. Zhang N, Kezunovic M (1995) Coordinating fuzzy ART neural networks to improve transmission line fault detection and classification. In: Proceedings of the IEEE general meeting, vol 1. pp 734–740

Index

W. Rebizant et al., *Digital Signal Processing in Power System Protection and Control,* Signals and Communication Technology,
DOI: 10.1007/978-0-85729-802-7,